Laboratory Exercises in

Microbiology

twelfth edition

Laboratory Exercises in
Microbiology

Nathan W. Rigel
Javier A. Izquierdo
HOFSTRA UNIVERSITY

LABORATORY EXERCISES IN MICROBIOLOGY, TWELFTH EDITION

Published by McGraw Hill LLC, 1325 Avenue of the Americas, New York, NY 10019. Copyright ©2023 by McGraw Hill LLC. All rights reserved. Printed in the United States of America. Previous editions ©2017, 2014, and 2011. No part of this publication may be reproduced or distributed in any form or by any means, or stored in a database or retrieval system, without the prior written consent of McGraw Hill LLC, including, but not limited to, in any network or other electronic storage or transmission, or broadcast for distance learning.

Some ancillaries, including electronic and print components, may not be available to customers outside the United States.

This book is printed on acid-free paper.

2 3 4 5 6 7 8 9 LKV 27 26 25 24

ISBN 978-1-264-77566-8 (bound edition)
MHID 1-264-77566-0 (bound edition)

Portfolio Manager: *Lauren Vondra*
Product Developer: *Darlene Schueller*
Marketing Manager: *Tami Hodge*
Content Project Managers: *Jeni McAtee, Rachael Hillebrand*
Buyer: *Sandy Ludovissy*
Designer: *David W. Hash*
Content Licensing Specialist: *Beth Cray*
Cover Image: *National Institute of Allergy and Infectious Diseases (NIAID)*
Compositor: *MPS Limited*

All credits appearing on page or at the end of the book are considered to be an extension of the copyright page.

The Internet addresses listed in the text were accurate at the time of publication. The inclusion of a website does not indicate an endorsement by the authors or McGraw Hill LLC, and McGraw Hill LLC does not guarantee the accuracy of the information presented at these sites.

Some of the laboratory experiments included in this text may be hazardous if materials are handled improperly or if procedures are conducted incorrectly. Safety precautions are necessary when you are working with chemicals, glass test tubes, hot water baths, sharp instruments, and the like, or for any procedures that generally require caution. Your school may have set regulations regarding safety procedures that your instructor will explain to you. Should you have any problems with materials or procedures, please ask your instructor for help.

mheducation.com/highered

Contents

About the Authors vii
Rules of Conduct and General Safety
  in the Laboratory viii
Tools for Your Success xii
Guided Tour through a Lab Exercise xiii
What's New in the Twelfth Edition xvi
Acknowledgments xx

PART ONE
Microscopic Techniques

1. Bright-Field Light Microscopy and Microscopic Measurement of Organisms 3
2. The Hanging Drop Slide and Bacterial Motility 15
3. Dark-Field Light Microscopy 19
4. Phase-Contrast Microscopy 25

PART TWO
Bacterial Cell Biology

5. Aseptic Technique 33
6. Negative Staining 43
7. Simple Staining 49
8. Gram Stain 57
9. Acid-Fast Staining Procedure 67
10. Endospore Staining 73
11. Capsule Staining 81
12. Flagella Staining 87

PART THREE
Basic Culture Techniques

13. Preparation of Microbiology Media and Equipment 95
14. The Spread-Plate Technique 103
15. The Streak-Plate Technique 109
16. The Pour-Plate Technique 117
17. Cultivation of Anaerobic Bacteria 123
18. Determination of Bacterial Numbers 133

PART FOUR
Microbial Biochemistry

19. Carbohydrates I: Fermentation and β-Galactosidase Activity 145
20. Carbohydrates II: Triple Sugar Iron Agar Test 151
21. Carbohydrates III: Starch Hydrolysis 157
22. Lipid Hydrolysis 161
23. Proteins I: The IMViC Tests 165
24. Proteins II: Gelatin and Casein Hydrolysis 173
25. Proteins III: Catalase Activity 181
26. Proteins IV: Oxidase Test 185
27. Proteins V: Urease Activity 191
28. Proteins VI: Lysine and Ornithine Decarboxylase Tests 197
29. Proteins VII: Phenylalanine Deamination 203
30. Proteins VIII: Dissimilatory Nitrate Reduction 209

PART FIVE
Environmental Factors Affecting Growth of Microorganisms

31. Temperature 217
32. pH 225
33. Osmotic Pressure 231
34. The Effects of Chemical Agents on Bacteria I: Disinfectants and Other Antimicrobial Products 237
35. The Effects of Chemical Agents on Bacteria II: Antibiotics 243
36. Bacterial Growth Curve 251

PART SIX
Environmental and Food Microbiology

37. Monitoring Water for Coliforms 265
38. Enumeration of Soil Microorganisms 273

39. Winogradsky Columns	279
40. Bioluminescence	287
41. Plate Counts and Quality Assessment of Milk	293

PART SEVEN
Medical Microbiology

42. Staphylococci	301
43. Pneumococcus	309
44. Streptococci	315
45. Neisseriae	323
46. Normal Human Flora	329

PART EIGHT
Eukaryotic Microbiology

47. Fungi I: Yeasts (*Ascomycota*)	337
48. Fungi II: *Zygomycota* (*Rhizopus*), *Ascomycota* (*Penicillium*), and *Basidiomycota* (*Agaricus*)	345

PART NINE
Microbial Genetics and Genomics

49. Isolation and Purification of Genomic DNA from Yeast and Bacteria	355
50. 16S rRNA Gene PCR and Sequencing	363
51. Mutations	369
52. Transformation	375
53. Conjugation	381
54. Generalized Transduction	387

Appendix A Dilutions with Sample Problems   A-1
Appendix B Reagents, Solutions, Stains, and Tests   A-11
Appendix C Culture Media   A-15
Appendix D Sources and Maintenance of Microbiological Stock Cultures   A-23
Appendix E Laboratory Technical Skills and Laboratory Thinking Skills   A-25
Appendix F Summary of Universal Precautions and Laboratory Safety Procedures   A-27
Appendix G Biosafety Level (BSL)   A-29
Appendix H pH and pH Indicators   A-31
Index   I-1

About the Authors

Nathan W. Rigel

Javier A. Izquierdo

Nathan W. Rigel is an associate professor in the Department of Biology at Hofstra University. Since starting his lab in 2013, Dr. Rigel's research has focused on the various protein trafficking pathways of *Acinetobacter*. While at Hofstra, Dr. Rigel has taught courses in microbiology, cell and molecular biology, bioinformatics, and bacterial genetics. Dr. Rigel completed his undergraduate studies in Microbiology at the Pennsylvania State University. He then moved south and earned his PhD in Microbiology and Immunology at the University of North Carolina at Chapel Hill while studying the accessory SecA2 protein export system found in mycobacteria. In 2009, Dr. Rigel returned north to study assembly of outer membrane proteins in Gram-negative bacteria as a postdoctoral research associate in the Department of Molecular Biology at Princeton University. Originally from a small town in central Pennsylvania, Dr. Rigel now lives in Queens with his wife and son. He can be reached at nathan.w.rigel@hofstra.edu.

Javier A. Izquierdo is an associate professor at Hofstra University, where he started his lab in 2014. Dr. Izquierdo received his BS in Biology at Case Western Reserve University. He earned his PhD in Microbiology from the University of Massachusetts Amherst, where he studied soil microbial ecology, and where he developed a passion for teaching microbiology thanks to many wonderful mentors. He went on to Dartmouth College for his postdoctoral work on the microbial ecology of cellulose degradation by anaerobic thermophiles. He then served as Biofuels Program Manager at the Research Triangle Institute in North Carolina, but his interests in teaching and mentoring research students brought him back to academia. His research group at Hofstra continues to work on understanding the ecology and physiology of cellulolytic microbes, as well as studying root-associated microbiomes of plants growing under stressful conditions. At Hofstra, he has taught courses in general microbiology, genomics and bioinformatics, and environmental microbiology. Born and raised in Venezuela, he now lives on Long Island with his wife and daughters. He can be reached at javier.a.izquierdo@hofstra.edu.

Rules of Conduct and General Safety in the Laboratory

Many of the microorganisms used in this course may be pathogenic for humans and animals. As a result, certain rules are necessary to avoid the possibility of infecting yourself or other people. Anyone who chooses to disregard these rules or exhibits carelessness that endangers others may be subject to immediate dismissal from the laboratory. If doubt arises as to the procedure involved in handling infectious material, consult your instructor.

The American Society for Microbiology, through its Office of Education and Training, has adopted the following on laboratory safety. Each point is considered essential for every introductory microbiology laboratory, regardless of its emphasis.

A student successfully completing basic microbiology will demonstrate the ability to explain and practice safe

1. **Microbiological procedures,** including
   a. reporting all spills and broken glassware to the instructor and receiving instructions for cleanup
   b. methods for aseptic transfer
   c. minimizing or containing the production of aerosols and describing the hazards associated with aerosols
   d. washing hands prior to and at the conclusion of laboratories, as well as at any time contamination is suspected
   e. never eating or drinking in the laboratory
   f. using universal precautions
   g. disinfecting lab benches prior to and at the conclusion of each lab session
   h. identification and proper disposal of different types of waste, including the use of special receptables for infectious materials and used glass.
   i. never applying cosmetics, including contact lenses, or placing objects (fingers, pencils) in the mouth or touching the face
   j. reading (and signing, if applicable) a laboratory safety agreement indicating that the student has read and understands the safety rules of the laboratory
   k. good lab practice, including returning materials to proper locations, proper care and handling of equipment, and keeping the benchtop clear of extraneous materials
   l. avoiding contamination of benches, floor, and wastebaskets.

2. **Protective procedures,** including
   a. tying long hair back, wearing personal protective equipment (eye protection, coats, closed shoes; glasses may be preferred to contact lenses), and using such equipment in appropriate situations
   b. always using appropriate pipetting devices

3. **Emergency procedures,** including
   a. locating and properly using emergency equipment (eye-wash stations, first-aid kits, fire extinguishers, chemical safety showers, telephones, and emergency numbers)
   b. reporting all injuries immediately to the instructor
   c. following proper steps in the event of an emergency

In addition, institutions where microbiology laboratories are taught will

1. train faculty and staff in proper waste stream management
2. provide and maintain necessary safety equipment and information resources
3. train faculty, staff, and students in the use of safety equipment and procedures
4. train faculty and staff in the use of Material Safety Data Sheet (MSDS). The Workplace Hazardous Materials Information System (WHMIS) requires that all hazardous substances, including microorganisms, be labeled in a specific manner. In addition, there must be a MSDS available to accompany each hazardous substance. Material Safety Data Sheets are now supplied with every chemical sold by supply houses. The person in charge of the microbiology laboratory should ensure that adherence to this law is enforced.

All laboratory work can be done more effectively and efficiently if the subject matter is understood before coming to the laboratory. To accomplish this, read the experiment several times before the laboratory begins. Know how each exercise is to be done and what principle it is intended to convey. Also, read the appropriate sections in your textbook that pertain to the experiment being performed; this will save you much time and effort during the actual laboratory period.

All laboratory experiments will begin with a brief discussion by your instructor of what is to be done, the location of the materials, and other important information. Feel free to ask questions if you do not understand the instructor or the principle involved.

Much of the work in the laboratory is designed to be carried out in groups or with a partner. This is to aid in coverage of subject matter, to save time and expense, and to encourage discussion of data and results.

For more information about the microorganisms that you will use in each lab and their biosafety levels (BSL), please refer to *BSL Recommendations for Select Agents—Supplement to CDC/NIH Biosafety in Microbiological and Biomedical Laboratories*, sixth edition (BMBL). This document is available online at https://www.cdc.gov/labs/BMBL.html.

Instructors: Student Success Starts with You

Tools to enhance your unique voice

Want to build your own course? No problem. Prefer to use an OLC-aligned, prebuilt course? Easy. Want to make changes throughout the semester? Sure. And you'll save time with Connect's auto-grading too.

65% Less Time Grading

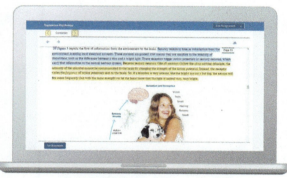

Laptop: McGraw Hill; Woman/dog: George Doyle/Getty Images

Study made personal

Incorporate adaptive study resources like SmartBook® 2.0 into your course and help your students be better prepared in less time. Learn more about the powerful personalized learning experience available in SmartBook 2.0 at **www.mheducation.com/highered/connect/smartbook**

Affordable solutions, added value

Make technology work for you with LMS integration for single sign-on access, mobile access to the digital textbook, and reports to quickly show you how each of your students is doing. And with our Inclusive Access program you can provide all these tools at a discount to your students. Ask your McGraw Hill representative for more information.

Padlock: Jobalou/Getty Images

Solutions for your challenges

A product isn't a solution. Real solutions are affordable, reliable, and come with training and ongoing support when you need it and how you want it. Visit **www.supportateverystep.com** for videos and resources both you and your students can use throughout the semester.

Checkmark: Jobalou/Getty Images

Students: Get Learning that Fits You

Effective tools for efficient studying

Connect is designed to help you be more productive with simple, flexible, intuitive tools that maximize your study time and meet your individual learning needs. Get learning that works for you with Connect.

Study anytime, anywhere

Download the free ReadAnywhere app and access your online eBook, SmartBook 2.0, or Adaptive Learning Assignments when it's convenient, even if you're offline. And since the app automatically syncs with your Connect account, all of your work is available every time you open it. Find out more at **www.mheducation.com/readanywhere**

> "I really liked this app—it made it easy to study when you don't have your textbook in front of you."
>
> - Jordan Cunningham, Eastern Washington University

Everything you need in one place

Your Connect course has everything you need—whether reading on your digital eBook or completing assignments for class, Connect makes it easy to get your work done.

Learning for everyone

McGraw Hill works directly with Accessibility Services Departments and faculty to meet the learning needs of all students. Please contact your Accessibility Services Office and ask them to email accessibility@mheducation.com, or visit **www.mheducation.com/about/accessibility** for more information.

Tools for Your Success

VIRTUAL LABS AND LAB SIMULATIONS

While the biological sciences are hands-on disciplines, instructors are now often being asked to deliver some of their lab components online, as full online replacements, supplements to prepare for in-person labs, or make-up labs.

These simulations help each student learn the practical and conceptual skills needed, then check for understanding and provide feedback. With adaptive pre-lab and post-lab assessment available, instructors can customize each assignment.

From the instructor's perspective, these simulations may be used in the lecture environment to help students visualize complex scientific processes, such as DNA technology or Gram staining, while at the same time providing a valuable connection between the lecture and lab environments.

RELEVANCY MODULES

Connect® offers a series of Relevancy Modules with auto-graded assessment content that can be assigned to help facilitate student-centered learning and put relevancy upfront for students. Each module consists of an overview of scientific concepts, videos, and in-depth application of those concepts. Ensure students are retaining information with auto-graded assessment questions that correlate to each module. Modules include: Scientific Thinking in Everyday Life, Fermentation and the Making of Beer, Microbes and Cancer, Vaccines: Your Best Defense, Global Health: Impact of Infectious Disease, COVID-19: The Rise of a Global Pandemic, Antibiotic Resistance, Emerging and Re-Emerging Infectious Diseases, and Biotechnology: The Helpful Side of Microorganisms.

TEGRITY

Tegrity in Connect is a tool that makes class time available 24/7 by automatically capturing every lecture. With a simple one-click start-and-stop process, you capture all computer screens and corresponding audio in a format that is easy to search, frame by frame. Students can replay any part of any class with easy-to-use, browser-based viewing on a PC, Mac, or other mobile device.

Educators know that the more students can see, hear, and experience class resources, the better they learn. Tegrity's unique search feature helps students efficiently find what they need, when they need it, across an entire semester of class recordings. Help turn your students' study time into learning moments immediately supported by your lecture.

McGraw Hill Create® is a self-service website that allows you to create custom course materials using McGraw Hill's comprehensive, cross-disciplinary content and digital products.

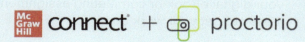

Remote Proctoring & Browser-Locking Capabilities

New remote proctoring and browser-locking capabilities, hosted by Proctorio within Connect, provide control of the assessment environment by enabling security options and verifying the identity of the student. Seamlessly integrated within Connect, these services allow instructors to control students' assessment experience by restricting browser activity, recording students' activity, and verifying students are doing their own work. Instant and detailed reporting gives instructors an at-a-glance view of potential academic integrity concerns, thereby avoiding personal bias and supporting evidence-based claims.

Guided Tour through a Lab Exercise

⚠️ SAFETY CONSIDERATIONS

This laboratory manual includes many of the safety precautionary measures established by the Centers for Disease Control and Prevention (CDC), Atlanta, Georgia; the Occupational Safety and Health Administration (OSHA); and the Environmental Protection Agency (EPA). A safety considerations box is included for each exercise to help both the instructor and the student prepare themselves for the possibility of accidents and discuss precautionary procedures.

LEARNING OUTCOMES

Each exercise has a set of learning outcomes that define the specific goals of the laboratory session. It is to the student's advantage to read through this list before coming to class. Upon completion of the exercise, the student should be able to meet all of the outcomes for that exercise.

SUGGESTED READING IN TEXTBOOK

These cross-references have been designed to save the students' time. By referring the student to sections, paragraphs, tables, and figures within the *Prescott Microbiology* textbook, unnecessary duplication is avoided.

PRONUNCIATION GUIDE

This section contains the phonetic pronunciations for all organisms used in the exercise. If students take the time to sound out new and unfamiliar terms and say them aloud several times, they will learn to use the vocabulary of microbiologists.

LEGAL CONSIDERATIONS

A special section on legal considerations has been added to those experiments where either the course director or the laboratory instructor should check with the appropriate college or university officials (legal department) to see if it is considered an invasion of a student's privacy to ask students to perform.

MATERIALS

To aid in the preparation of all exercises, each procedure contains a list of the required cultures, media, reagents, and other equipment necessary to complete the exercise in the allocated lab time. Appendices B and C provide recipes for reagents, stains, and culture media. Appendix D describes the maintenance of microorganisms and supply sources.

MEDICAL APPLICATION

Many students using this laboratory manual are either in one of the allied health disciplines, such as nursing, or in a preprofessional program such as pre-med, pre-PA, pre-dental, or pre-vet and need to know the clinical relevance of each exercise performed. To satisfy this need, a medical application section is included for the medically oriented exercises. Medical applications describe a specific application of the purpose of the exercise.

WHY ARE THE FOLLOWING BACTERIA, SLIDES, OR OTHER MICROORGANISMS USED IN THIS EXPERIMENT?

The authors have chosen specific viruses, bacteria, fungi, protozoa, algae, and various prepared slides for each exercise. This microbial material has been selected based on cost, ease of growth, availability, reliability, and most important, the ability to produce the desired experimental results. In addition, this section provides biochemical, morphological, and taxonomic information about the microorganism(s) that the student should find helpful when performing the experiment.

PRINCIPLES

This section precedes each exercise and/or procedure; it contains a brief discussion of the microbiological principles, concepts, and techniques that underlie the experimental procedures being performed in the exercise.

HELPFUL HINTS

Additional information on what to watch out for and what can go wrong, and helpful tidbits to make the experiment work properly.

PROCEDURE

Explicit instructions are augmented by diagrams to aid students in executing the experiment as well as interpreting the results. Where applicable, actual results are shown so that the student can see what should be obtained.

LABORATORY REPORT

Various pedagogical techniques are used for recording the obtained results. This part of the exercise can be turned in to the instructor for checking or grading.

ASSESSMENT
CRITICAL THINKING AND LEARNING OUTCOMES REVIEW

By definition, critical thinking is the art of analyzing and evaluating thinking with a view to improve it. Microbiologists and students of microbiology need to be clear as to the purpose at hand and the question at issue.

Critical thinking and learning outcomes review questions are located at the end of each laboratory report. These were written so that students can test their understanding of the concepts and techniques presented in each exercise and so that the instructor can determine the student's ability to utilize critical thinking and understand the experimental concepts and techniques.

NEW ART AND IMAGES

Illustrations and photos have been updated to improve the clarity of protocols, to tie more closely with *Prescott's Microbiology*, and to present the material in a more contemporary style.

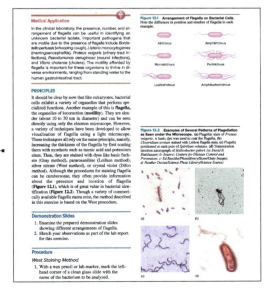

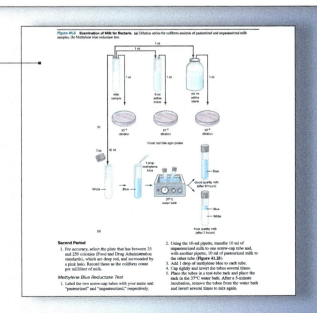

What's New in the Twelfth Edition

In revising the manual for the twelfth edition, feedback has been received from different sources. In particular, we have made extensive changes to enhance each exercise by updating the text material, updating the art where necessary, adding new and improved figures and illustrations, increasing student involvement, and cross-referencing all aspects of the manual with the changes that were made in the twelfth edition of *Prescott's Microbiology*. All changes were made to enhance clarity, provide more emphasis, and reduce redundancy. In each exercise, the student is continually encouraged to develop a constant awareness of the presence and ubiquity of microorganisms and their relationship to humans.

NEW EXERCISES AND REMOVED MATERIAL

Four new exercises have been added that cover Winogradsky Columns (Exercise 39), 16S rRNA Gene PCR and Sequencing (Exercise 50), Mutations (Exercise 51), and Generalized Transduction (Exercise 54). These exercises enhance and expand the scope of Part 6 (Environmental and Food Microbiology) and Part 9 (Microbial Genetics and Genomics), introducing students to new techniques while reinforcing concepts and techniques introduced in other exercises. A few exercises have been removed from this edition to reduce redundancy or due to limited accessibility of reagents. Exercises from previous editions on the API 20E System, the EnteroPluri-*Test* System, Using the Bergey's Manual, Hand Washing, and Hemagglutination Reactions have been removed. Other exercises such as General Unknown have been integrated into new exercises.

UPDATED LEARNING OUTCOMES

Colleges and universities throughout the country are now requiring science courses to identify Student Learning Outcomes in various assessment instruments. These Learning Outcomes are designed as a guide for students to understand the minimum material they are expected to learn from each exercise within the manual. In the twelfth edition of *Laboratory Exercises in Microbiology*, each exercise begins with specific Learning Outcomes that represent each major concept, the skills, and the theories each student should master. At the end of each laboratory exercise, an Assessment of Critical Thinking and Learning Outcomes Review are presented on one page. This assessment serves as a check to help students confirm their understanding of the learning outcomes in each exercise.

CONTENT CHANGES TO THE TWELFTH EDITION BY PART AND EXERCISE

Each exercise has been thoroughly reviewed and has undergone revision. There are also a number of new exercises. The highlights include the following:

PART ONE (Microscopic Techniques)

Exercise 1—Bright-Field Light Microscopy and Microscopic Measurement of Organisms. The title of this exercise has been changed and instructions for calibrating the ocular micrometer have been expanded with more details.

Exercise 3—Dark-Field Light Microscopy. The title of this exercise has been changed and new photomicrographs have been added to provide more examples of the uses of dark-field microscopy.

Exercise 4—Phase-Contrast Microscopy. The title of this exercise has been changed and new photomicrographs have been added to better depict the uses of phase contrast microscopy.

PART TWO (Bacterial Cell Biology)

Exercise 5—Aseptic Technique. This exercise is now the lead for this section and introduces students to aseptic technique before using live cultures for the subsequent staining exercises.

Exercise 6—Negative Staining. The Principles section has been updated to better introduce students to the applications of this technique. Diagrams have been updated to explain the mode of action of the dyes used in this exercise by way of their chemical structure.

Exercise 7—Simple Staining. The title of this exercise has been simplified. The morphology and procedural diagrams have been updated, as well as the photomicrographs.

Exercise 8—Gram Stain. The procedure to stain *Hyphomicrobium* has been removed. New figures depicting the structure of Gram-positive and Gram-negative cell walls have been added, as well as new photomicrographs.

Exercise 9—Acid-Fast Staining Procedure. The title of this exercise has been changed to indicate that this exercise will only cover the Ziehl-Neelsen procedure, and not the Kinyoun procedure.

Exercise 10—Endospore Staining. The title of this exercise has been changed. A new figure depicting the life cycle of endospore-forming bacteria has been added, as well as new photomicrographs to better depict the diversity of spores students may be able to observe.

Exercise 11—Capsule Staining. Figures with explanatory diagrams and a photomicrograph of capsule staining have all been updated to improve clarity.

Exercise 12—Flagella Staining. The title of this exercise has been changed to indicate that this exercise will only cover the West staining method, and not the Difco procedure. A new figure depicting different arrangements of flagella has been added, as well as new photomicrographs.

PART THREE (Basic Culture Techniques)

Exercise 13—Preparation of Microbiology Media and Equipment. The title has been changed to better represent the focus of this exercise. The introductory text in the Principles section and explanatory figures have been updated for clarity.

Exercise 14— The Spread-Plate Technique. Procedural diagrams and figures have been updated for clarity.

Exercise 15—The Streak-Plate Technique. The title has been changed to better represent the focus of this exercise. The principles of selective and differential media are now introduced in the main Principles section with new images.

Exercise 16— The Pour-Plate Technique. Procedural diagrams have been updated for clarity.

Exercise 17—Cultivation of Anaerobic Bacteria. The Wright's tube procedure has been removed. Diagrams and figures have been updated for clarity. The lab exercises have been modified to compare three organisms using all the techniques.

Exercise 18—Determination of Bacterial Numbers. Procedural diagrams have been updated and new photos of spectrophotometers have been added.

PART FOUR (Microbial Biochemistry)

Exercise 19—Carbohydrates I: Fermentation and β-Galactosidase Activity. The sugar differentiation disk experiment has been removed. New photos of inoculated tubes have been added. *Salmonella enterica* has been chosen to replace *Salmonella cholerae-suis*.

Exercise 20—Carbohydrates II: Triple Sugar Iron Agar Test. Photos and diagrams explaining the interpretation of TSIA tubes have been updated and condensed into one figure. Examples of possible results with specific organisms have been reformatted in a new figure for clarity.

Exercise 21—Carbohydrates III: Starch Hydrolysis. A new photo of inoculated media has been added.

Exercise 22—Lipid Hydrolysis. A new photo of inoculated media has been added.

Exercise 23—Proteins I: The IMViC Tests. A previous exercise on hydrogen sulfide and motility has been integrated into this exercise. A more extensive explanation of the principles of these tests has been provided. Photos for each test have been updated.

Exercise 24—Proteins II: Gelatin and Casein Hydrolysis. This exercise now integrates two previous exercises. Photos of inoculated media have been updated for clarity.

Exercise 26—Proteins IV: Oxidase Test. This exercise focuses only on the oxidase test and not on bioluminescence. A new figure with results from strip tests has been added.

Exercise 27—Proteins V: Urease Activity. A photo of inoculated media has been updated for clarity. *Salmonella enterica* has been chosen to replace *Salmonella cholerae-suis*.

Exercise 28—Proteins VI: Lysine and Ornithine Decarboxylase Tests. A photo of inoculated media has been updated for clarity.

Exercise 29—Proteins VII: Phenylalanine Deamination. Figures have been updated for clarity, including a new photo of inoculated media.

Exercise 30—Proteins VIII: Dissimilatory Nitrate Reduction. Diagrams and figures have been updated for clarity.

PART FIVE (Environmental Factors Affecting Growth of Microorganisms)

Exercise 31—Temperature. *Serratia marcescens* has been added to this exercise, and *Bacillus subtilis* has been chosen to replace *Bacillus globisporus*. The experimental setup focuses now on a temperature range from 4°C to 70°C.

Exercise 32—pH. *Pseudomonas aeruginosa* has been added to the list of organisms used in this exercise.

Exercise 33—Osmotic Pressure. Explanatory diagrams have been updated and text discussing water activity has been removed.

Exercise 35—The Effects of Chemical Agents on Bacteria II: Antibiotics. Photos and explanatory diagrams have been updated for clarity. The EnteroPluri-*Test* is no longer part of this exercise.

Exercise 36—Bacterial Growth Curve. The title of this exercise has been changed to indicate that only the classical growth curve method will be used, and not the two-hour method. Figures have been updated for clarity.

PART SIX (Environmental and Food Microbiology)

Exercise 37—Monitoring Water for Coliforms. The title of this exercise has been changed to indicate that the Colilert®-18 test has been removed. The experimental diagram of the MPN portion of the exercise has been updated for clarity.

Exercise 38—Enumeration of Soil Microorganisms. The experimental diagram of the MPN portion of the exercise has been updated for clarity.

Exercise 39—Winogradsky Columns. This is a new exercise that covers the assembly of Winogradsky columns, the enrichment of microorganisms growing at different levels of the columns and their microscopic observation.

Exercise 40—Bioluminescence. This is an extensively revised exercise that covers what previously was covered in the oxidase exercise, and has been expanded with a larger focus on quorum sensing and additional exploratory experiments.

Exercise 41—Plate Counts and Quality Assessment of Milk. The title of this exercise has been changed to describe the exercise more accurately. Figures and procedural diagrams have been updated for clarity.

PART SEVEN (Medical Microbiology)

Exercise 42—Staphylococci. Figures of inoculated media and microscopy have been extensively updated. The Staphaurex test has been removed, while previous experiments on DNase and coagulase activity have been integrated into this exercise.

Exercise 43—Pneumococcus. The procedure has been extensively revised, including the removal of throat swabs and the rapid strep test. Photos of media and microscopy have been updated.

Exercise 44—Streptococci. Figures of inoculated media and microscopy have been extensively updated.

Exercise 45—Neisseriae. Microscopy and a diagram for the candle jar incubation have been updated.

Exercise 46—Normal Human Flora. This exercise now integrates previous exercises on the isolation of human flora and on unknown identification. The exercise now focuses on skin microflora and expands the previous scope to include the cultivation, differentiation, and identification of *Staphylococcus* and *Micrococcus*.

PART EIGHT (Eukaryotic Microbiology)

Exercise 47—Fungi I: Yeasts (*Ascomycota*). A new photomicrograph has been added.

Exercise 48—Fungi II: Zygomycota (*Rhizopus*), Ascomycota (*Penicillium*), and Basidiomycota (*Agaricus*). Figures of colony morphology, experimental setup, and microscopy have been extensively updated.

PART NINE (Microbial Genetics and Genomics)

Exercise 49—Isolation and Purification of Genomic DNA from Yeast and Bacteria. This exercise now combines two exercises from the previous edition on the DNA isolation from yeast and from *E. coli*. The procedure used on *E. coli* is now used on two additional Gram-negative organisms: *Serratia marcescens* and *Enterobacter aerogenes*. If feasible, this exercise is designed to use the extracted DNA for 16S rRNA gene sequencing as part of Exercise 50.

Exercise 50—16S rRNA Gene PCR and Sequencing. This is a new exercise that introduces students to the polymerase chain reaction (PCR), verifying the success of their reactions through agarose gel electrophoresis, and very basic analysis of a DNA sequencing reaction. This exercise can use the DNA obtained from Exercise 49 or on other DNA samples, and it introduces electrophoresis and sequencing for subsequent exercises.

Exercise 51—Mutations. This is a new exercise where students will select for mutations in the *rpoB* gene in *E. coli*, introducing them to the role of spontaneous mutations in antibiotic resistance. The exercise builds on culture and molecular skills developed in previous exercises.

Exercise 52—Transformation. This exercise has been extensively updated to use *E. coli* and naturally-competent *Acinetobacter baylyi*, instead of *B. subtilis*, and to introduce students to the electroporation technique. The figure discussed in the Principles section has been updated to include plasmid DNA.

Exercise 53—Conjugation. This exercise has been updated to use different Hfr and F⁻ *E. coli* strains, using a leucine and threonine auxotroph resistant to streptomycin as the recipient and a leucine and threonine prototroph as the donor. A new figure depicting Hfr conjugation has been added to the Principles section.

Exercise 54—Generalized Transduction. This is a new exercise where students will use phage P1 to move a kanamycin resistance cassette from one *E. coli* strain to another.

APPENDICES

New media recipes have been added and the list of microorganisms used has been updated. In addition, tables for transmission absorbance and identification charts have been removed.

Acknowledgments

The authors wish to thank the team at McGraw Hill who made this twelfth edition possible. In particular, we would like to thank Darlene Schueller, Lauren Vondra, Tami Hodge, David Hash, Beth Cray, and Jeni McAtee for all their unwavering support throughout this process. Thanks to Beth Baugh and Rajesh Negi for their help as well. The authors also wish to thank Drs. Jo Willey, Kaethe Sandman, and Dorothy Wood, authors of *Prescott's Microbiology,* for their insightful feedback and discussions throughout the development of this edition. The authors are indebted to Dr. Chris Boyko who has been instrumental in coordinating lab needs and media preparation for our labs. A special thanks goes to our teaching assistants Arjun, Casey, Cheyenne, Fatima, Harshani, Julianna, Katie, Lauren, Megan, Suzie, and Tyler for the excitement and the energy they have brought to the microbiology teaching lab. Lastly, the authors would like to sincerely thank the many students who have taken the General Microbiology course at Hofstra University and enriched our experience as instructors.

The authors have also benefitted greatly from the reviews by many colleagues. Thank you to the following individuals for providing a detailed review.

Michael S. Carter
  Salisbury University
Stella Doyungan
  Texas A&M University-Corpus Christi
Jeffrey C. Hoyt
  Paradise Valley Community College
Jeffrey A. Isaacson
  Nebraska Wesleyan University
Carrie Kinsey
  Indiana University Kokomo
Wendy Trzyna
  Marshall University
Rachel Zufferey
  St. John's University

Nathan Rigel would like to thank his wife Lisa and son Kieran for the much-needed encouragement in tackling this project. All first-time authors should be so lucky!

Javier Izquierdo would like to thank his wife Sara and his daughters Anna and Silvia for always sharing his excitement for microbiology and for all their support throughout the writing of this book.

PART 1
Microscopic Techniques

Microbiologists employ a variety of light microscopes in their work: bright-field, dark-field, phase-contrast, differential interference contrast (DIC), and fluorescence are most commonly used. In fact, the same microscope may be a combination of types: bright-field and phase-contrast, or phase-contrast and fluorescence. You will use these microscopes and the principles of microscopy extensively in this course as you study the form, structure, staining characteristics, and motility of different microorganisms. Therefore, proficiency in using the different microscopes is essential to all aspects of microbiology and must be mastered at the very beginning of a microbiology course. The next four exercises have been designed to accomplish this major objective.

After completing Exercise 1, you will be able to demonstrate the ability to use a bright-field light microscope. This will meet the following American Society for Microbiology Core Curriculum skills:

- correctly setting up and focusing the microscope
- proper handling, cleaning, and storage of the microscope
- correct use of all lenses
- recording microscopic observations

JGI/Daniel Grill/Tetra Images, LLC/Alamy Stock Photo

Our ability to understand the microbial world has been revolutionized by the use of the microscope. Ever since Antonie van Leeuwenhoek reported the first observations of bacteria in the mid-1600s, the microscope has become an essential tool to understand the physiology, ecology, and diversity of microbes. Because different microscopes and microscopy techniques have different applications, it is crucially important to become very familiar with these tools.

EXERCISE 1

Bright-Field Light Microscopy and Microscopic Measurement of Organisms

SAFETY CONSIDERATIONS

- Slides and coverslips are glass. Be careful with them. Do not cut yourself when using them.
- The coverslips are very thin and easily broken.
- Dispose of any broken glass in the appropriately labeled container.
- If your microscope has an automatic stop, do not use it as the stage micrometer is too thick to allow it to function properly. It may result in a shattered or broken slide or lens.

MATERIALS

- compound microscope
- lens paper and lens cleaner
- immersion oil
- prepared stained slides of several types of bacteria (rods, cocci, spirilla), fungi, algae, and protozoa
- glass slides
- coverslips
- dropper with bulb
- newspaper or cut-out letter e's
- tweezers
- ocular micrometer
- stage micrometer
- safety glasses
- disposable gloves
- lab coat

LEARNING OUTCOMES

Upon completion of this exercise, students will demonstrate the ability to

1. Identify all the parts of a compound microscope
2. Correctly use the microscope—especially the oil immersion lens
3. Make use of a wet-mount preparation
4. Distinguish how microorganisms can be measured under the light microscope
5. Calibrate an ocular micrometer
6. Perform some measurements on different microorganisms

SUGGESTED READING IN TEXTBOOK

1. There Are Several Types of Light Microscopes, section 2.2; see also figures 2.1–2.5.
2. See table 2.2.

Medical Application

In the clinical laboratory, natural cell size, arrangement, and motility are important characteristics in the identification and characterization of a bacterial pathogen.

Why Are Prepared Slides Used in This Exercise?

Because this is a microbiology course and most of the microorganisms studied are bacteria, this is an excellent place to introduce the student to the three basic bacterial shapes: cocci, rods, and spirilla. By gaining expertise in using the bright-field light microscope, the student should be able to observe these three bacterial shapes by the end of the lab period. In addition, the student will gain an appreciation for the small size and arrangement of prokaryotic cell structure.

One major objective of this exercise is for the student to understand how microorganisms can be measured under the light microscope and to perform some measurements on different microorganisms. By making measurements on prepared slides of various bacteria, fungi, algae, and protozoa, the student will gain an appreciation for the size of different microorganisms discussed throughout both the lecture and laboratory portions of this course.

PRINCIPLES

The success of a student in a microbiology laboratory is directly related to the competent use of the compound bright-field light microscope. It is called a bright-field microscope because it forms a dark image against a brighter background. The **bright-field light microscope** is an instrument that magnifies images using two lens systems. It is used to examine both stained and unstained specimens. Initial magnification occurs in the **objective lens.** Most microscopes have at least three objective lenses on a rotating base, and each lens may be rotated into alignment with the **eyepiece** or **ocular lens,** in which the final magnification occurs. The objective lenses are identified as the **low-power, high-dry,** and **oil immersion objectives.** Each objective is also designated by other terms. These terms give either the **linear magnification** or the **focal length.** The latter is about equal to or greater than the **working distance** between the specimen and the tip of the objective lens. For example, the low-power objective is also called the **10×,** or **16 millimeter (mm), objective;** the high-dry is called the **40×,** or **4 mm, objective;** and the oil immersion is called the **90×, 100×,** or **1.8 mm objective.** As the magnification increases, the size of the lens at the tip of the objective becomes progressively smaller and admits less light. This is one of the reasons that changes in position of the **substage condenser** and **iris diaphragm** are required when using different objectives if the specimens viewed are to be seen distinctly. The condenser focuses the light on a small area above the stage, and the iris diaphragm controls the amount of light that enters the condenser. It is important to remember that, as a rule, one must increase the light as one increases the magnification. When the oil immersion lens is used, immersion oil fills the space between the objective and the specimen. Because immersion oil has the same **refractive index** as glass, the loss of light is minimized (**Figure 1.1**). The **eyepiece,** or **ocular,** at the top of the tube magnifies the image formed by the objective lens. As a result, the total magnification seen by the observer is obtained by multiplying the magnification of the objective lens by the magnification of the ocular, or eyepiece. For example, when using the 10× ocular and the 40× objective, total magnification is $10 \times 40 = 400$ times.

Procedure for Basic Microscopy: Proper Use of the Microscope

1. Always carry the microscope with two hands. Place it on the desk with the open part away from you.
2. Clean all of the microscope's lenses only with lens paper and lens cleaner if necessary. Do not use paper towels or Kimwipes; they can scratch the lenses. Do not remove the oculars or any other parts from the body of the microscope.
3. Cut a lowercase *e* from a newspaper or other printed page. Prepare a wet-mount as illustrated in **Figure 1.2.** Place the glass slide on the stage of the microscope and secure it firmly using stage clips. If your microscope has a mechanical stage device, place the slide securely in it. Move the slide until the letter *e* is over the opening in the stage.

Figure 1.2 Preparation of a Wet-Mount Slide. (**a**) Add a drop of water to a slide. (**b**) Place the specimen (letter *e*) in the water. (**c**) Place the edge of a coverslip on the slide so that it touches the edge of the water. (**d**) Slowly lower the coverslip to prevent forming and trapping air bubbles. Note the safety gloves on the hands.

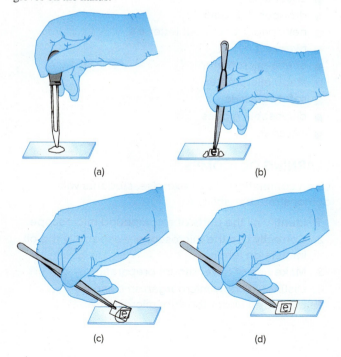

Figure 1.1 The Oil Immersion Objective. An oil immersion objective lens operating in air and with immersion oil. Light rays that must pass through air are bent (refracted), and many do not enter the objective lens. The immersion oil prevents the loss of light rays.

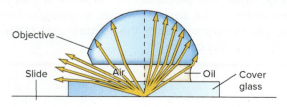

4. With the low-power objective in position, lower the objective until the tip is within 5 mm of the slide. Be sure that you lower the objective while looking at the microscope from the side.
5. Look into the microscope and slowly raise the objective by turning the coarse adjustment knob counterclockwise until the object comes into view. Once the specimen is in view, use the fine adjustment knob to focus the desired image.
6. Open and close the diaphragm, and lower and raise the condenser, noting what effect these actions have on the appearance of the object being viewed. Usually the microscope is used with the substage condenser in its topmost position. The diaphragm should be open and then closed down until just a slight increase in contrast is observed (**Table 1.1**).
7. Repeat steps 4 through 6 with the stained bacteria that are provided (**Figure 1.3a–d**) and use the oil immersion lens to examine these slides. The directions for using this lens are as follows: First, locate the stained area with the low-power objective, then turn the oil immersion lens into the oil and focus with the fine adjustment. An alternate procedure is to get the focus very sharp under high power, then move the revolving nosepiece until you are halfway between the high-power and oil immersion objectives. Place a small drop of immersion oil in the center of the illuminated area on the slide. Continue revolving

Figure 1.3 Examples of Bacterial Shapes as Seen with the Bright-Field Light Microscope. (a) *Staphylococcus aureus* cocci; singular, coccus (×1,000). (b) *Bacillus subtilis* rods or bacilli; singular, bacillus (×1,000). (c) Spirilla; singular, spirillum (*Spirillum volutans*; ×1,000). (d) Numerous small spirilla (*Rhodospirillum rubrum*; ×1,000). (a–d: ©Javier Izquierdo/McGraw Hill)

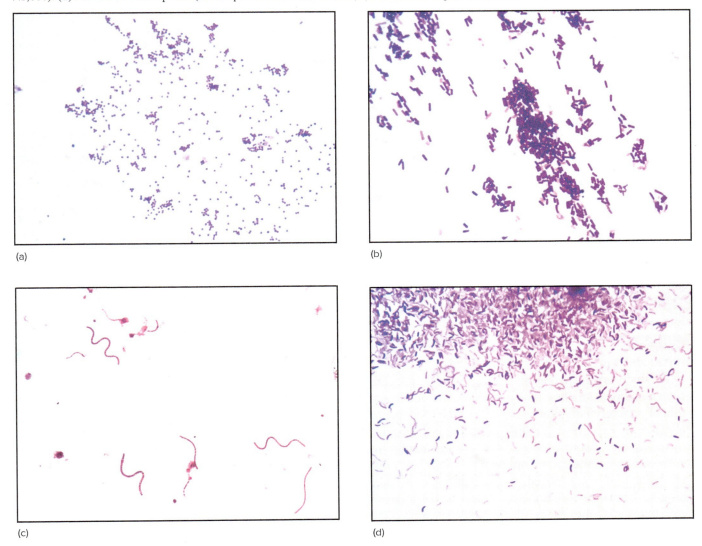

Table 1.1 Troubleshooting the Bright-Field Light Microscope

Common Problem	Possible Correction
No light passing through the ocular	■ Check to ensure that the microscope is completely plugged into a good receptacle ■ Check to ensure that the power switch to the microscope is turned on ■ Make sure the objective is locked or clicked in place ■ Make sure the iris diaphragm is open
Insufficient light passing through the ocular	■ Raise the condenser as high as possible ■ Open the iris diaphragm completely ■ Make sure the objective is locked or clicked in place
Lint, dust, eyelashes interfering with view	■ Clean ocular with lens paper and cleaner
Particles seem to move in hazy visual field	■ Air bubbles in immersion oil; add more oil or ensure that oil immersion objective is in the oil ■ Make sure that the high-dry objective is not being used with oil ■ Make sure a temporary coverslip is not being used with oil. Oil causes the coverslip to float since the coverslip sticks to the oil and not the slide, making viewing very hazy or impossible

the nosepiece until the oil immersion objective clicks into place. The lens will now be immersed in oil. Sharpen the focus with the fine adjustment knob. Draw a few of the bacteria in the spaces provided.

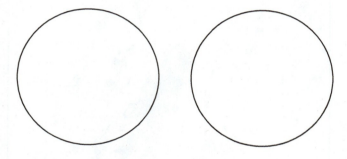

8. After you are finished with the microscope, place the low-power objective in line with the ocular, lower the tube to its lowest position, clean the oil from the oil immersion lens with lens paper and lens cleaner, and return the microscope to its proper storage place.

PRINCIPLES OF MICROSCOPIC MEASUREMENT

It frequently is necessary to accurately measure the size of the microorganism one is viewing. For example, size determinations are often indispensable in the identification of a bacterial unknown. The size of microorganisms is generally expressed in metric units and is determined by the use of a microscope equipped with an ocular micrometer. An **ocular micrometer** is a small glass disk on which uniformly spaced lines of unknown distance, ranging from 0 to 100, are etched. The ocular micrometer is inserted into the ocular of the microscope and then calibrated against a **stage micrometer,** which has uniformly spaced lines of known distance etched on it. The stage micrometer is usually divided into 0.01 mm and 0.1 mm graduations. The ocular micrometer is calibrated using the stage micrometer by aligning the images at the left edge of the scales.

The dimensions of microorganisms in dried, fixed, or stained smears tend to be reduced as much as 10 to 20% from the dimensions of the living microorganisms. Consequently, if the actual dimensions of a microorganism are required, measurements should be made in a wet-mount.

Procedure

Calibrating an Ocular Micrometer

1. If you were to observe the ocular micrometer without the stage micrometer in place, it would appear as shown in **Figure 1.4a**. In this manner, the stage micrometer would appear as illustrated in **Figure 1.4b**.
2. When in place, the two micrometers appear as shown in **Figure 1.4c**. Turn the ocular in the body tube until the lines of the ocular micrometer are parallel with those of the stage micrometer (**Figure 1.4d**). Match the lines at the left edges of the two micrometers by moving the stage micrometer.
3. Calculate the actual distance in millimeters between the lines of the ocular micrometer by observing how many spaces of the stage micrometer are included within a given number of spaces on the ocular micrometer. You will get the greatest accuracy in calibration if you use more ocular micrometer spaces to match with stage micrometer lines.

Figure 1.4 Calibrating an Ocular Micrometer.
(Light microscope: Science Photo Library/Getty Images)

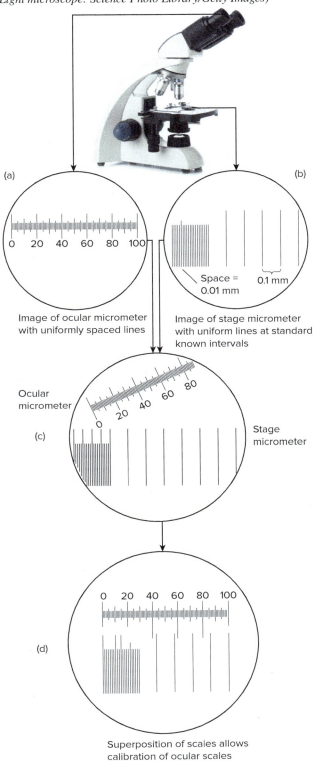

(a) Image of ocular micrometer with uniformly spaced lines

(b) Image of stage micrometer with uniform lines at standard known intervals
Space = 0.01 mm; 0.1 mm

(c) Ocular micrometer / Stage micrometer

(d) Superposition of scales allows calibration of ocular scales (10 ocular units = 0.07 mm)

Because the smallest space on the stage micrometer equals 0.01 mm or 10 μm (**Figure 1.4b**), you can calibrate the ocular micrometer using the following:

10 spaces on the ocular micrometer = Y spaces on the stage micrometer.

Since the smallest space on a stage micrometer = 0.01 mm, then

10 spaces on the ocular micrometer = Y spaces on the stage micrometer × 0.01 mm, and 1 space on the ocular micrometer = Y spaces on the stage micrometer $\dfrac{\times\, 0.01\text{ mm}}{10}$

For example, if 10 spaces on the ocular micrometer = 6 spaces on the stage micrometer, then

1 ocular space × $\dfrac{6 \times 0.01\text{ mm}}{10}$,

1 ocular space = 0.006 mm or 6.0 μm.

This numerical value holds only for the specific objective-ocular lens combination used and may vary with different microscopes.

4. Calibrate for each of the objectives on your microscope and record in the table below. Show all calculations in a separate sheet of paper and discuss with your instructor.

Low power (10×)	1 ocular space = _____ mm
High-dry power (40×)	1 ocular space = _____ mm
Oil immersion (90×)	1 ocular space = _____ mm

5. After your ocular micrometer has been calibrated, determine the dimensions of the prepared slides provided by the instructor and record your observations in the lab report.

The distance between the lines of an ocular micrometer is an arbitrary measurement that has meaning only if the ocular micrometer is calibrated for the specific objective being used. If it is necessary to insert an ocular micrometer in your eyepiece (ocular), ask your instructor whether it is to be inserted below the bottom lens or placed between the two lenses. Make sure that the etched graduations are on the upper surface of the glass disk that you are inserting. With stained preparations such as Gram-stained bacteria, the bacteria may measure smaller than they normally are if the only stained portion of the cell is the cytoplasm (Gram-negative bacteria), whereas those whose walls are stained (Gram-positive bacteria) will measure closer to their actual size.

HELPFUL HINTS

- Forcing the fine or coarse adjustment knobs on the microscope beyond their gentle stopping points can render the microscope useless.
- The lower the magnification, the less light should be directed upon the object.
- The fine adjustment knob on the microscope should be centered prior to use to allow for maximum adjustment in either direction.
- If a slide is inadvertently placed upside down on the microscope stage, you will have no difficulty focusing the object under low and high power. However, when progressing to oil immersion, you will find it impossible to bring the object into focus.
- Slides should always be placed on and removed from the stage when the low-power (4× or 10×) objective is in place. Removing a slide when the higher objectives are in position may scratch the lenses.
- A note about wearing eyeglasses. A microscope can be focused; therefore, it is capable of correcting for near- or farsightedness. Individuals who wear eyeglasses that correct for near- or farsightedness do not have to wear their glasses. The microscope cannot correct for astigmatism; thus these individuals must wear their glasses. If eyeglasses are worn, they should not touch the oculars for proper viewing. If you touch the oculars with your glasses, they may scratch either the glasses or the oculars.
- Because lens cleaner can be harmful to objectives, be sure not to use too much cleaner or leave it on too long.

Parts of the Light Microscope and Their Function

Microscope Part	Function
Abbe condenser	Collects and concentrates light upward through the object on the stage. Contains an iris diaphragm
Arm	Provides support between the tube and the base; used for carrying the microscope
Base	Holds the light source and supports the rest of the microscope
Coarse adjustment	This knob, depending on the type of microscope, allows either the stage or the nosepiece to be slowly raised or lowered to provide initial focusing
Fine adjustment	Once the object is in focus with the course adjustment knob, the fine adjustment knob allows for very precise focusing
Illuminator	Provides the source of light and is located in the base
Iris diaphragm	Is used to adjust the amount of light entering the condenser by moving a lever
Rotating nosepiece	Holds the objective lenses, which are moved and rotated by means of the knurled ring
Stage	The fixed platform that holds slides and has a center hole to allow light to pass through and enter the lenses. A mechanical stage allows easy movement by means of adjustment controls in two (horizontal and vertical) directions of the slide
Objective lenses	Usually three lenses of different powers (10×, 40×, and 90 or 100×); used to magnify the object on the stage
Ocular lenses	The eyepiece lenses through which the object on the stage is viewed. There may be one (monocular) or two (binocular); they usually provide additional magnification of 5 or 10 times.

Laboratory Report 1

Name: _____

Date: _____

Lab Section: _____

Bright-Field Light Microscopy and Microscopic Measurement of Organisms

Parts of a Compound Microscope

1. Your microscope may have all or most of the features described below and illustrated in Figure 2.3 in your textbook. By studying this figure and reading your textbook, label the compound microscope in Figure LR1.1 on the next page. Locate the indicated parts of your microscope and answer the following questions.

    a. What is the magnification stamped on the housing of the oculars on your microscope? _____

    b. What are the magnifications of each of the objectives on your microscope? _____

    c. Calculate the total magnification for each ocular/objective combination on your microscope.

Ocular	×	Objective	=	Total Magnification
_____		_____		_____
_____		_____		_____
_____		_____		_____
_____		_____		_____

    d. List the magnification and numerical aperture for each objective on your microscope.

Magnification of Objective	Numerical Aperture (NA)
_____	_____
_____	_____
_____	_____
_____	_____

    e. With some compound microscopes, loosening a lock screw allows you to rotate the body tube 180°. What is the advantage of being able to rotate the body tube? _____

    f. Note the horizontal and vertical scales on the mechanical stage. What is the function of these scales? _____

    g. Where is the diaphragm on your microscope located? _____

    How can you regulate the diaphragm? _____

    h. Locate the substage condenser on your microscope. What is its function, and how can it be regulated? _____

    i. Can the light intensity of your microscope be regulated? Explain. _____

Figure LR1.1 Modern Bright-Field Compound Microscope. *(James Redfearn/McGraw Hill)*

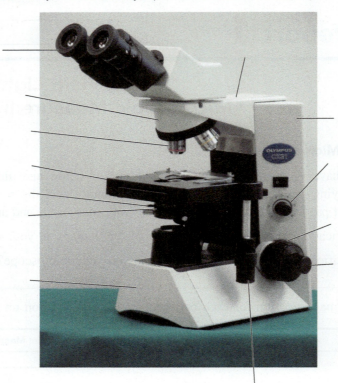

Microscopic Measurement of Microorganisms

2. After your ocular micrometer has been calibrated, determine the dimensions of the prepared slides of the following microorganisms.

Microorganism	Length	Width	Magnification
Bacterium name _____	_____	_____	_____
Fungus name _____	_____	_____	_____
Alga name _____	_____	_____	_____
Protozoan name _____	_____	_____	_____

10   Microscopic Techniques

3. Draw and label, as completely as possible, the microorganisms that you measured.

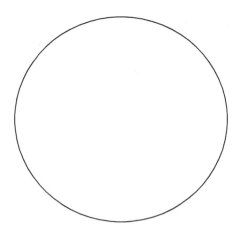

Genus and species: _____

Magnification: ×_____

Genus and species: _____

Magnification: ×_____

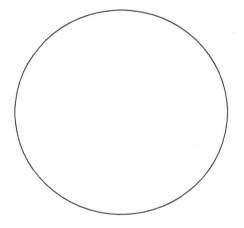

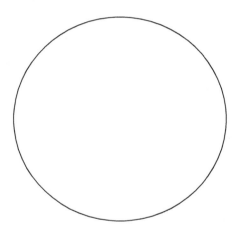

Genus and species: _____

Magnification: ×_____

Genus and species: _____

Magnification: ×_____

ASSESSMENT
Critical Thinking and Learning Outcomes Review

1. Differentiate between the linear magnification and the focal length of an objective.

2. Why is the low-power objective placed in position when the microscope is stored or carried?

3. Why is oil necessary when using the 90× to 100× objective?

4. What is the function of the iris diaphragm? The substage condenser?

5. What is meant by the limit of resolution?

6. How can you increase the bulb life of your microscope if the bulb's voltage is regulated by a rheostat?

7. In general, at what position should you keep your microscope's substage condenser lens?

8. What are the bacterial shapes you observed?

9. How can you increase the resolution on your microscope?

10. In microbiology, what is the most commonly used objective? Explain your answer.

11. In microbiology, what is the most commonly used ocular? Explain your answer.

12. If 5× instead of 10× oculars were used in your microscope with the same objectives, what magnifications would be achieved?

13. Why is it necessary to calibrate the ocular micrometer with each objective?

14. In the prepared slides, which organism was the largest?

15. When identifying microorganisms, why should a wet-mount be used when making measurements?

16. What is a stage micrometer?

17. Complete the following for the 10× objective:
    a. _____ ocular micrometer divisions = _____ stage micrometer divisions
    b. _____ ocular micrometer divisions = 1 stage micrometer division = _____ mm
    c. One ocular micrometer division = _____ stage micrometer divisions = _____ mm

18. Complete the following on units of measurement:

	Unit	Abbreviation	Value
a.	1 centimeter	_____	10^{-2} meter
b.	1 millimeter	mm	_____
c.	_____	μm	10^{-6} meter
d.	1 nanometer	_____	10^{-9} meter
e.	1 angstrom	_____	10^{-10} meter

19. In summary, what is the major purpose of this exercise?

EXERCISE 2: The Hanging Drop Slide and Bacterial Motility

SAFETY PRECAUTIONS

- Be careful with the Bunsen burner flame.
- Slides and coverslips are glass, and it is possible to cut yourself when using them.
- Dispose of any broken glass in the appropriately labeled container.
- Discard contaminated depression slides in a container with disinfectant or in a biohazard container.

MATERIALS

- suggested bacterial strains:
  24- to 48-hour tryptic soy broth cultures of *Pseudomonas aeruginosa* (small, motile bacillus), *Bacillus subtilis* (large, motile bacillus), and *Spirillum volutans* (large, motile spiral bacterium)
- microscope or phase-contrast microscope
- lens paper and lens cleaner
- immersion oil
- clean depression slides and coverslips
- petroleum jelly (Vaseline®)
- inoculating loop
- toothpicks
- Bunsen burner
- methylene blue dye
- container for biohazard waste
- safety glasses
- disposable gloves
- lab coat

LEARNING OUTCOMES

Upon completion of this exercise, students will demonstrate the ability to

1. Make a hanging drop slide in order to observe living bacteria
2. Differentiate between the three bacterial species used in this exercise on the basis of size, shape, arrangement, and motility

SUGGESTED READING IN TEXTBOOK

1. External Structures Are Used for Attachment and Motility, section 3.8; see also figures 3.37–3.40.
2. Bacteria Move in Response to Environmental Conditions, section 3.9; see also figures 3.41–3.44.

Pronunciation Guide

- *Bacillus subtilis* (bah-SIL-lus sub-TIL-us)
- *Pseudomonas aeruginosa* (soo-do-MO-nas a-ruh-jin-OH-sah)
- *Spirillum volutans* (spy-RIL-lum VOL-u-tans)

Why Are the Following Bacteria Used in This Exercise?

The major objectives of this exercise are to allow students to gain expertise in making hanging drop slides and observing the motility of living bacteria. Motility is also an important characteristic used to identify microorganisms. To accomplish these objectives, the authors have chosen three bacteria that are easy to culture and vary in size, shape, arrangement of flagella, and types of motion. Specifically, *Pseudomonas aeruginosa* is a straight or slightly curved rod (1.5 to 3.0 μm in length) that exhibits high motility by way of a polar flagellum; *Bacillus subtilis* is a large (3.0 to 5.0 μm in length), rod-shaped, and straight bacillus that moves by peritrichous flagella; and *Spirillum volutans* is a rigid, helical cell (14 to 60 μm in length) that is highly motile since it contains large bipolar tufts of flagella having a long wavelength and about one helical turn. *P. aeruginosa* is widely distributed in nature and may be a saprophytic or opportunistic animal pathogen. *B. subtilis* is found in a wide range of habitats and is a model organism for the study of Gram positive bacteria. *S. volutans* occurs in stagnant freshwater environments.

PRINCIPLES

Many bacteria display no motion and are termed **nonmotile.** However, in an aqueous environment, these same bacteria appear to be moving erratically. This erratic movement is due to **Brownian movement.** Brownian movement results from the random motion of the water molecules bombarding the bacteria and causing them to move.

True **motility** (self-propulsion) has been recognized in other bacteria and involves several different mechanisms: swimming by flagella (**flagellar motion**), **swarming** by flagella, **spirochete motility,** and **gliding motility.**

The above types of motility or nonmotility can be observed over a long period in a hanging drop slide. Hanging drop slides are also useful in observing the general shape of living bacteria, binary fission, and the arrangement of bacterial cells when they associate together (*see Figure 1.3*). A ring of petroleum jelly or Vaseline around the edge of the coverslip keeps the slide from drying out.

Procedure

1. With a toothpick, spread a small ring of petroleum jelly around the concavity of a depression slide (**Figure 2.1a**). Do not use too much jelly.
2. After thoroughly mixing one of the cultures, use the inoculating loop to aseptically place a small drop of one of the bacterial suspensions in the center of a coverslip (**Figure 2.1b**).
3. If desired, a loopful of methylene blue may be mixed with a loopful of bacteria to increase contrast in the hanging drop procedure.
4. Lower the depression slide, with the concavity facing down, onto the coverslip so that the drop protrudes into the center of the concavity of the slide (**Figure 2.1c**). Press gently to form a seal.
5. Turn the hanging drop slide over (**Figure 2.1d**) and place on the stage of the microscope so that the drop is over the light hole.
6. Examine the drop by first locating its edge under low power and focusing on the drop. Switch to the high-dry objective and then, using immersion oil, to the 90 to 100× objective. In order to see the bacteria clearly, close the diaphragm as much as possible for increased contrast. Note bacterial shape, size, arrangement, and motility as you complete the report for Exercise 2. Be careful to distinguish between motility and Brownian movement.

Disposal

Discard your coverslips and any contaminated slides in a container with disinfectant solution.

HELPFUL HINTS

- Always make sure the specimen is on the top side of the slide.
- Particular care must be taken to avoid breaking the coverslip since it is more vulnerable when supported only around its edges.
- With depression slides, the added thickness of the slide and coverslip may preclude the use of the oil immersion objective with some microscopes.
- If your microscope is equipped with an automatic stop, it may be necessary to bring the image into focus by using the coarse adjustment knob.
- In order to determine true motility, look for a directional movement that is several times the long dimension of the bacterium. Movement can also occur in any direction in the same field of view.
- Ignore Brownian movement, which is caused by the inherent thermal motion of molecules against the bacteria, causing them to appear to vibrate. True motility is always directional.
- Always examine a hanging drop slide immediately, because motility decreases rapidly with time.

Figure 2.1 Preparation of a Hanging Drop Slide.

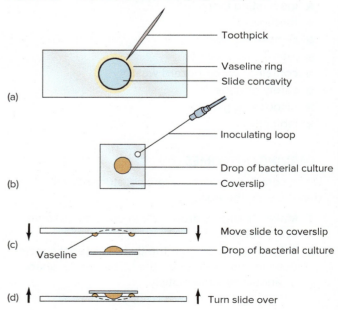

Name: _____

Date: _____

Lab Section: _____

Laboratory Report 2

The Hanging Drop Slide and Bacterial Motility

1. Examine the hanging drop slide and complete the following table with respect to the size, shape, and motility of the different bacteria.

Bacterium	Size	Shape	True Motility or Brownian Movement	Cell Arrangement
B. subtilis	_____	_____	_____	_____
P. aeruginosa	_____	_____	_____	_____
S. volutans	_____	_____	_____	_____

2. Draw a representative field for each bacterium.

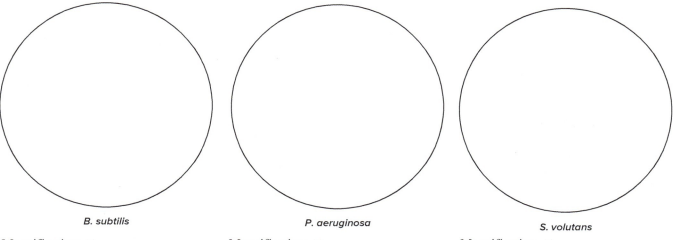

    B. subtilis   P. aeruginosa   S. volutans

Magnification: ×_____   Magnification: ×_____   Magnification: ×_____

3. In summary, what is the major purpose of this exercise?

ASSESSMENT
Critical Thinking and Learning Outcomes Review

1. Why are unstained bacteria more difficult to observe than stained bacteria?

2. What are some reasons for making a hanging drop slide?

3. Describe the following types of bacterial movement:

    a. Brownian movement

    b. flagellar motion

    c. gliding motion

4. Why do you have to reduce the amount of light with the diaphragm in order to see bacteria in a hanging drop slide?

5. Can the hanging drop slide be used to examine other microorganisms? Explain which ones.

6. Which of the bacteria exhibited true motility on the slides?

7. How does true motility differ from Brownian movement?

EXERCISE 3: Dark-Field Light Microscopy

SAFETY CONSIDERATIONS

- Gently scrape the gum line or gingival sulcus with a flat toothpick so that you obtain a small amount of surface scrapings and not lacerated gum tissue or impacted food.
- Slides and coverslips are glass. Do not cut yourself when using them.
- Dispose of any broken glass in the appropriately labeled container.
- Do not throw used toothpicks in the wastebasket. Place them in the appropriate container for disposal.

MATERIALS

- dark-field light microscope
- flat toothpicks
- lens paper and lens cleaner
- immersion oil
- slides and coverslips
- prepared slides (e.g., *Treponema pallidum, Volvox, Paramecium*)
- tweezers
- container for biohazard waste
- safety glasses
- disposable gloves
- lab coat

LEARNING OUTCOMES

Upon completion of this exercise, students will demonstrate the ability to

1. Explain the principles behind dark-field microscopy
2. Correctly use the dark-field microscope
3. Make a wet-mount and examine it for spirochetes with the dark-field microscope

SUGGESTED READING IN TEXTBOOK

1. There Are Several Types of Light Microscopes, section 2.2; see also figures 2.6 and 2.7.

Pronunciation Guide

- *Treponema denticola* (trep-o-NE-mah dent-A-cola)
- *T. pallidum* (T. pal-LA-dum)
- *Volvox* (vol-vox)
- *Paramecium* (par-A-mes-ium)

Why Is the Following Bacterium Used in This Exercise?

Treponema denticola often is a part of the normal microbiota of the oral mucosa; thus, this spirochete is readily available and does not have to be cultured. Most species stain poorly if at all with Gram's or Giemsa's methods and are best observed with dark-field or phase-contrast microscopy. Thus *T. denticola* is an excellent specimen to observe when practicing the use of a dark-field microscope, and also allows the student to continue practicing the wet-mount preparation. *T. denticola* is a slender, helical cell, 6 to 16 μm in length. In a wet-mount, the bacteria show both rotational and translational movements due to two or three periplasmic flagella inserted at each end of the protoplasmic cylinder. Young cells rotate rapidly on their axes. Thus by using *T. denticola,* you are also able to observe bacterial motility.

Medical Application

Treponema pallidum is the bacterium that causes the disease syphilis. The diagnosis of syphilis is relatively easy and rapid using specific antibody tests against the antigens of *T. pallidum*. Spirochetes can be observed by dark-field microscopy (or immunofluorescence) in fresh discharge from lesions, but only when these observations are made immediately since *T. pallidum* rarely survives transport to the laboratory.

PRINCIPLES

The compound microscope may be fitted with a dark-field condenser that has a numerical aperture (resolving power) greater than the objective. The condenser also contains a dark-field stop. The compound microscope now becomes a **dark-field microscope.** Light passing through the specimen is diffracted and enters the objective lens, whereas undiffracted light does not, resulting in a bright image against a dark background (**Figures 3.1** and **3.2**). Since light objects against a dark background are seen more clearly by the eye than the reverse, dark-field microscopy is useful in observing unstained living microorganisms, microorganisms that are difficult to stain, internal structures in larger eukaryotic microorganisms, and spirochetes (**Figure 3.2**), which are poorly defined by bright-field microscopy.

Procedure

1. Place a drop of immersion oil directly on the dark-field condenser lens.
2. Position one of the prepared slides so that the specimen is directly over the light opening.
3. Raise the dark-field condenser with the height control until the oil on the condenser lens just touches the slide.
4. Lock the 10× objective into position. Focus with the coarse and fine adjustment knobs until the spirochetes come into sharp focus. Do the same with the 40× objective.
5. Use the oil immersion objective lens to observe the spirochetes. Draw several of these cells in the space provided in the report for Exercise 3.
6. Nonpathogenic spirochetes (*T. denticola*) may be part of the normal microbiota of the oral mucosa. To make a wet-mount of these, gently scrape your gum line with a flat toothpick. Stir the scrapings into a drop of water on a slide. Gently lower a coverslip (*see Figure 1.2*) to prevent trapping air bubbles. Examine with the dark-field microscope and draw several spirochetes in the space provided in the report for Exercise 3. Alternatively, your instructor may also provide you with a culture of *T. denticola*.

Disposal

Discard all toothpicks in the beaker of disinfectant. Do the same for all slides and coverslips. Clean the prepared slides and return them to the proper slide box.

HELPFUL HINTS

- It is good practice to always clean the condenser lens before placing a drop of oil on it.
- Make sure the prepared slide is placed right side up (coverslip up) on the stage.
- If you have trouble focusing with the oil immersion lens, ask for help from your instructor.
- Always make sure that the substage condenser diaphragm is wide open for adequate illumination of the specimen.

Figure 3.1 Dark-Field Microscopy. The simplest way to convert a microscope to dark-field microscopy is to place a dark-field stop (star diaphragm, insert) underneath the condenser lens system. The condenser then produces a hollow cone of light so that only light entering the objective comes from the specimen.

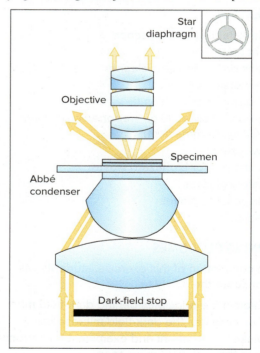

Figure 3.2 Examples of Microorganisms Seen with Dark-Field Microscopy. (a) A spirochete, *Treponema pallidum* (×500). (b) A green alga, *Volvox*. (a: Centers for Disease Control and Prevention; b: Stephen Durr)

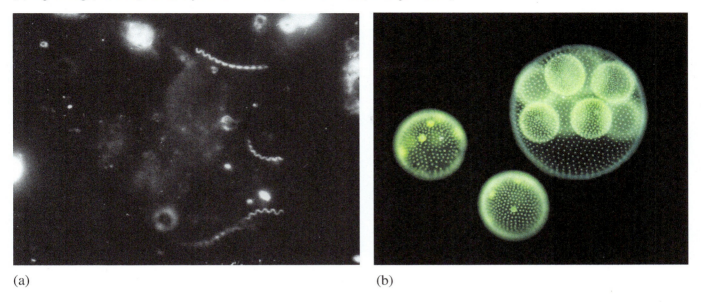

(a)   (b)

NOTES

Laboratory Report 3

Name: _____

Date: _____

Lab Section: _____

Dark-Field Light Microscopy

1. Drawing of spirochetes from a prepared slide.   Drawing of spirochetes from a wet-mount.

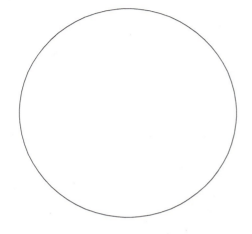
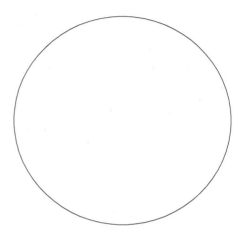

Magnification: ×_____   Magnification: ×_____

Genus and species: _____   Genus and species: _____

Shape: _____   Shape: _____

2. Label the following parts of a dark-field microscope.

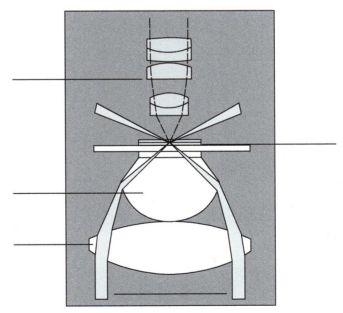

23

ASSESSMENT
Critical Thinking and Learning Outcomes Review

1. What is the principle behind dark-field microscopy?

2. When would you use the dark-field microscope?

3. Why is the field dark and the specimen bright when a dark-field microscope is used to examine a specimen?

4. Differentiate between bright-field and dark-field microscopy.

5. What is the function of the Abbé condenser in dark-field microscopy?

6. What is the function of the dark-field stop?

7. In dark-field microscopy, why is a drop of oil placed directly on the condenser lens?

EXERCISE 4 Phase-Contrast Microscopy

SAFETY CONSIDERATIONS

- Be careful with the glass slides and coverslips.
- Dispose of the slides and coverslips, the used Pasteur pipettes, and pond water properly when finished.

MATERIALS

- pond water
- phase-contrast light microscope
- microscope slides and coverslips
- Pasteur pipette
- methyl cellulose
- lens paper and lens cleaner
- prepared protist slides (at your instructor's discretion)
- container for biohazard waste
- safety glasses
- disposable gloves
- lab coat

LEARNING OUTCOMES

Upon completing this exercise, students will demonstrate the ability to

1. Explain the basic principles behind phase-contrast microscopy
2. Correctly use the phase-contrast microscope
3. Make a wet-mount of pond water and classify the microorganisms that are present

SUGGESTED READING IN TEXTBOOK

1. There Are Several Types of Light Microscopes, section 2.2; see also Figures 2.3, 2.6, 2.8, and 2.10.

Pronunciation Guide

- *Volvox* (vol-vox)
- *Paramecium* (par-A-mes-ium)

Why Are the Following Protists and Pond Water Used in This Exercise?

Most microorganisms and their organelles are colorless and often difficult to see by ordinary bright-field or dark-field microscopy. In fact, most bacterial organelles are so small that they are impossible to see by these techniques. Phase-contrast microscopy permits the observation of otherwise indistinct, living, unstained microbes and their associated subcellular structures. Thus by using prepared slides of protists such as *Paramecium* and *Volvox,* you will gain expertise in using the phase-contrast microscope.

Pond water is usually teeming with microbes. By using the phase-contrast microscope and slowing down the many microorganisms with methyl cellulose, you will be able to observe the internal structures of protists such as *Paramecium* and *Volvox.*

PRINCIPLES

The internal organelles of transparent microbes are often impossible to see by ordinary bright-field or dark-field microscopy. Microorganisms and their organelles are only visible when they absorb, reflect, refract, or diffract more light than their environment. The **phase-contrast microscope** permits the observation of otherwise invisible living, unstained microorganisms (**Figure 4.1***a–c*).

In phase-contrast microscopy, the condenser has an annular diaphragm, which produces a hollow cone of light. The objective has a glass disk (the phase plate) with a thin film of transparent material deposited on it, which accentuates phase changes produced in the specimen (**Figure 4.2**). This phase change is observed in the specimen as a difference in **light intensity.** Phase plates may either restrict (positive phase plate) the diffracted light relative to the undiffracted light, producing **dark-phase contrast (Figure 4.3),** or advance (negative phase plate) the undiffracted light relative to the directed light, producing **bright-phase contrast.**

Phase-contrast microscopy is especially useful for studying microbial motility, determining the shape of living cells, and detecting internal structures such as endospores and inclusion bodies. These are clearly visible because they have refractive indices markedly different from that of water. Phase-contrast microscopes are also widely used to study various eukaryotic cells.

Procedure

1. Make a wet-mount of pond water. Add a drop of methyl cellulose to slow the swimming of the microorganisms.
2. Place the slide on the stage of the phase-contrast microscope so that the specimen is over the light hole.
3. Rotate the 10× objective into place.
4. Rotate into position the annular diaphragm that corresponds to the 10× objective. It is absolutely

Figure 4.1 Some Examples of Microorganisms Seen with Phase-Contrast Microscopy. (**a**) The cyanobacterium *Nostoc caeruleum*. (**b**) *Paramecium caudatum*. (**c**) *Euglena*. *(a, c: Stephen Durr; b: James Redfearn/McGraw Hill)*

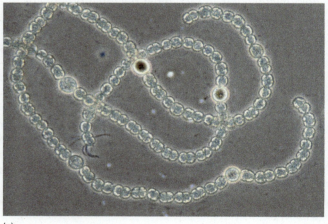

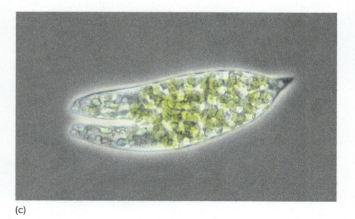

Figure 4.2 Phase-Contrast Microscopy. The optics of a dark-phase contrast microscope.

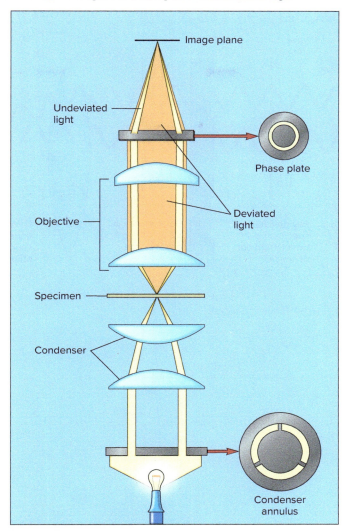

necessary that the cone of light produced by the annular diaphragm below the condenser be centered exactly with the phase plate of the objective. Consequently, there are three different annular diaphragms that match the phase plates of the three different phase objectives (10×, 40×, and 90× or 100×). The substage unit beneath the condenser contains a disk that can be rotated in order to position the correct annular diaphragm.

5. Focus with the 10× objective and observe the microorganisms.
6. Rotate the nosepiece and annular diaphragm into the proper position for observation with the 40× objective.
7. Do the same with the oil immersion lens.
8. In the report for Exercise 4, sketch several of the microorganisms that you have observed.
9. Repeat steps 2–8 for prepared slides of *Paramecium* and *Volvox*, as well as any additional slides provided to you by your instructor.

HELPFUL HINTS

- Make sure the specimen is directly over the light hole in the stage of the microscope.
- The phase elements must be properly aligned. Misalignment is the major pitfall that beginning students encounter in phase-contrast microscopy.
- If your microscope is not properly aligned, ask your instructor for help.
- When you use an oil immersion objective, add immersion oil to the top of the condenser as well as to the top of the cover glass.
- Halo effects are normal with phase microscopy.

Figure 4.3 The Production of Contrast in a Phase-Contrast Microscope. This diagram illustrates the behavior of deviated and undeviated (i.e., undiffracted) light rays (shown in yellow). Because the light rays tend to cancel each other out (far right), the image of the specimen will be dark against a light background (*see Figure 4.1a–c*).

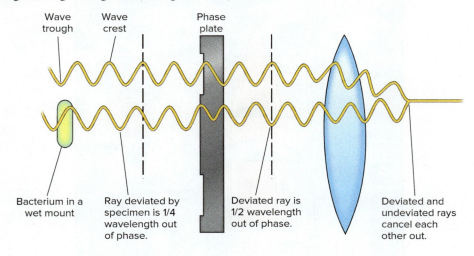

Laboratory Report 4

Name: _____

Date: _____

Lab Section: _____

Phase-Contrast Microscopy

1. Some typical microorganisms in pond water as seen with the phase-contrast light microscope.

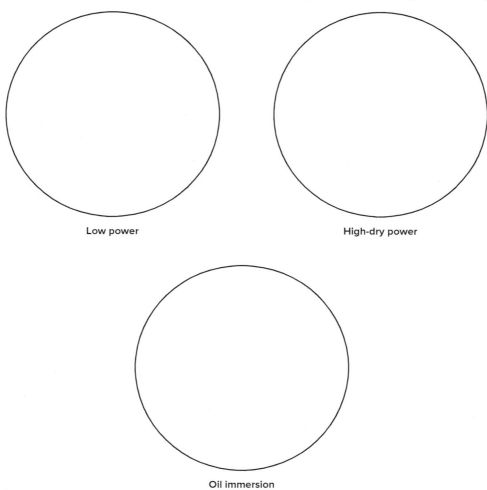

Low power

High-dry power

Oil immersion

2. Drawings of *Paramecium, Volvox,* or another organism as seen with the phase-contrast microscope.

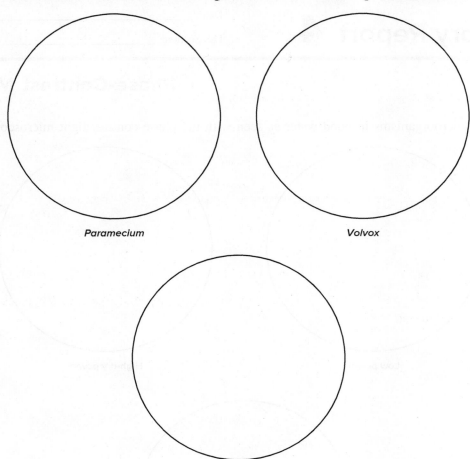

30   Microscopic Techniques

ASSESSMENT
Critical Thinking and Learning Outcomes Review

1. In the phase-contrast microscope, what does the annular diaphragm do?

2. Under what circumstances is it appropriate to use the phase-contrast microscope?

3. Explain how the phase plate works in a phase-contrast microscope that produces bright objects with respect to the background.

4. What happens to the phase of diffracted light in comparison to undiffracted light in a phase-contrast microscope?

5. Compare and contrast the optics of a phase-contrast microscope to the optics of a bright-field microscope. How are they similar? How do they differ?

6. What is the difference between a bright-phase-contrast and a dark-phase-contrast microscope?

7. Is it possible to see the nucleus of a protozoan using a phase-contrast microscope? What about the nucleoid of a bacterium? Explain your answer.

PART 2
Bacterial Cell Biology

Living microbes sometimes are studied by bright-field or phase-contrast microscopy. These techniques are useful for observing traits like motility. The first exercise in Part Two provides an opportunity for observing bacteria alive and unstained.

Since living microbes are generally colorless, staining is necessary in order to make them readily visible. This is particularly important for determining cellular morphology, as well as observing structural features that can aid in microbial identification. The exercises in Part Two of the lab manual have been designed to give students expertise in staining and slide preparation, an appreciation for differences in microbial cell morphology, and experience in staining some specialized bacterial structures such as endospores, capsules, and flagella.

After completing the exercises in Part Two, you will be able to demonstrate how to properly prepare slides for examination under the microscope. This will meet the following American Society for Microbiology Core Curriculum skills:

- cleaning and disposal of slides
- preparing smears from solid and liquid cultures
- performing wet-mounts and/or hanging drop preparations
- performing Gram stains

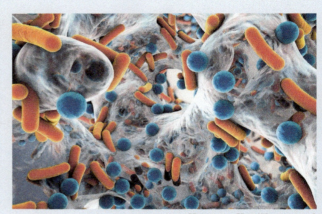

Kateryna Kon/Shutterstock

Bacteria come in a wide variety of shapes and sizes. In recent years, advances in microscopy and culture techniques have made it possible to examine the cellular morphology of an ever growing number of microbes. We now appreciate that bacteria are more than simple bags of cytoplasm. For example, they contain internal structures akin to the cytoskeleton found in eukaryotes. Understanding the molecular nature of these structures is the basis of bacterial cell biology, an exciting frontier in microbiological research.

EXERCISE 5

Aseptic Technique

SAFETY CONSIDERATIONS

- To fill a pipette, always use a bulb or other mechanical device.
- If the pipette is contaminated, immediately place it in the proper container.
- Do not carry sterile pipettes through the lab.
- Be careful of the Bunsen burner flames and red-hot inoculating loops and needles.

MATERIALS

- suggested bacterial strains (provided as broth, slant, and plate cultures):
  Serratia marcescens or *Micrococcus roseus*
- vortex mixer
- inoculating loop
- inoculating needle
- Bunsen burner
- pipette with pipettor
- tryptic soy broth tubes
- tryptic soy agar slants
- tryptic soy agar deeps
- wax pencil or lab marker
- container for biohazard waste
- safety glasses
- disposable gloves
- lab coat

LEARNING OUTCOMES

Upon completion of this exercise, students will demonstrate the ability to

1. Correctly use a pipette
2. Correctly use an inoculating loop and needle
3. Use aseptic technique to remove and transfer bacteria for subculturing
4. Explain the reasoning behind pure culture preparations
5. Describe how bacterial cultures can be maintained

SUGGESTED READING IN TEXTBOOK

1. Laboratory Culture of Cellular Microbes Requires Media and Conditions That Mimic the Normal Habitat of a Microbe, section 7.7; see also Table 7.3.

Pronunciation Guide

- *Micrococcus roseus* (mi-kro-KOK-us ros-E-us)
- *Serratia marcescens* (se-RA-she-ah mar-SES-sens)

Why Are the Following Bacteria Used in This Exercise?

Serratia marcescens is found in water, soil, and food, and is widely distributed in nature. It is easy to maintain in the laboratory, is facultatively anaerobic, is chemoorganotrophic, and has both a respiratory and a fermentative type of metabolism. It grows very quickly in tryptic soy broth or agar and often produces the red pigment prodigiosin when grown between 20°C and 37°C. Like most species of *Micrococcus*, *M. roseus* is a common inhabitant of human skin. In contrast to *S. marcescens*, it grows much slower, taking 3 days of incubation at 30°C for colonies to appear on agar plates. Later in this lab manual, we will study the microbial inhabitants of human skin in more detail.

PRINCIPLES

Before handling bacteria in clinical or research labs, all aspiring microbiologists must first become proficient in manipulating the instruments used for culture transfer. A **pipette** is an instrument often used to transfer aliquots of culture, to prepare serial dilutions of microorganisms, and to dispense chemical reagents. The **blow-out pipette** (also called a **serological pipette**) can come in glass or disposable plastic (**Figure 5.1a**). With the blow-out pipette, the final few drops of liquid must be emptied in order to deliver the correct volume.

Figure 5.1 **Transferring Liquid by Pipetting.** (**a**) A blow-out (serological) pipette. (**b**) A plastic pump. The pump is attached to the pipette and the wheel turns to move fluid either into or out of the pipette. (**c**) A pipette bulb. While pressing the *A* valve, squeeze the bulb, *B*, and it will collapse. To draw fluid into the pipette, press the *S* valve; to release fluid, press the *E* valve. (**d**) Electronically powered pipetting device for dispensing liquid. Pressing the top button will draw liquid in while pressing the bottom button will force liquid out. (**e**) An adjustable-volume micropipettor has a dial that allows a person to set the volume to be dispersed, and a disposable tip that can be disposed of after use. Depressing the plunger on the top of the micropipettor allows one to either fill or empty the fluid in the tip.

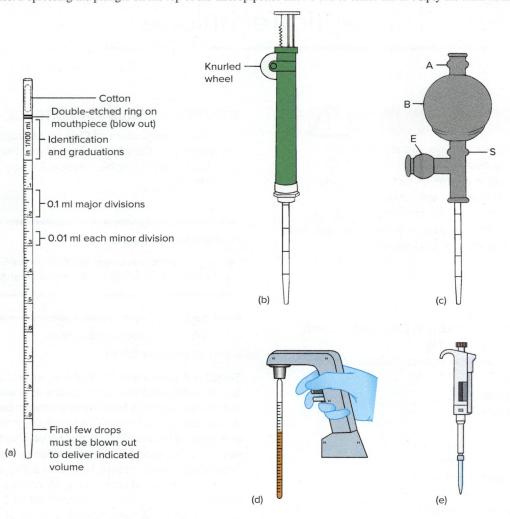

To fill a pipette, use a bulb or other mechanical device (**Figure 5.1**). Draw the desired amount of fluid into the pipette. The volume is read at the bottom of the meniscus.

Often the mouth end of a pipette is carefully plugged with a small piece of cotton before sterilization. This helps prevent cross-contamination of the bulb or mechanical device of the pipette. The disposable plastic tips used with micropipettors are equipped with filters for this purpose.

When handling sterile pipettes, be very careful to avoid contaminating them. If you think your pipette is contaminated but aren't sure, don't risk it. Dispose of it and use a fresh pipette to avoid contaminating your cultures and ruining your experiment. Sterile glass pipettes are usually stored in a pipette can tip first. The bottom of the can should contain a wad of paper or cotton to protect the pipette tips from breakage. If the top of the can sticks while it is being put on or taken off, a twisting motion will often unstick or free it. After the pipettes have been loaded into the cans, they must be autoclaved and dried. As an alternative, individually-wrapped, single-use plastic can be used instead of glass pipettes. Such pipettes are convenient and do not need to be autoclaved in advance; however, their use will generate significant plastic waste that will need to be decontaminated prior to disposal.

Figure 5.2 Microbiological Transfer Instruments. (a) Inoculating needle, and (b) inoculating loop.

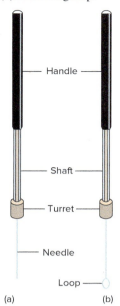

To correctly use the pipette, remove it from a pipette can or its wrapping. Do not put the pipette down before it is used because this will risk contamination. Slowly draw culture up into the pipette, and then slowly expel the culture into its intended vessel. After a pipette has been used to transfer culture, it should immediately be placed tip down in a container of a disinfectant or disposed of in a biohazard container.

In addition to pipettes, **inoculating needles** and **loops** can be used to transfer microorganisms between types of culture media. This process of transferring microorganisms without contamination is called **aseptic technique**. As shown in **Figure 5.2**, both needles and loops may consist of handles, a shaft, and a turret, which holds a nickel-chromium or platinum wire. If the wire is straight, it is an inoculating needle; if a loop is present, it is an inoculating loop. Before using either, the end of the wire must be sterilized by passing it slowly through the tip of the flame from a Bunsen burner. When done correctly, all parts of the wire will turn red with heat. The needle or loop should then be used before it becomes contaminated. After you have finished using an inoculating loop or needle, it should be thoroughly flame-sterilized.

Microorganisms are transferred from one culture medium to another by **subculturing**, using aseptic technique (**Figure 5.3**). **Asepsis** means free from sepsis, a toxic condition resulting from the presence of microorganisms. This aseptic technique is a fundamentally important skill that all microbiologists must master. As such, it will be used in most of the remaining exercises in this manual. Since microorganisms are always present in the laboratory, if aseptic technique is not followed, there is a good possibility that **external contamination** will result and will interfere with your experiment. Proper aseptic technique also protects you from **contamination** with the cultures you are handling.

Procedure for Culture Transfer Instruments and Techniques

Pipetting

1. Proper pipetting will be demonstrated in the laboratory by the instructor. After the demonstration, practice using pipettes with some distilled water and the bulbs or mechanical devices provided.

Aseptic Technique

1. Using a wax pencil or lab marker, label the tube or plate to be inoculated with the date, your name, and the name of the test microorganism (**Figure 5.3a**).
2. Gently mix the primary culture tube in order to put the bacteria into a uniform suspension (**Figure 5.3b**). The tube can be tapped to create a vortex that will suspend the microorganisms, or if a vortex mixer is available, it can be used.
3. Place the stock culture tube and the tube to be inoculated in the palm of one hand and secure with the thumb. The tubes are then separated to form a V in the hand (**Figure 5.3c**). They should be held at an angle so that the open ends are not vertical and directly exposed to airborne laboratory contaminants.
4. Using the other hand, flame the inoculating loop or needle over a Bunsen burner from the handle to the tip to avoid introducing aerosolized material into the air until the wire becomes red-hot (**Figure 5.3d**). Alternatively, a benchtop incinerator can also be used (**Figure 5.4**).
5. Using the same hand that is holding the inoculating loop, remove the caps from the two tubes, hold them between your fingers, and briefly flame the necks of the tubes over a Bunsen burner (**Figure 5.3e**) by passing them through the flame. **Do not allow the tubes to become red-hot.**

Aseptic Technique

Figure 5.3 Aseptic Technique for Bacterial Removal and Subculturing. Notice the safety gloves on the hands. If plastic culture tubes are used instead of glass, do not flame them during steps (**e**) or (**j**).

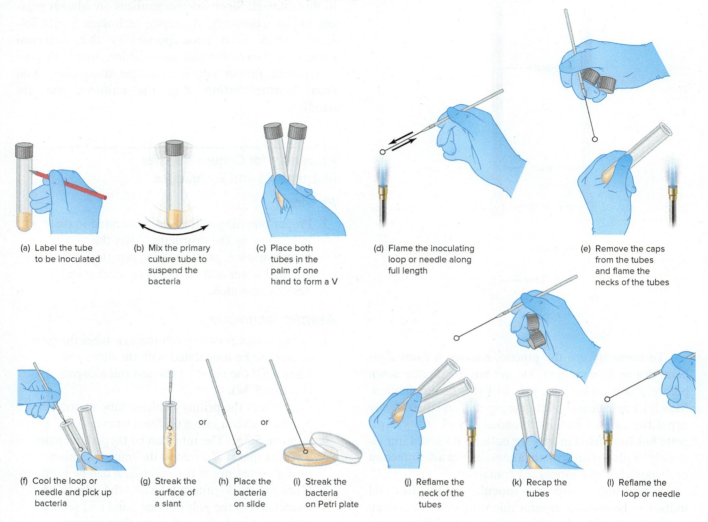

6. Cool the hot loop in the broth culture (**Figure 5.3f**). With the sterile inoculating loop, transfer 1 loopful of culture from the stock culture tube into the new broth tube. At this point, one could also streak the surface of a slant (**Figure 5.3g**), transfer the bacteria to the surface of a slide (**Figure 5.3h**), or streak the bacteria onto the surface of a Petri plate (**Figure 5.3i**). When picking up bacteria from a slant, cool the hot loop or needle by holding it against the top of the slant until it stops "hissing."
7. Reflame the neck of the tube (**Figure 5.3j**).
8. Recap the tube (**Figure 5.3k**).
9. Reflame or sterilize the loop or needle (**Figure 5.3l**).

Using aseptic technique, perform the following transfers: plate to broth, slant to deep (**Figure 5.5a**), and broth to slant (**Figure 5.5b**).

10. Incubate your tubes at 30°C for 24 to 48 hours (or one week at room temperature). Afterwards, examine all of the tubes for the presence of bacterial growth. Growth is indicated by turbidity (cloudiness) in the broth culture (**Figure 5.6**), and the appearance of flocculant material on the slant and along the line of inoculation in the agar deep. Record your results in the lab report for this exercise.

Disposal

When you are finished with all of the pipettes and tubes, discard them in the designated place for sterilization and disposal.

Figure 5.4 A Benchtop Incinerator. This oven sterilizes needles, loops, and culture tube mouths in 5 seconds at optimum sterilizing temperature of 871°C (1,600°F). (*James Redfearn/McGraw Hill*)

Figure 5.6 Some Typical Growth Patterns in Broth Media. (**a**) Growth is diffuse and present throughout the tube. (**b**) Growth is restricted to the surface of the tube, near the air–liquid interface. (**c**) All growth is collected at the bottom of the tube in a pellet of cells. (**d**) Growth is located just below the air–liquid interface, but does not extend deeper toward the bottom of the tube. (**e**) Growth of large, puffy clusters of cells occurs midway down the tube.

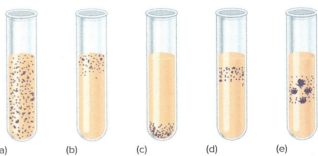

(a) (b) (c) (d) (e)

Figure 5.5 Transferring Techniques. (**a**) Stab technique for transferring bacteria. Notice that the inoculating needle is moved into the tube without touching the walls of the tube, and the needle penetrates the medium to two-thirds of its depth. (**b**) Technique for streaking the surface of a slant with a loop.

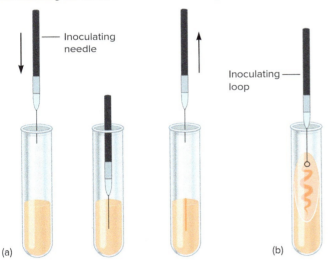

HELPFUL HINTS

- Consider the material contained within the pipette contaminated if it is drawn up in the pipette and the liquid touches the cotton.
- Always check the loop size to see that it is approximately 3 mm in diameter, because a significantly larger or smaller loop often fails to hold liquids properly during transfer.
- When pipetting, always position your eyes so that they are horizontal with the top of the fluid column in the pipette; this avoids errors that can occur from misalignment of the meniscus with the graduated line on the pipette. Hold the pipette vertical and use your forefinger to control the flow.
- Be sure to flame and cool needles between all inoculations to avoid incidental cross-contamination of cultures.

Aseptic Technique

NOTES

Laboratory Report 5

Name: _____

Date: _____

Lab Section: _____

Aseptic Technique

1. Examine the pure stock cultures for bacterial distribution and color of growth. Record your results by drawing exactly what you observed and completing the table.

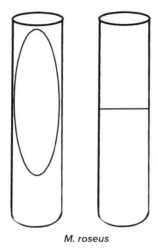
M. roseus

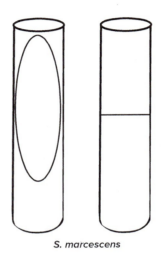
S. marcescens

Type of Culture	Description of Observed Growth
Tryptic soy agar deep	
Tryptic soy agar slant	
Tryptic soy broth	

ASSESSMENT
Critical Thinking and Learning Outcomes Review

1. Describe how to use a serological pipette.

2. What is the purpose of flaming the neck of a culture tube during aseptic technique?

3. What is the purpose of subculturing?

4. In subculturing, when would you use an inoculating loop versus an inoculating needle?

5. How is it possible to contaminate a subculture?

6. How would you determine whether culture media given to you by the laboratory instructor are sterile before you use them?

7. What are some signs of growth in a liquid medium?

8. Why is it best to use an inoculating loop instead of a needle to make the transfers from a broth culture?

Aseptic Technique   41

9. How do agar slants differ from agar deeps?

10. Imagine instead of using *M. roseus* and *S. marcescens*, you were given cultures of a strict aerobe and a strict anaerobe. After following the procedures outlined in this exercise, would the growth patterns of these microbes be the same or different? Explain your answer.

11. What are some signs of growth in a solid medium?

12. In summary, what is the major purpose of this exercise?

EXERCISE 6: Negative Staining

SAFETY CONSIDERATIONS

- Be careful with the Bunsen burner flame.
- The stains used in this experiment will not wash out of your clothing, so make sure to wear a lab coat.
- When preparing a negative stain smear, push the top slide away from the end of the slide you are holding (Figure 6.3).
- Slides should always be discarded in a container with disinfectant or in a biohazard container.

MATERIALS

- suggested microbes:
  Escherichia coli, *Staphylococcus aureus*, and *Sacchromyces cerevisiae*
- Solutions of the following stains: Nigrosin, India ink, Eosin Blue, and Congo Red
- microscope slides
- inoculating loop
- immersion oil
- microscope
- lens paper and lens cleaner
- wax pencil or lab marker
- Bunsen burner
- container for biohazard waste
- safety glasses
- disposable gloves
- lab coat

LEARNING OUTCOMES

Upon completion of this exercise, students will demonstrate the ability to

1. Explain the reason for the negative staining procedure
2. Stain different microbes using the negative staining procedure
3. Prepare slides using the thin smear technique

SUGGESTED READING IN TEXTBOOK

1. Staining Specimens Helps to Visualize and Identify Microbes, section 2.3; see also figures 2.16–2.18.

Pronunciation Guide

- *Escherichia coli* (esh-er-I-ke-a KOH-lie)
- *Staphylococcus aureus* (staf-iloh-KOK-us ORE-ee-us)
- *Saccharomyces cerevisiae* (sakah-ro-MI-seez ser-ah-VEES-ee-eye)

Why Are the Following Microbes Used in This Exercise?

Escherichia coli, Staphylococcus aureus, and *Sacchromyces cerevisiae* are three of the most widely studied microbes. Students will gain familiarity with these model organisms throughout the exercises in this lab manual.

PRINCIPLES

A major disadvantage with many staining techniques is that they can distort cellular morphology. In addition, some cells are resistant to simple staining techniques due to the structure of their cell envelope. A classic example of this is the mycobacteria, which have a thick, waxy cell envelope that resists many stains. To circumvent these issues, negative staining techniques can be utilized.

In the microbiology lab, **negative staining** is achieved by mixing a liquid suspension of cells with an acidic stain or dye, called a **chromogen,** and then spreading out the mixture on the surface of a slide to form a thin microbial film. Commonly used negative stains include Nigrosin, India ink, Eosin Blue, and Congo Red. These stains will not penetrate and stain microbial cells due to repulsion between the negative charge of the stains and the negatively charged cell surface. Instead, these stains either produce a deposit around the microbe or produce a dark background so that the microbe appears as unstained cells with a clear area around them (**Figures 6.1** and **6.2**).

Procedure

1. With a wax pencil or lab marker, label the left-hand corner of each glass slide with the name of the microbe being used.
2. For each microbe, use a sterilized inoculating loop to transfer several loopfuls of each culture to one end of the appropriately labeled slide (**Figure 6.3a**).
3. Add 1 to 2 loopfuls of Nigrosin, India ink, Congo Red, or Eosin Blue solution to the culture (**Figure 6.3b**) and mix thoroughly. Be careful not to add too much dye. The drop should be about 0.5 cm in diameter.
4. Spread the mixture over the slide using a second slide. The second slide should be held at a 45° angle so that the microbe/stain solution spreads across its edge (**Figure 6.3c**). The resulting microbe-stain smear will be thinner farther away from the initial spot of dye. (**Figure 6.3d**). This is known as a **thin smear.**
5. Allow the smear to air dry completely (**Figure 6.3e**). Do not heat-fix these slides!
6. With the low-power objective, find an area of the smear that is of the optimal thickness for observation. Start with the most transparent part of the slide to find your bearings and work your way to an area that is slightly opaque.
7. Use the oil immersion lens to observe the smear, and draw each microbe in the lab report for Exercise 6.
8. Clean the oil immersion lens carefully and dispose of your slides as directed by your instructor.

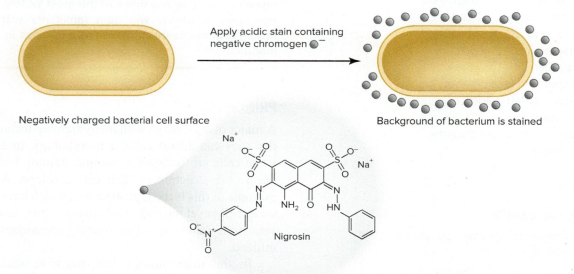

Figure 6.1 **Chemistry of Acidic Dyes.** (a) Acidic dyes have a negatively charged chromogen that is repelled by the negatively charged bacterial cell surface and thus the background is colorized while the bacterium remains transparent. (b) Structures of the commonly used acidic dye Nigrosin used for negative staining.

Figure 6.2 Negative Staining. (**a**) India Ink Stain of the bacterium *Bacillus anthracis*. (**b**) India ink stain of the fungus *Cryptococcus neoformans*. Notice the dark background around the clear yeast cells. *(a: Lansing Prescott; b: Source: Dr. Leanor Haley/CDC)*

(a)

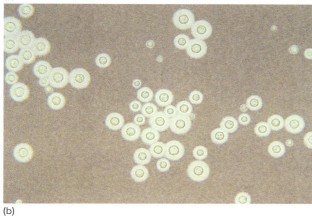
(b)

Figure 6.3 Negative Staining Procedure and Thin Smear Preparation. By pushing the mixture of bacterial culture and stain across the length of the slide, a thin film will form. After allowing time to air dry, the slide can then be visualized under the microscope.

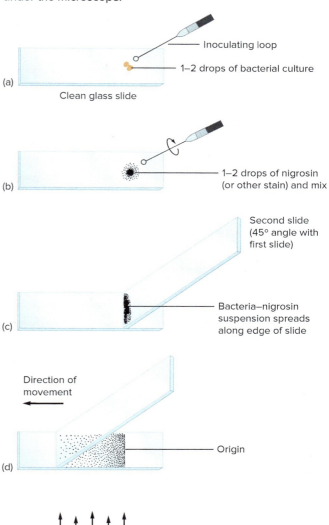

HELPFUL HINTS

- For a successful thin smear, the slides must be absolutely clean and free from oil and grease—including fingerprints.
- If an inconsistent smear is obtained, it is better to prepare a new slide than to search unsuccessfully for an appropriate area on a poorly stained slide.
- Do not use too much stain; only a small drop of stain is necessary.
- Prepare a smear that consists of a thin layer of cells without clumps.
- View the thinner or clearer portions of the film first before moving to thicker areas.
- In the negative staining procedure, heat fixing is not done; therefore, keep in mind that the microbes are not necessarily killed and the slides should be disposed of in a biohazard container.

Negative Staining

NOTES

Laboratory Report 6

Name: _____
Date: _____
Lab Section: _____

Negative Staining

1. Draw a representative field of your microscopic observation as seen with the oil immersion lens.

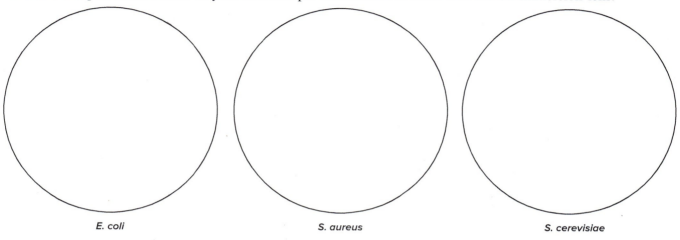

E. coli   S. aureus   S. cerevisiae

Magnification: ×_____

2. Describe the microscopic appearance of *E. coli, S. aureus,* and *S. cerevisiae.*

    a. *E. coli*

    b. *S. aureus*

    c. *S. cerevisiae*

3. In summary, what is the major purpose of this exercise?

ASSESSMENT
Critical Thinking and Learning Outcomes Review

1. In what situation is negative staining used?

2. Name three stains that can be used for negative staining.
    a.
    b.
    c.

3. Why do the bacteria remain unstained in the negative staining procedure?

4. What is an advantage of negative staining?

5. Why didn't you heat-fix the bacterial suspension before staining?

6. Why is negative staining also called either indirect or background staining?

7. When streaking with the second slide, why must it be held at a 45° angle?

EXERCISE 7 Simple Staining

SAFETY CONSIDERATIONS

- Always use a slide holder or clothespin to hold glass slides when heat-fixing them.
- Never touch a hot slide until it cools.
- If a glass slide is held in the flame too long, it can shatter.
- If the stains used in this experiment get on your clothing, they will not wash out.

MATERIALS

- suggested bacterial strains:
  tryptic soy broth and agar slants of *Escherichia coli* and *Staphylococcus aureus*
- microscope
- microscope slides
- paper towels
- inoculating loop and needle
- sterile distilled water
- Bunsen burner
- methylene blue
- crystal violet
- carbolfuchsin
- wax pencil or lab marker
- immersion oil
- lens paper and lens cleaner
- slide holder or clothespin
- container for biohazard waste
- safety glasses
- disposable gloves
- lab coat

LEARNING OUTCOMES

Upon completion of this exercise, students will demonstrate the ability to

1. Learn the proper procedure for preparing a bacterial smear
2. Do several simple staining procedures

SUGGESTED READING IN TEXTBOOK

1. Staining Specimens Helps to Visualize and Identify Microbes, section 2.3; see also figure 2.16.
2. Bacteria Are Diverse but Share Some Common Features, section 3.2; see also figures 3.1 and 3.2.

Pronunciation Guide

- *Escherichia coli* (esh-er-I-ke-a KOH-lie)
- *Staphylococcus aureus* (staf-iloh-KOK-us ORE-ee-us)

Why Are the Following Bacteria Used in This Exercise?

The major objective and outcome of this exercise is to allow students to gain expertise in smear preparation and simple staining. To accomplish this outcome, you will use two of the mostly studied bacteria: *Escherichia coli* and *Staphylococcus aureus*. *E. coli* is a Gram-negative rod (0.5 × 2 μm) that is characterized as a facultative anaerobe. Due to the ease in which it can be manipulated, *E. coli* has served as one of the primary model organisms in microbiology labs. Another commonly used model bacterium is *S. aureus*, a Gram-positive coccus (~1 μm in diameter). Staphylococci get their name based on their appearance in grape-like clusters when viewed under the microscope. Both *E. coli* and *S. aureus* will be used frequently in the experiments throughout this lab manual. In the wild, both *E. coli* and *S. aureus* can cause a variety of infections, including food-borne gastrointestinal diseases.

PRINCIPLES

While negative staining is satisfactory when making simple observations on bacterial morphology and size, more specific stains are necessary if bacterial detail is to be observed **(Table 7.1)**. One way of achieving this detail involves smear preparation and simple staining. A **bacterial smear** is a dried preparation of bacterial

Table 7.1	Comparison of Positive and Negative Stains	
	Positive Staining	**Negative Staining**
Appearance of cell	Colored by dye	Clear and colorless
Background	Not stained (generally white)	Stained (dark gray or black)

Dyes employed	Basic dyes:
	  Crystal violet
	  Methylene blue
	  Safranin
	  Malachite green
		Acidic dyes:
		  Nigrosin
		  India ink

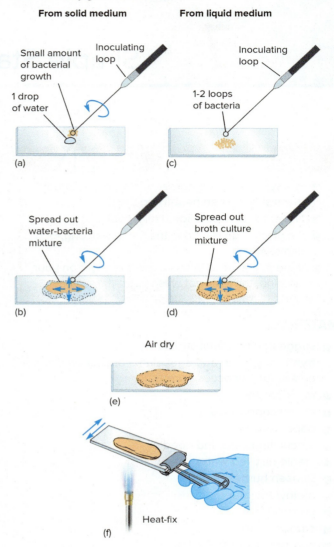

Figure 7.1 Bacterial Smear Preparation. Smears can be prepared from bacteria grown on agar plates or in liquid culture. Note the safety glove on the hand.

cells on a glass slide. In a bacterial smear that has been properly processed, (1) the bacteria are evenly spread out on the slide in such a concentration that they are adequately separated from one another, (2) the bacteria are not washed off the slide during staining, and (3) bacterial form is not distorted.

In making a smear, bacteria from either a broth culture or an agar slant or plate may be used. If a slant or plate is used, a *small* amount of bacterial growth is transferred to a drop of sterile distilled water on a glass slide (**Figure 7.1a**) and mixed. The mixture is then spread out evenly over a large area on the slide (**Figure 7.1b**).

One of the most common errors in smear preparation from agar cultures is the use of too large an inoculum. This invariably results in the occurrence of large aggregates of bacteria piled on top of one another.

If the medium is liquid, place one or two loops of the medium directly on the slide (**Figure 7.1c**) and spread the bacteria over a large area (**Figure 7.1d**). Allow the slide to air dry at room temperature (**Figure 7.1e**).

After the smear is dry, the next step is to attach the bacteria to the slide by **heat-fixing.** This is accomplished by gentle heating (**Figure 7.1f**), passing the slide a couple of times through the hot portion of the flame of a Bunsen burner. Most bacteria can be **fixed** to the slide and killed in this way without serious distortion of cell structure.

The use of a single stain or dye (called a **chromogen**) to create contrast between the bacteria and the background is referred to as **simple staining.** Its chief value lies in its simplicity and ease of use. Simple staining is often employed when information about cell shape, size, and arrangement of bacterial and archaeal cells is desired. In this procedure, one places the heat-fixed slide on a staining rack, covers the smear with a small amount of the desired stain for the proper amount of time, washes the stain off with water for a few seconds, and, finally, blots it dry. Basic dyes such as **crystal violet** (20 to 30 seconds staining time), **carbolfuchsin** (5 to 10 seconds staining time), or **methylene blue** (1 minute staining time) are often used. These basic dyes bind to bacterial cells by ionic interactions. They have positively charged chemical groups and bind to negatively charged molecules like nucleic acids, many proteins, and the surfaces of bacterial cells (**Figure 7.2**). Once bacteria have been properly stained, it is usually an easy matter to discern their overall shape. For future reference, the most common shapes are presented in **Figure 7.3.**

Procedure

Smear Preparation

1. Using a wax pencil or lab marker, label each of four slides with the name of the cultures of *E. coli* and *S. aureus*.
2. For both broth cultures, shake the culture tube and, with an inoculating loop, aseptically transfer

Figure 7.2 Chemistry of Basic Dyes. (**a**) Basic dyes have a positively charged chromogen that forms an ionic bond with the negatively charged bacterial cell and thus colorizes the bacterium. (**b**) Structure of representative basic dyes.

Negatively charged bacterial cell
Apply basic stain containing positive chromogen +
Bacterium is stained
Cl⁻
Crystal Violet

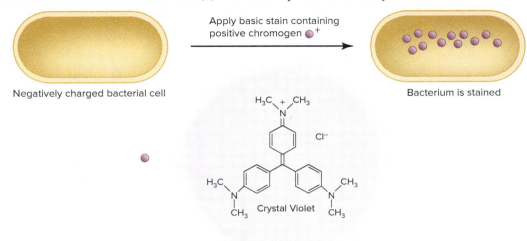

Figure 7.3 Common Bacterial Shapes. Note that the same cellular shape can be arranged differently. Examining cellular morphology can be useful for identifying an unknown bacterial isolate.

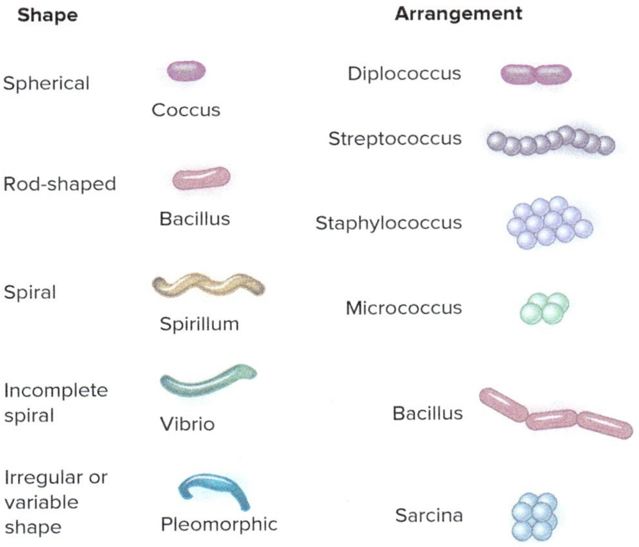

Shape
Arrangement
Spherical
Coccus
Diplococcus
Streptococcus
Rod-shaped
Bacillus
Staphylococcus
Spiral
Spirillum
Micrococcus
Incomplete spiral
Vibrio
Bacillus
Irregular or variable shape
Pleomorphic
Sarcina

1 to 2 loopfuls of bacteria to the center of the slide. Spread this out to about a ½-inch area. When preparing a smear from a slant or plate, place a loopful of sterile distilled water in the center of the slide. With the inoculating loop, aseptically pick up a *very small* amount of culture and mix into the drop of water. Spread this out as above. Four slides should be prepared: one for each broth culture and one for each plate/slant culture.

3. Allow the slides to air dry for at least 10 minutes.
4. While holding the microscope slide with a clothespin or slide holder, pass the slide through a Bunsen burner flame three times to heat-fix and kill the bacteria.
5. The smear is now heat-fixed.

Simple Staining

1. Place the fixed smears on a staining loop or rack over a sink or other suitable waste receptacle (**Figure 7.4*a***).
2. Wearing gloves, stain your slides with methylene blue for 1 minute; carbolfuchsin for 10 seconds; or crystal violet for 20 to 30 seconds.
3. Wash stain off slide with water for a few seconds (**Figure 7.4*b***).
4. Blot slides dry with paper towels (**Figure 7.4*c***). Be careful not to rub the smear when drying the slide because this will remove the stained bacteria.
5. Examine each slide under the oil immersion lens and complete the report for this exercise.
6. Coordinate your efforts with your labmates so that you have a set of slides with each of the three stains. If time allows, you can cover bacterial smears for varying lengths of time with a given stain in order to get a feel for how reactive they are. This will allow you to assess the impact of overstaining or understaining a slide preparation. See **Figure 7.5*a–c*** for examples of bacteria stained with crystal violet.

Disposal

When you are finished with all of the glass slides, discard them as directed by your instructor.

Figure 7.4 Simple Staining Procedure. Note the safety gloves on the hands and use of a slide holder.

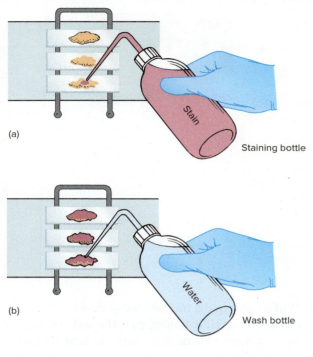

Figure 7.5 Simple Staining. Bacteria stained with simple stains. Shown here are examples of typical (**a**) bacilli, (**b**) cocci, and (**c**) spirals. *(a: Source: Centers for Disease Control and Prevention; b: Kallayanee Naloka/Shutterstock; c: Ed Reschke/Photolibrary/Stone/Getty Images)*

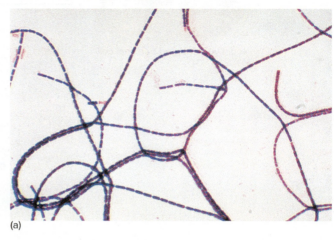

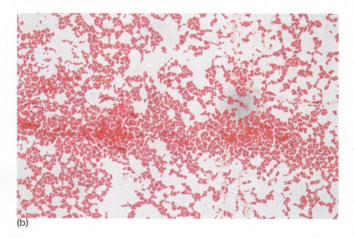

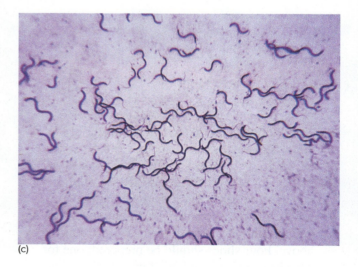

HELPFUL HINTS

- When heat-fixing a smear, always make sure that the smear is on the top of the slide as you pass it through the flame.
- Bacteria growing on solid media tend to cling to each other and must be dispersed sufficiently by diluting with water. If this is not done, the smear will be too thick and uneven.
- Be careful not to use too much cell paste in making the smear. It is easy to ruin your results by using too many bacteria.
- Always wait until the slide is dry before heat-fixing.
- Fixing smears with an open flame may create artifacts.
- The inoculating loop must be relatively cool before inserting it into any broth. If the loop is too hot, it will spatter the broth and suspend bacteria into the air.
- Always flame the inoculating loop after using it and before setting it down.
- When rinsing with water, direct the stream of water so that it runs gently over the smear.

Laboratory Report 7

Name: _____

Date: _____

Lab Section: _____

Simple Staining

1. For each slide, draw a representative field to document the cellular morphology of *E. coli* and *S. aureus*. Then, complete the table for the simple staining procedure.

	E. coli	S. aureus
Bacterium		
Magnification		
Stain		
Cell shape/grouping		
Cell color		
Background color		
Culture source		

2. In summary, what was the major purpose of this exercise?

NOTES

ASSESSMENT
Critical Thinking and Learning Outcomes Review

1. What are the two purposes of heat fixation?
    a.

    b.

2. What is the purpose of simple staining?

3. What is a chromogen?

4. Why are basic dyes more successful in staining bacteria than acidic dyes?

Simple Staining

5. Name three basic stains not used in this exercise.

   a.

   b.

   c.

6. Why is time an important factor in simple staining?

7. How would you define a properly prepared bacterial smear?

8. Why should you avoid using too much liquid culture when preparing a smear?

EXERCISE 8 Gram Stain

SAFETY CONSIDERATIONS

- Be careful with the Bunsen burner flame.
- Ethanol is flammable so keep it away from the flame.
- If the stains used in this experiment get on your clothing, they will not wash out.
- Discard slides in a biohazard container.
- Hold all slides with forceps or a clothespin when heat-fixing.
- Crystal violet, safranin, and iodine can cause irritation to the eyes, respiratory system, and skin.
- Wear suitable protective gloves, lab coat, and safety glasses.

MATERIALS

- suggested bacterial strains:
  broth and plate cultures of *Staphylococcus aureus*, *Escherichia coli*, and a mixture of *S. aureus* and *E. coli*.
- solutions of Crystal Violet, Gram's iodine, 95% ethanol, and safranin
- Bismark brown stain (for color-blind students)
- clean glass slides
- inoculating loop
- Bunsen burner
- paper towels
- microscope
- lens paper and lens cleaner
- immersion oil
- slide warmer
- staining rack
- container for biohazard waste
- safety glasses
- disposable gloves
- lab coat

LEARNING OUTCOMES

Upon completion of this exercise, students will demonstrate the ability to

1. Explain the biochemistry underlying the Gram stain
2. Summarize the theoretical basis for differential staining procedures
3. Perform a satisfactory Gram stain
4. Differentiate a mixture of bacteria into Gram-positive and Gram-negative cells

SUGGESTED READING IN TEXTBOOK

1. Staining Specimens Helps to Visualize and Identify Microbes, section 2.3; see also figures 2.17 and 2.18.
2. The Cell Envelope Often Includes Layers Outside the Cell Wall, section 3.6; see also figures 3.21 and 3.23.

Pronunciation Guide

- *Escherichia coli* (esh-er-I-ke-a KOH-lie)
- *Staphylococcus aureus* (staf-il-oh-KOK-us ORE-ee-us)

Why Are the Following Bacteria Used in This Exercise?

The major objective of this exercise is to enable you to correctly use the Gram stain to differentiate a mixture of bacteria into Gram-positive and Gram-negative cells. The classical standards for this differentiation are *Staphylococcus aureus* and *Escherichia coli*. *S. aureus* cells are cocci of about 1 μm in diameter, and are grouped in clusters that resemble bunches of grapes. This bacterium is Gram-positive, nonmotile, and does not form spores. Species of *Staphylococcus*, including *S. aureus*, are common inhabitants of the skin of warm-blooded vertebrates. *S. aureus* can also cause skin and wound infections. Increasingly, these infections are caused by strains of *S. aureus* that are resistant to antibiotics (e.g., MRSA and VRSA). In contrast, *E. coli* cells are short bacilli, 2 to 6 μm in length, occurring singly or in short chains. This bacterium is Gram-negative and serves as the workhorse for much of modern molecular biology research. *E. coli* occurs as part of the normal flora in the intestinal tract of warm-blooded animals. As with *S. aureus*, there are many strains of *E. coli* that can cause infections, particularly in the gastrointestinal and urinary tracts.

Medical Application

Gram staining is one of the simplest and most useful tests in the clinical microbiology laboratory. It is the differential staining procedure most commonly used for the direct examination of specimens and bacterial colonies because it has a broad staining spectrum. The staining spectrum includes almost all bacteria. The significant exceptions include *Treponema*, *Mycoplasma*, *Chlamydia*, and *Rickettsia*, which are too small to visualize by light microscopy or lack a cell wall. Another exception are members of the genus *Mycobacterium*, which take up the Gram stain reagents poorly due to their thick, waxy cell envelope.

PRINCIPLES

Simple staining depends on the fact that bacteria differ chemically from their surroundings and thus can be stained to contrast with their environment. Due to differences in their subcellular structures, bacteria also differ from one another chemically and physically. In the laboratory, this means that different bacteria can potentially react differently to a given staining procedure. This is the principle of **differential staining.** Differential staining can distinguish between types of bacteria based on their underlying chemical and physical properties.

The **Gram stain (Figure 8.1)** is named after Christian Gram, a Danish scientist and physician. This technique is among the most useful and widely employed differential stain in bacteriology. It divides most bacteria (but not archaea) into two groups—**Gram-positive** and **Gram-negative.** These two groups are defined by structural differences in their cell envelopes. The cell envelope of a bacterium includes everything beyond the cytoplasm. In Gram-positive bacteria (**Figure 8.2a**), this includes the plasma (or cytoplasmic) membrane and a thick cell wall, usually composed of a polymer called peptidoglycan. In Gram-negative bacteria (**Figure 8.2b**), the cell envelope is composed of the inner (plasma) membrane, the outer membrane, and a cell wall that is much thinner than its counterpart in Gram-positives.

Figure 8.1 Gram Stain. Light micrograph of a typical Gram stain. (**a**) Gram-positive cocci in clusters. (**b**) Gram-negative bacilli in short chains. (*Toeytoey2530/iStock/Getty Images*)

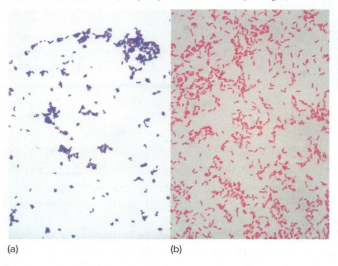

(a)   (b)

The plasma membranes of both Gram-positive and Gram-negative bacteria are composed of phospholipid bilayers. Notably, the outer membrane is an asymmetric bilayer made of phospholipids on the inner leaflet and lipopolysaccharide (LPS) on the outer leaflet.

The basis of the Gram stain is that Gram-positive bacteria retain the color of the primary dye (crystal violet) whereas the Gram-negative bacteria lose the primary dye when washed in a decolorizing solution. Gram-negatives then take on the color of the second dye (the counterstain safranin).

The first step in the procedure involves staining with the dye crystal violet. This is the **primary stain.** It is followed by treatment with an iodine solution, which functions as a **mordant;** that is, it increases the interaction between the bacterial cell and the dye by causing the dye to form large complexes in the peptidoglycan meshwork of the bacterial wall. Because the peptidoglycan layer in Gram-positive bacteria is thicker, these complexes become trapped in the Gram-positive cell wall but can be easily washed out of a Gram-negative cell wall. The smear is then decolorized by washing with an agent such as 95% ethanol. Thus, Gram-positive bacteria retain the crystal violet-iodine complex when washed with the decolorizer, whereas Gram-negative bacteria lose their crystal violet-iodine complex and become colorless. Finally, the smear is **counterstained** with a basic dye, different in color from crystal violet. This counterstain is usually safranin. The safranin will stain the colorless, Gram-negative bacteria pink but importantly does not alter the dark purple color of the Gram-positive bacteria.

Figure 8.2 Schematic of Bacterial Cell Envelopes. Cartoon representation of a typical Gram-positive (**a**) and Gram-negative (**b**) bacterial cell envelope.

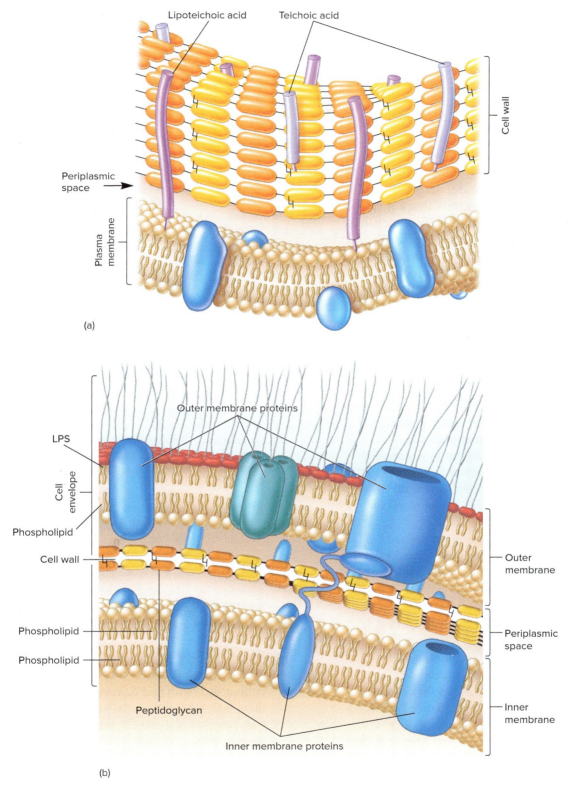

Gram Stain 59

The end result is that Gram-positive bacteria are deep purple in color and Gram-negative bacteria are pinkish to red in color (**Figure 8.1**).

The Gram stain does not always yield clear results. Species will differ from one another in regard to the ease with which the crystal violet-iodine complex is removed by ethanol. Gram-positive cultures may often appear Gram-negative if they get too old. Thus it is always best to Gram stain young cultures rather than older ones. Furthermore, some bacterial species are **Gram variable.** That is, some cells in the same culture will appear Gram-positive and some, Gram-negative. Therefore one should always be certain to run Gram stains on several cultures under carefully controlled conditions in order to make certain that a given bacterial strain is truly Gram-positive or Gram-negative.

Indistinct Gram-stain results can be confirmed by a simple test using KOH. Place a drop of 10% KOH on a clean glass slide and mix with a loopful of bacterial paste. Wait 30 seconds, then pull the loop slowly through the suspension and up and away from the slide. A Gram-negative organism will produce a gooey string; a Gram-positive organism remains fluid.

Procedure for Traditional Gram-Stain Technique

1. Prepare heat-fixed smears of *E. coli*, *S. aureus*, and the mixture of *E. coli* and *S. aureus*.
2. Place the slides on the staining rack.
3. Cover the smears with crystal violet and let stand for 30 seconds (**Figure 8.3a**).
4. Rinse with water for 5 seconds (**Figure 8.3b**).
5. Cover with Gram's iodine and let stand for 1 minute (**Figure 8.3c**).
6. Rinse with water for 5 seconds (**Figure 8.3d**).
7. Decolorize with 95% ethanol for 15 to 30 seconds. Do not decolorize too long. Add the decolorizer drop by drop until the crystal violet fails to wash from the slide (**Figure 8.3e**).
8. Rinse with water for 5 seconds (**Figure 8.3f**).
9. Counterstain with safranin for about 1 minute (**Figure 8.3g**). Safranin preparations vary considerably in strength, and different staining times may be required for each batch of stain. (If you are color-blind, use Bismark brown stain rather than safranin.)
10. Rinse with water for 5 seconds (**Figure 8.3h**).
11. Blot dry with paper towels (**Figure 8.3i**) and examine under oil immersion. Gram-positive organisms stain blue to purple; Gram-negative organisms stain pink to red (**Figure 8.4**). There is no need to place a coverslip on the stained smear.

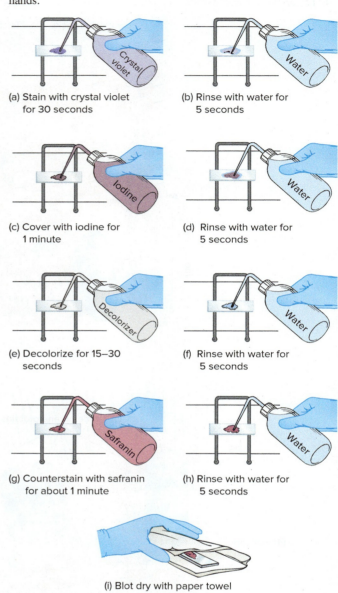

Figure 8.3 Gram-Stain Procedure. Note the gloves on the hands.

(a) Stain with crystal violet for 30 seconds
(b) Rinse with water for 5 seconds
(c) Cover with iodine for 1 minute
(d) Rinse with water for 5 seconds
(e) Decolorize for 15–30 seconds
(f) Rinse with water for 5 seconds
(g) Counterstain with safranin for about 1 minute
(h) Rinse with water for 5 seconds
(i) Blot dry with paper towel

See Figure 8.1 for an example of Gram-positive and Gram-negative bacteria.

Control Procedure

1. Prepare two heat-fixed slides of the mixed culture of *E. coli* and *S. aureus*.
2. Stain one slide with crystal violet only (steps 3 to 6).
3. Carry the second slide through the decolorizing process (steps 3 to 8).
4. Examine these two slides and compare with the mixed culture slide that was carried all the way through the staining procedure (steps 1 to 10). Your observations should help you understand how the Gram stain works.

Figure 8.4 Steps in the Gram-Stain Procedure and State of Bacteria. Notice the color changes that occur at each step in the Gram-staining process and the chemical reactions that occur in the cell wall.

	Steps in Staining	State of Bacteria	Chemical Reaction in Cell Wall	
	Step 1: Crystal violet (primary stain)	Cells stain purple.	Gram-positive	Gram-negative
			Both cell walls affix the dye	
	Step 2: Gram's iodine (mordant)	Cells remain purple.	Dye complex trapped in cell wall	No effect of iodine
	Step 3: Ethanol (decolorizer)	Gram-positive cells remain purple; Gram-negative cells become colorless.	Dye complex remains trapped in cell wall	Outer membrane weakened; wall loses dye
	Step 4: Safranin (counterstain)	Gram-positive cells remain purple; Gram-negative cells appear pink.	Pink dye masked by violet	Pink dye stains the colorless cell

Whenever you are preparing slides from an unknown culture, it is good practice to include known cultures or controls. It is very important that controls be included in each staining run, preferably on the same slide using *Staphylococcus aureus* and *Escherichia coli*. You can accomplish this by drawing two small circles on a slide using a lab marker. Make a smear inside one circle with your unknown isolate, and a smear of the control organism inside the other circle. When performing the Gram stain on a clinical specimen, particularly when the results will be used as a guide to identify the cause of an infection and design a course of treatment, such controls ensure that the iodine solution is providing proper mordant activity and that decolorization was performed properly.

Disposal

When you are finished with all of the glass slides, discard them in the designated place for sterilization and disposal.

> **HELPFUL HINTS**
>
> - Don't make your smears too thick; thick smears will require more time to decolorize than thin smears.
> - Decolorization has occurred when the solution flows colorlessly from the slide. If you cannot tell accurately when the solution becomes colorless, try decolorizing for about 30 seconds.
> - Some common sources of Gram-staining errors are the following:
>   - ☐ The inoculating loop was too hot.
>   - ☐ Excessive heat was used during the heat-fixing procedure.
>   - ☐ Decolorizing alcohol was left on the slide for too long.
> - Do not blot the slide vigorously with paper towels or you may rub off the Gram stain.
> - If the Gram's iodine is yellow or pale in color, it has lost its potency and should be discarded.
> - If the cultures used to prepare the smear are older than 24 hours, some Gram-positive bacteria will appear Gram-negative.
> - If too much decolorizing agent is used or applied for more than 30 seconds, a lot of the crystal violet-iodine complex will be removed from the Gram-positive cell walls.
> - Make sure the iris diaphragm is fully open on the microscope.

NOTES

Laboratory Report 8

Name: _____

Date: _____

Lab Section: _____

Gram Stain

1. Draw the Gram-stained bacteria (from broth cultures) in the following circles.

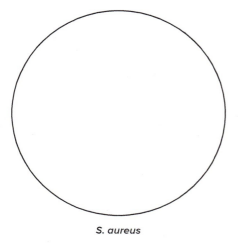 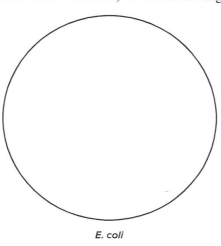 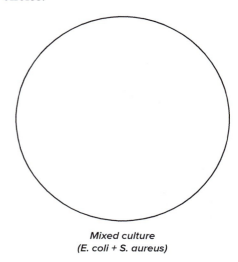

      S. aureus   E. coli   Mixed culture (E. coli + S. aureus)

2. Draw the control Gram-stain results in the following circles.

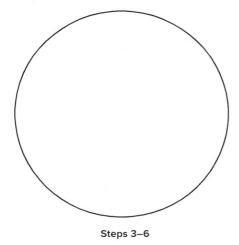 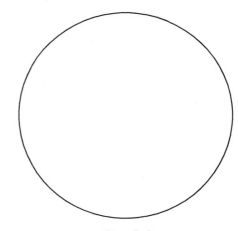

        Steps 3–6   Steps 3–8

Bacterial color _____   _____

3. Draw the Gram-stained bacteria (from plates) in the following circles.

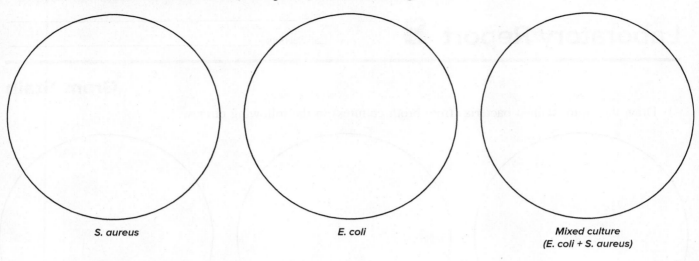

S. aureus

E. coli

Mixed culture
(E. coli + S. aureus)

ASSESSMENT
Critical Thinking and Learning Outcomes Review

1. What is the difference between a simple and a differential stain?

2. Name the reagent used and state the purpose of each of the following in the Gram stain:
    a. primary stain

    b. counterstain

    c. decolorizer

    d. mordant

3. What common mistakes can lead to poor results in the Gram stain? Explain your answer.

4. Would you expect a week-old culture of *S. aureus* to show the same Gram reaction as an overnight culture? Why or why not?

5. How is it possible that *S. aureus* can be a normal part of your skin microbiota, yet in some cases it can cause severe infections?

6. What is meant by "Gram variable"?

7. What parts of the bacterial cell are most involved with Gram staining? Explain your answer.

EXERCISE 9 Acid-Fast Staining Procedure

SAFETY CONSIDERATIONS

- A volatile and flammable liquid (acid-alcohol) is used in this experiment. Do not use near an open flame.
- If the carbolfuchsin or methylene blue gets on your clothing, it will not wash out.
- NOTE: When carbolfuchsin is heated, phenol is driven off.
- Phenol is poisonous and caustic, so always use a chemical hood with the exhaust fan on for the hot plate or boiling water bath setup and wear gloves and safety glasses.
- Discard slides in a container with disinfectant or in a biohazard container.
- Mycobacteria should be handled in a safety cabinet to prevent dissemination in case the human pathogen *Mycobacterium tuberculosis* occurs among the cultures.
- Infected material should be disinfected by heat because mycobacteria are relatively resistant to chemical disinfectants.

MATERIALS

- tryptic soy broth culture (18–24 hours) of *Escherichia coli* and nutrient agar slant culture of *Mycobacterium smegmatis* or *Mycobacterium phlei* —5-day-old cultures
- Ziehl's carbolfuchsin: carbolfuchsin prepared with either Tergitol No. 4 (a drop per 30 ml of carbolfuchsin) or Triton-X (2 drops per 100 ml of carbolfuchsin). Tergitol No. 4 and Triton-X act as detergents, emulsifiers, and wetting agents.
- alkaline methylene blue
- acid-alcohol
- clean glass slides
- commercial slides showing acid-fast *Mycobacterium tuberculosis*
- inoculating loop
- hot plate
- microscope
- paper towels
- lens paper and lens cleaner
- immersion oil
- staining racks
- 1-ml pipettes with pipettor
- container for biohazard waste
- safety glasses
- disposable gloves
- lab coat

LEARNING OUTCOMES

Upon completion of this exercise, students will demonstrate the ability to

1. Understand the biochemical basis of the acid-fast stain
2. Perform an acid-fast stain
3. Differentiate bacteria into acid-fast and non-acid-fast groups

SUGGESTED READING IN TEXTBOOK

1. Staining Helps to Visualize and Identify Microbes, section 2.3; see also figures 2.16–2.18.
2. Order *Mycobacteriales* Includes Important Human Pathogens, section 22.1; see also figures 22.5–22.6.
3. Bacteria Can Be Transmitted by Airborne Routes, section 38.1, Mycobacterium Infections.
4. Direct Contact Diseases Can Be Caused by Bacteria, section 38.3, Mycobacterial Skin Infections.

Pronunciation Guide

- *Escherichia coli* (esh-er-I-ke-a KOH-lie)
- *Mycobacterium phlei* (mi-ko-bak-TE-re-um FLEH-ee)
- *M. smegmatis* (M. smeg-MEH-tis)
- *M. tuberculosis* (M. too-ber-ku-LO-sis)
- *Nocardia* (no-KAR-dee-ah)

Why Are the Following Bacteria Used in This Exercise?

One of the major objectives of this exercise is to give you expertise in acid-fast staining. To allow you to differentiate between acid-fast and non-acid-fast bacteria, the authors have chosen one of the cultures from Exercise 8, *Escherichia coli*. *E. coli* is a good example of a non-acid-fast bacterium. *Mycobacterium smegmatis* and *M. phlei* are nonpathogenic members of the genus *Mycobacterium*. These bacteria are straight or slightly curved rods, acid-fast at some stage of growth, and not readily stained by Gram's method. They are also nonmotile, nonsporing, without capsules, and slow growers. The mycobacteria are widely distributed in soil and water; some species are obligate parasites and pathogens of vertebrates.

Medical Application

In the clinical laboratory, the acid-fast stain is important in identifying bacteria in the genus *Mycobacterium;* specifically, *M. leprae* (leprosy) and *M. tuberculosis* (tuberculosis). Acid-fast staining of sputum often directs the treatment, allowing the physician or nurse practitioner to administer drugs such as isoniazid and rifampin, which are specific for TB infections, long before the bacteria could be cultured in the clinical laboratory. This differential stain is also used to identify members of the aerobic actinomycete genus *Nocardia*—specifically, the opportunistic pathogens *N. brasiliensis* and *N. asteroides,* which cause the lung disease nocardiosis.

PRINCIPLES

A few species of bacteria in the genera *Mycobacterium* and *Nocardia* do not readily stain with simple stains. However, these microorganisms can be stained by heating them with carbolfuchsin. The heat drives the stain into the cells. Once the microorganisms have taken up the carbolfuchsin, they are not easily decolorized by acid-alcohol, and hence are termed **acid-fast.** Paul Ehrlich developed the acid-fast stain in 1882 while working with the tubercle bacillus *Mycobacterium tuberculosis*. This acid-fastness is due to the high lipid content (**mycolic acid**) in the cell wall of these microorganisms, which prevents dyes from readily binding to the cells. The **Ziehl-Neelsen acid-fast staining procedure** (developed by Franz Ziehl, a German bacteriologist, and Friedrich Neelsen, a German pathologist, in the late 1800s) is a very useful differential staining technique that makes use of this difference in retention of carbolfuchsin. Acid-fast microorganisms

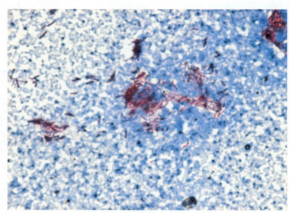

Figure 9.1 Stain of *Mycobacterium* Acid-Fast Rods. In this photomicrograph, *Mycobacterium smegmatis* stains red and the background bacteria blue. *(Auburn University Photographic Services/McGraw Hill)*

will retain this dye and appear red (**Figure 9.1**). Microorganisms that are not acid-fast, termed **non-acid-fast,** will appear blue due to the counterstaining with methylene blue after they have been decolorized by the acid-alcohol.

Procedure

1. Prepare a smear consisting of a mixture of *E. coli* and *M. smegmatis*.
2. Allow the smear to air dry and then heat-fix (*see Figure 7.1*).
3. Place the slide on a hot plate that is within a chemical hood (with the exhaust fan on), and cover the smear with a piece of paper towel that has been cut to the same size as the microscope slide. Saturate the paper with Ziehl's carbolfuchsin (**Figure 9.2a**). Heat for 3 to 5 minutes. Do not allow the slide to dry out, and avoid excess flooding. Also, prevent boiling by adjusting the hot plate to a proper temperature. A boiling water bath with a staining rack or loop held 1 to 2 inches above the water surface also works well. Instead of using a hot plate to heat-drive the carbolfuchsin into the bacteria, an alternate procedure is to cover the heat-fixed slide with a piece of paper towel. Soak the towel with the carbolfuchsin and heat, well above a Bunsen burner flame.
4. Remove the slide, let it cool, and rinse with water for 30 seconds (**Figure 9.2b**).
5. Decolorize by adding acid-alcohol drop by drop until the slide remains only slightly pink.

This requires 10 to 30 seconds and must be done carefully (**Figure 9.2c**).

6. Rinse with water for 5 seconds (**Figure 9.2d**).
7. Counterstain with alkaline methylene blue for about 2 minutes (**Figure 9.2e**).
8. Rinse with water for 30 seconds (**Figure 9.2f**).
9. Blot dry with paper towels (**Figure 9.2g**).
10. There is no need to place a coverslip on the stained smear. Examine the slide under oil immersion and record your results in the report for Exercise 8. Acid-fast organisms stain red; the background and other organisms stain blue or brown (**Figure 9.3**). See Figure 9.1 for an example of the acid-fast stain.
11. Examine the prepared slide of *Mycobacterium tuberculosis*.

Disposal

When you are finished with all of the glass slides, discard them in the designated place for sterilization and disposal.

Figure 9.2 **Acid-Fast Staining Procedure.** Note the safety gloves on the hands.

(a) Saturate with carbolfuchsin and heat for 5 minutes in an exhaust hood

(b) Cool and rinse with water for 30 seconds

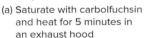
(c) Decolorize with acid-alcohol until pink (10–30 seconds)

(d) Rinse with water for 5 seconds

(e) Counterstain with methylene blue for about 2 minutes

(f) Rinse with water for 30 seconds

(g) Blot dry with paper towels

Figure 9.3 **The Ziehl-Neelsen Acid-Fast Stain.** Appearance of bacteria in a fixed smear at each step in this acid-fast procedure.

Steps in Staining	Acid-Fast	Non-Acid-Fast
Bacteria prior to staining are transparent.		
Carbolfuchsin (primary stain) stains the bacteria reddish. Steam enhances penetration of the primary stain into the bacteria.		
Acid-alcohol (decolorizing agent) removes stain from non-acid-fast bacteria.		
Methylene blue (counterstain) stains non-acid-fast bacteria blue.		

HELPFUL HINTS

- Light (diaphragm and condenser adjustments) is critical in the ability to distinguish acid-fast-stained microorganisms in sputum or other viscous background materials.
- If the bacteria are not adhering to the slide, it may be possible to mix the bacteria with sheep serum or egg albumen during smear preparation. This will help the bacteria adhere to the slide.
- As in the Gram stain, not all acid-fast microorganisms are consistent in their reactions. Generally, young cultures of microorganisms may not be as acid-fast as older ones since they have not yet accumulated enough mycolic acid.

NOTES

Laboratory Report 9

Name: _____

Date: _____

Lab Section: _____

Acid-Fast Staining Procedure

1. Complete the following table with respect to the acid-fast stain and draw representative specimens.

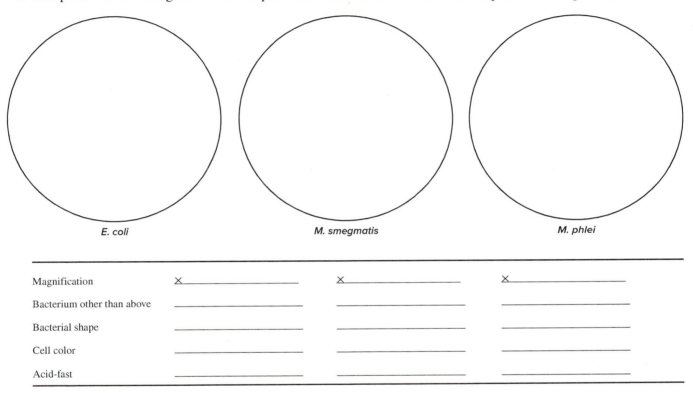

	E. coli	M. smegmatis	M. phlei
Magnification	×_____	×_____	×_____
Bacterium other than above	_____	_____	_____
Bacterial shape	_____	_____	_____
Cell color	_____	_____	_____
Acid-fast	_____	_____	_____

2. Are you satisfied with your results? _____ If not, what can you do to improve your technique the next time you prepare an acid-fast stain from a broth culture?

3. In summary, what is the major purpose of this exercise?

ASSESSMENT
Critical Thinking and Learning Outcomes Review

1. What is the purpose of the heat during the acid-fast staining procedure?

2. What is the function of the counterstain in the acid-fast staining procedure?

3. Are mycobacteria considered Gram-positive or Gram-negative? Explain your answer.

4. For what diseases would you use an acid-fast stain?

5. What cell wall component is responsible for the acid-fast property of mycobacteria?

6. Is a Gram stain an adequate substitute for an acid-fast stain? Why or why not?

EXERCISE 10 Endospore Staining

SAFETY CONSIDERATIONS

- Be careful with the Bunsen burner flame and boiling water bath.
- If either malachite green or safranin gets on your clothes, it will not wash out.
- Wear safety glasses.
- Discard slides in a container with disinfectant or in a biohazard container.

MATERIALS

- suggested bacterial strains:
  24- to 48-hour nutrient agar slant cultures of *Bacillus megaterium* and *Bacillus macerans,* and old (more than 48 hours) thioglycollate cultures of *Clostridium acetobutylicum* and *Bacillus circulans*
- clean glass slides
- microscope
- immersion oil
- wax pencil or lab marker
- inoculating loop
- hot plate or boiling water bath with staining rack or loop
- 5% malachite green solution
- safranin
- paper towels
- lens paper and lens cleaner
- slide warmer
- forceps
- Petri plates
- container for biohazard waste
- safety glasses
- disposable gloves
- lab coat

LEARNING OUTCOMES

Upon completion of this exercise, students will demonstrate the ability to

1. Explain the biochemistry underlying endospore staining
2. Perform an endospore stain
3. Differentiate between bacterial endospore and vegetative cell forms

SUGGESTED READING IN TEXTBOOK

1. Staining Helps to Visualize and Identify Microbes, section 2.3.
2. Bacterial Endospores Are a Survival Strategy, section 3.10; see also figures 3.45–3.47.
3. Direct Contact Diseases Can be Caused by Bacteria, section 38.3; see also figure 38.26.
4. Zoonotic Diseases Arise from Human-Animal Interactions, section 38.5, anthrax; see also figure 38.33.
5. Phylum Firmicutes, Class *Bacilli:* Aerobic Endospore-Forming Bacteria, section 22.2; see also figure 22.13.
6. Phylum Firmicutes, Class *Clostridia:* Anaerobic Endospore-Forming Bacteria, section 22.3 and figure 22.24.

Pronunciation Guide

- *Bacillus megaterium* (bah-SIL-us meg-AH-ter-ee-um)
- *B. macerans* (ma-ser-ANS)
- *B. circulans* (sir-KOO-lanz)
- *Clostridium acetobutylicum* (klos-STRID-ee-um ah-SE-to-buty-licum)

Why Are the Following Bacteria Used in This Exercise?

Because the major objective of this exercise is to provide experience in endospore staining, the authors have chosen several bacteria that vary in the size and shape of their endospores. *Bacillus megaterium* is a cylindrical to oval or pear-shaped cell about 1.2 to 1.5 μm in diameter and 2 to 5 μm long; it tends to occur in short, twisted chains. The spores are central and vary from short oval to elongate. Spores occur in soil. *Bacillus macerans* is an elongated cell 0.5 to 0.7 μm wide and 2.5 to 5 μm in length with terminal spores. Spores are relatively scarce in soil. *Bacillus circulans* is an elongated cell 2 to 5 μm in length and 0.5 to 0.7 μm wide. In most strains, the spore is terminal to subterminal; it is central in a spindle-shaped sporangium if the bacillus is short. The spores are found in soil. *Clostridium acetobutylicum* is a straight or slightly curved rod, 2.4 to 7.6 μm in length and 0.5 to 1.7 μm wide, with rounded ends. The cells occur singly, in pairs, in short chains, and occasionally as long filaments. They are motile with peritrichous flagella. Spores are oval and eccentric to subterminal and are found in soil and animal feces.

Medical Application

Only a few bacteria produce endospores. Those of medical importance include *Bacillus anthracis* (anthrax), *Clostridium tetani* (tetanus), *C. botulinum* (botulism), *C. difficile* (pseudomembranous colitis), and *C. perfringens* (gas gangrene). In the clinical laboratory, the location and size of endospores vary with the species; thus they are often of value in identifying bacteria.

PRINCIPLES

Bacteria in some Gram-positive bacterial genera such as *Bacillus* and *Clostridium* produce quite a resistant structure capable of surviving for long periods in an unfavorable environment and then giving rise to a new bacterial cell (**Figure 10.1**). This structure is called an **endospore** since it develops within the parent (mother) bacterial cell. Endospores are spherical to elliptical in shape and may be either smaller or larger than the parent bacterial cell. Endospore position within the cell is characteristic and may be central, subterminal, or terminal.

Endospores do not stain easily, but, once stained, they strongly resist decolorization. This property is the basis of the **Schaeffer-Fulton** (Alice B. Schaeffer and MacDonald Fulton were microbiologists at Middlebury College, Vermont, in the 1930s) or **Wirtz-Conklin method** (Robert Wirtz and Marie E. Conklin were bacteriologists in the early 1900s) of staining endospores. The endospores are stained with malachite green. Heat is used to provide stain penetration. The rest of the cell is then decolorized and counterstained a light red with safranin (**Figure 10.2**).

Figure 10.1 The Life Cycle of Endospore-Forming Bacteria.

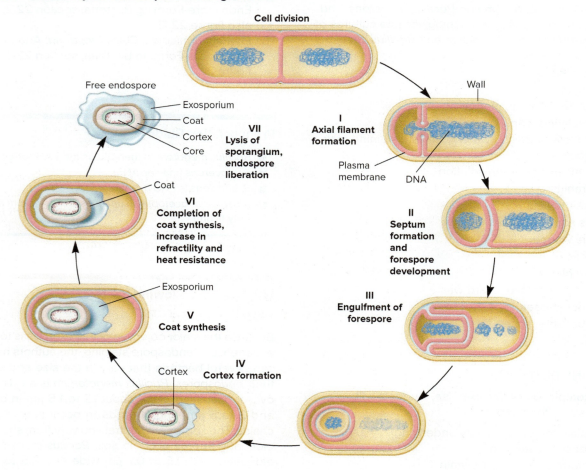

74 Bacterial Cell Biology

Figure 10.2 Endospore Staining. Color changes that occur at each step in the endospore staining procedure.

Reagent	Endospore Former	Non-Endospore-Former
None (heat-fixed smear)		
Malachite green and heat		
Water		
Safranin		

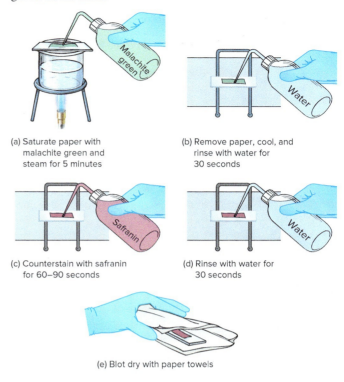

Figure 10.3 Endospore Staining Procedure. Note the safety gloves on the hands.

(a) Saturate paper with malachite green and steam for 5 minutes
(b) Remove paper, cool, and rinse with water for 30 seconds
(c) Counterstain with safranin for 60–90 seconds
(d) Rinse with water for 30 seconds
(e) Blot dry with paper towels

Two procedures for endospore staining are presented in this exercise. The first is the classical Schaeffer-Fulton or Wirtz-Conklin procedure and the second is a microwave procedure.

Procedure

Schaeffer-Fulton Procedure

1. With a wax pencil or lab marker, place the names of the respective bacteria on the edge of four clean glass slides.
2. As shown in Figure 5.3, aseptically transfer one species of bacterium with an inoculating loop to each of the respective slides, air dry (or use a slide warmer), and heat-fix.
3. Place the slide to be stained on a hot plate or boiling water bath equipped with a staining loop or rack. Cover the smear with a paper towel that has been cut the same size as the microscope slide.
4. Perform this step with proper ventilation (or under a hood), safety glasses, and gloves. Soak the paper with the malachite green staining solution. Gently heat on the hot plate (just until the stain steams) for 5 to 6 minutes after the malachite green solution begins to steam. Replace the malachite green solution as it evaporates so that the paper remains saturated during heating (**Figure 10.3a**). Do not allow the slide to become dry.
5. Remove the paper using forceps, allow the slide to cool, and rinse the slide with water for 30 seconds (**Figure 10.3b**).
6. Counterstain with safranin for 60 to 90 seconds (**Figure 10.3c**).
7. Rinse the slide with water for 30 seconds to remove excess water (**Figure 10.3d**).
8. Blot dry with paper towels (**Figure 10.3e**) and examine under oil immersion. A coverslip is not necessary. The spores, both endospores and free spores, stain green; vegetative cells stain red. Draw the bacteria in the space provided in the report for Exercise 10. See **Figure 10.4a–d** for examples of endospore staining.

Microwave Procedure

1. With a wax pencil or lab marker, place the names of the respective bacteria on the edge of four clean glass slides.
2. As shown in Figure 5.3, aseptically transfer one species of bacterium with an inoculating loop to each of the respective slides, air dry (or use a slide warmer), and heat-fix.
3. Cut two pieces of paper towel to fit in an open Petri plate two layers thick. Saturate the toweling with tap water.

Figure 10.4 **Examples of Endospores.** (a) Central spores of *Bacillus* stained with malachite green (×1,000). Notice that the cells are rod shaped and straight, often arranged in pairs or chains, with rounded squared ends. The endospores are oval and not more than one spore per cell. (b) *Clostridium tetani* showing round, terminal spores that usually distend the cell (×1,000). Notice that the cells are rod shaped and are often arranged in pairs or short chains with rounded or sometimes pointed ends. (c) *Bacillus megaterium* showing short oval to elongate spores. (d) *Clostridium botulinum.* The blue structures are the spherical to ovoid/elongate endospores (×1,000). *(a: Source: Larry Stauffer/Oregon State Public/CDC; b and c: Lisa Burgess/McGraw Hill; d: Michael Abbey/Science Source)*

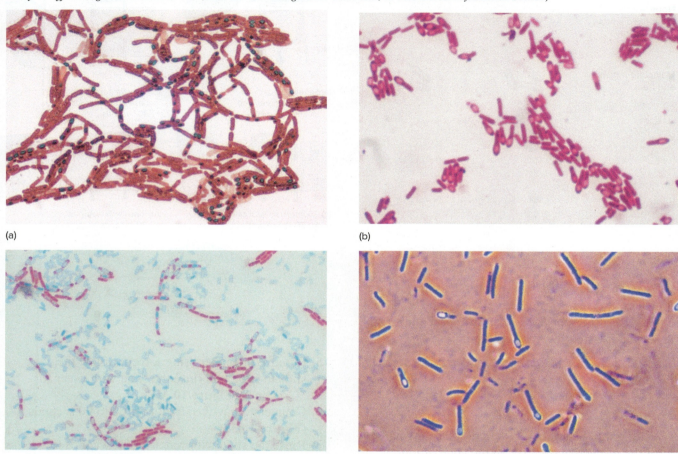

4. Place the heat-fixed slide on top of the two layers of paper towel and flood the slide with the 5% malachite green solution.
5. Place the open Petri plate(s) containing the slide(s) in a microwave oven for 30 seconds at high power.
6. Remove the Petri plate(s) and slide(s) and allow to cool. Rinse with water for 30 seconds.
7. Counterstain with safranin for 30 seconds.
8. Rinse briefly to remove excess safranin. Blot dry with paper towels and examine under oil immersion.

Disposal

When you are finished with all of the glass slides, discard them in the designated place for sterilization and disposal.

> **HELPFUL HINTS**
> - Do not boil the stain—always steam gently.
> - After steaming the slide, cool it before flooding it with cold water. If the slide is not cooled, it may shatter or crack when rinsed with cold water.
> - If you get any stain on your fingers, clean them with stain-removing cream.

Laboratory Report 10

Name: _____

Date: _____

Lab Section: _____

Endospore Staining

1. Make drawings and answer the questions for each of the bacterial endospore slides.

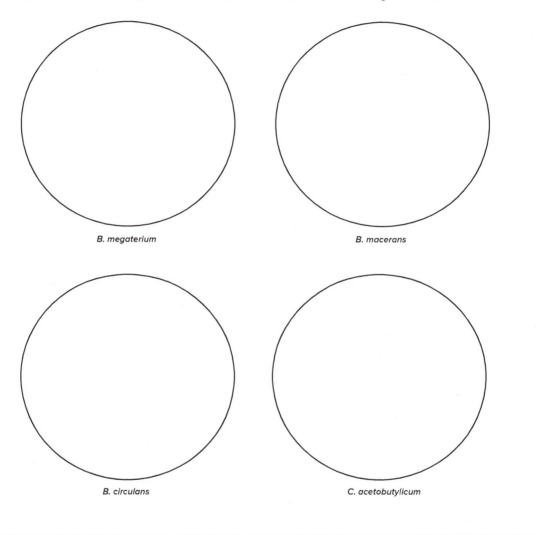

B. megaterium

B. macerans

B. circulans

C. acetobutylicum

Bacterium				
Magnification	×____	×____	×____	×____
Bacterium other than above				
Spore color				
Color of vegetative cell				
Location of endospore (central, terminal, subterminal)				

77

2. Are you satisfied with the results of your endospore stain? _____ If not, how can you improve your results the next time you prepare an endospore stain?

3. In summary, what is the major purpose of this exercise?

ASSESSMENT
Critical Thinking and Learning Outcomes Review

1. Why is heat necessary in order to stain endospores?

2. Where are endospores located within vegetative cells?

3. In the Schaeffer-Fulton endospore stain, what is the primary stain? The counterstain?

4. Name two disease-causing bacteria that produce endospores.
    a.

    b.

Endospore Staining

5. What is the function of an endospore?

6. Why are endospores so difficult to stain?

7. What do endospore stains have in common with the acid-fast stain?

Bacterial Cell Biology

EXERCISE 11 Capsule Staining

SAFETY CONSIDERATIONS

- Be careful with the Bunsen burner flame.
- If India ink, crystal violet, or safranin gets on your clothes, it will not wash out.
- Seventy percent ethyl alcohol is flammable—keep away from open flames.
- Discard slides in a container with disinfectant or in a biohazard container.

MATERIALS

- suggested bacterial strains:
  18-hour skim milk cultures of *Klebsiella pneumoniae* and *Alcaligenes denitrificans*
- Tyler's crystal violet (1% aqueous solution) or Gram's crystal violet (1% aqueous solution)
- 20% (w/v) solution of copper sulfate ($CuSO_4 \cdot 5H_2O$)
- microscope
- immersion oil
- lens paper and lens cleaner
- clean glass slides
- wax pencil or lab marker
- paper towels
- inoculating loop
- 70% ethyl alcohol
- India ink or Spot Test India ink ampules
- safranin stain
- container for the biohazard waste and container for the copper sulfate waste
- safety glasses
- disposable gloves
- lab coat

LEARNING OUTCOMES

Upon completion of this exercise, students will demonstrate the ability to

1. Explain the biochemistry of the capsule stain
2. Perform a capsule stain
3. Distinguish capsular material from the bacterial cell

SUGGESTED READING IN TEXTBOOK

1. The Cell Envelope Often Includes Layers Outside the Cell Wall, section 3.6; see also figure 3.28.

Pronunciation Guide

- *Alcaligenes denitrificans* (al-kah-LIJ-e-neez de-ni-tri-fi-KANS)
- *Klebsiella pneumoniae* (kleb-se-EL-lah nu-MO-ne-EYE)

Why Are the Following Bacteria Used in This Exercise?

One of the major objectives of this exercise is to provide you with experience in capsule staining. To help accomplish this objective, the authors have chosen one capsulated and one noncapsulated bacterium. *Klebsiella pneumoniae* is a nonmotile, capsulated rod, 0.6 to 6 μm in length, and is arranged singly, in pairs, or in short chains. Cells contain a large polysaccharide capsule and give rise to large mucoid colonies. *K. pneumoniae* occurs in human feces and clinical specimens, water, grain, fruits, and vegetables. *Alcaligenes denitrificans* occurs as a rod, a coccal rod, or a coccus; is 0.5 to 2.6 μm in length; and usually occurs singly in water and soil. It is motile with 1 to 4 peritrichous flagella. No capsule is present.

Medical Application

Many bacteria (e.g., *Bacillus anthracis* [anthrax], *Streptococcus mutans* [tooth decay], *Streptococcus pneumoniae* [pneumonia], and the fungus *Cryptococcus neoformans* [cryptococcosis, or fungal meningitis]) contain a gelatinous covering called a capsule. This capsule makes these microbes less vulnerable to phagocytosis. In the clinical laboratory, demonstrating the presence of a capsule is a means of diagnosis and determining the organism's virulence, the degree to which a pathogen can cause disease.

PRINCIPLES

Many bacteria and some fungi have a slimy layer surrounding them, which is usually referred to as a **capsule** (**Figure 11.1a**). The capsule's composition, as well as its thickness, varies between individual bacterial species. Polysaccharides, polypeptides, and glycoproteins have all been found in capsules. Often a pathogenic bacterium with a thick capsule will be more virulent than a strain with little or no capsule since the capsule protects the bacterium against the phagocytic activity of the host's phagocytic cells. However, one cannot always determine if a capsule is present by simple staining procedures, such as using negative staining and India ink. An unstained area around a bacterial cell may be due to the separation of the cell from the surrounding stain upon drying. Two convenient procedures for determining the presence of a capsule are Anthony's (E. E. Anthony, Jr., a bacteriologist at the University of Texas, Austin, in the 1930s) capsule staining method (**Figure 11.1b**) and the Graham and Evans (Florence L. Evans, a bacteriologist at the University of Illinois in the 1930s) procedure.

Anthony's procedure employs two reagents. The primary stain is crystal violet, which gives the bacterial cell and its capsular material a dark purple color. Unlike the cell, the capsule is nonionic and the primary stain cannot adhere. Copper sulfate is the decolorizing agent. It removes excess primary stain as well as color from the capsule. At the same time, the copper sulfate acts as a counterstain by being absorbed into the capsule and turning it a light blue. In this procedure, smears should not be heat-fixed since shrinkage is likely to occur and create a clear zone around the bacterium, which can be mistaken for a capsule.

Procedure

Capsule Staining (Anthony's Procedure)

1. With a wax pencil or lab marker, label the left-hand corner of a clean glass slide with the name of the bacterium that will be stained. Be sure to wear gloves.
2. As shown in Figure 5.3, aseptically transfer a loopful of culture with an inoculating loop to the slide. Allow the slide to air dry. Do not heat-fix! Heat-fixing can cause the bacterial cells to shrink and give a false appearance to the capsule.
3. Place the slide on a staining rack. Flood the slide with crystal violet and let stand for 4 to 7 minutes (**Figure 11.2a**).
4. With the slide over a proper waste container, gently wash off the crystal violet with 20% copper sulfate (**Figure 11.2b**). Do not wash the stain directly into the sink.
5. Blot dry with paper towels (**Figure 11.2c**).
6. Examine under oil immersion (a coverslip is not necessary) and draw the respective bacteria in the report for Exercise 11. Capsules appear as faint halos around dark cells.

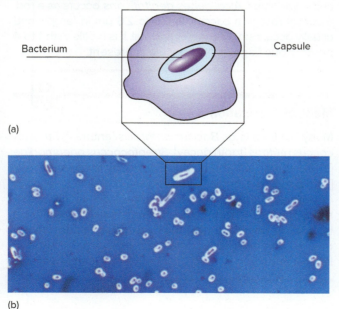

Figure 11.1 Anthony's Capsule Staining Method. (a) Drawing of a single bacterium, capsule, and background material. (b) *Klebsiella pneumoniae* capsules; light micrograph (×1,000). Capsules appear as white halos around blue backgrounds. (b: Lisa Burgess/McGraw Hill)

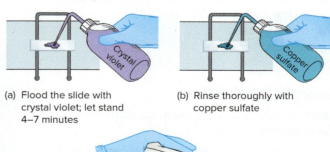

Figure 11.2 Capsule Staining Procedure. Note the safety gloves on the hands.

(a) Flood the slide with crystal violet; let stand 4–7 minutes
(b) Rinse thoroughly with copper sulfate
(c) Blot dry with paper towels

Modified Capsule Stain (Graham and Evans)

1. Thoroughly clean the slide to be used with a household cleanser and alcohol. Be sure to wear gloves.
2. Mix two loopfuls of culture with a small amount (1 to 2 drops) of India ink at one end of the slide.
3. Spread out the drop using a second slide in the same way one prepares a thin smear (*see Figure 6.3*).
4. Dry the smear.
5. Gently rinse with distilled water so that the bacteria do not wash off the slide.
6. Stain for 1 minute with Gram's crystal violet.
7. Rinse again with water.
8. Stain for 90 seconds with safranin stain.
9. Rinse with water and blot dry.
10. If a capsule is present, the pink to red bacteria are surrounded by a clear zone. The background is dark.

Disposal

When you are finished with all of the glass slides, discard them in the designated place for sterilization and disposal.

> **HELPFUL HINTS**
>
> - As with any materials stained with similar or identical colors, light adjustments under the microscope will be critical for optimal visualization of capsules.
> - Be sure to use a very small amount of India ink or the capsules will not be clearly visible

NOTES

Laboratory Report 11

Name: _____

Date: _____

Lab Section: _____

Capsule Staining

1. Fill in the table and make drawings of a representative field of each preparation as seen with the oil immersion lens.

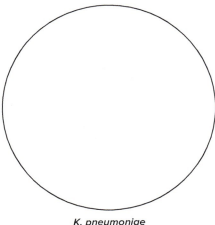
K. pneumoniae

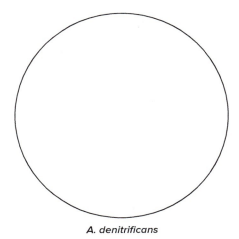
A. denitrificans

Magnification	×_____	×_____
Bacterium other than above	_____	_____
Capsule size (in μm)	_____	_____
Capsule color	_____	_____
Vegetative cell color	_____	_____

2. Are you satisfied with the results of your capsule stain? _____ If not, how can you improve your results the next time you do a capsule stain?

3. In summary, what is the major purpose of this exercise?

ASSESSMENT
Critical Thinking and Learning Outcomes Review

1. What three chemical substances have been identified in bacterial capsules?
   a.
   b.
   c.

2. What is the relationship between the presence of capsules and bacterial pathogenicity?

3. What is the dual function of copper sulfate in capsule staining?

4. What would happen in this procedure if the slide were heat-fixed?

5. How is the capsule stain used in clinical microbiology?

6. Name several bacteria that have capsules.

7. What is the function of a capsule?

EXERCISE 12 Flagella Staining

SAFETY CONSIDERATIONS

- Be careful with the boiling water bath and Bunsen burner flame.
- West stain solutions A and B are irritants; do not breathe vapors or get these solutions on your skin.
- Always prepare West solutions A and B while using a fume hood and wearing safety glasses, gloves, and a lab coat.
- Also, always use the hoods when applying heat to the staining solutions.
- Discard slides in a biohazard container when you are finished examining them.

MATERIALS

- suggested bacterial strains: *Alcaligenes faecalis* and *Pseudomonas aeruginosa*
- prepared demonstration slides of monotrichous flagella (*Pseudomonas aeruginosa*), lophotrichous flagella (*Helicobacter pylori*), peritrichous flagella (*Proteus vulgaris*), amphitrichous flagella (*Spirillum volutans*)
- wax pencil or lab marker
- inoculating loop
- acid-cleaned glass slides with frosted ends
- clean distilled water
- microscope
- immersion oil
- lens paper and lens cleaner
- boiling water bath
    250-ml beaker filled halfway with distilled water
    ring stand
    wire gauze pad
    Bunsen burner or hot plate
- Pasteur pipettes with pipettor
- West stain
    solution A
    solution B
- container for biohazard waste
- safety glasses
- disposable gloves
- lab coat

LEARNING OUTCOMES

Upon completion of this exercise, students will demonstrate the ability to

1. Explain the biochemical basis of flagella staining
2. Perform a flagella stain
3. Illustrate the different types of flagellar arrangement

SUGGESTED READING IN TEXTBOOK

1. Staining Specimens Helps to Visualize and Identify Microbes, section 2.3; see also figure 2.18.
2. External Structures Are Used for Attachment and Motility, section 3.8; see also figures 3.37–3.40.

Pronunciation Guide

- *Alcaligenes faecalis* (al-kah-LIJ-en-eez fee-KAL-iss)
- *Pseudomonas aeruginosa* (soo-do-MO-nas a-ruh-jin-OH-sah)

Why Are the Following Bacteria Used in This Exercise?

After this exercise you should be able to correctly stain bacteria to determine the presence of flagella and their arrangement. You will examine two Gram-negative bacteria that have different flagellar arrangements. *Alcaligenes faecalis* cells exist as cocci or coccobacilli that are 0.5 to 3 µm in length, usually occurring singly. Motility is provided by one or more peritrichous flagella. *A. faecalis* normally occurs in water and soil. *Pseudomonas aeruginosa* cells are rods 2 to 3 µm in length and 0.5 to 1 µm in width. They occur singly and in pairs with motility provided by polar flagella. *P. aeruginosa* is widely distributed in nature and can also cause severe infections in humans.

Medical Application

In the clinical laboratory, the presence, number, and arrangement of flagella can be useful in identifying an unknown bacterial isolate. Important pathogens that are motile due to the presence of flagella include *Bordetella pertussis* (whooping cough), *Listeria monocytogenes* (meningoencephalitis), *Proteus vulgaris* (urinary tract infections), *Pseudomonas aeruginosa* (wound infections), and *Vibrio cholerae* (cholera). The motility afforded by flagella is important for these organisms to thrive in diverse environments, ranging from standing water to the human gastrointestinal tract.

PRINCIPLES

It should be clear by now that like eukaryotes, bacterial cells exhibit a variety of organelles that perform specialized functions. Another example of this is **flagella**, the organelles of locomotion (**motility**). They are slender (about 10 to 30 nm in diameter) and can be seen directly using only the electron microscope. However, a variety of techniques have been developed to allow visualization of flagella using a light microscope. These techniques all rely on the same principle, namely, increasing the thickness of the flagella by first coating them with mordants such as tannic acid and potassium alum. Then, they are stained with dyes like basic fuchsin (Gray method), pararosaniline (Leifson method), silver nitrate (West method), or crystal violet (Difco method). Although the procedures for staining flagella can be cumbersome, they often provide information about the presence and location of flagella (**Figure 12.1**), which is of great value in bacterial identification (**Figure 12.2**). Though a variety of commerically available flagella stains exist, the method described in this exercise is based on the West procedure.

Demonstration Slides

1. Examine the prepared demonstration slides showing different arrangements of flagella.
2. Sketch your observations as part of the lab report for this exercise.

Procedure

West Staining Method

1. With a wax pencil or lab marker, mark the left-hand corner of a clean glass slide with the name of the bacterium to be analyzed.

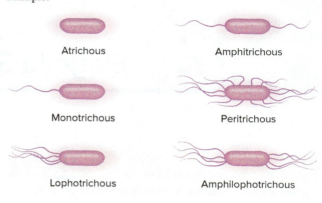

Figure 12.1 **Arrangement of Flagella on Bacterial Cells.** Note the difference in position and number of flagella in each example.

Atrichous
Amphitrichous
Monotrichous
Peritrichous
Lophotrichous
Amphilophotrichous

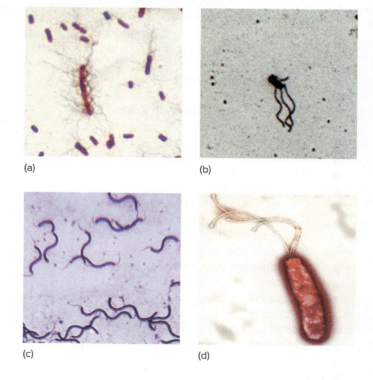

Figure 12.2 **Examples of Several Patterns of Flagellation as Seen under the Microscope.** (a) Flagellar stain of *Proteus vulgaris*. A basic dye was used to coat the flagella. (b) *Clostridium tertium* stained with Leifson flagella stain. (c) Flagella positioned at each pole of *Spirillum volutans*. (d) Transmission electron micrograph of *Helicobacter pylori*. (a: David B. Fankhauser; b: Source: Centers for Disease Control and Prevention; c: Ed Reschke/Photolibrary/Stone/Getty Images; d: Heather Davies/Science Photo Library/Science Source)

(a) (b) (c) (d)

2. Using aseptic technique, transfer the bacterium with an inoculating loop from the turbid liquid at the bottom of the culture to 3 small drops of distilled water in the center of a clean slide that has been carefully wiped off with clean lens paper. Gently spread the diluted bacterial suspension over a 3-cm area (**Figure 12.3a**).
3. Let the slide air dry for 15 minutes (**Figure 12.3b**).
4. Cover the dry smear with solution A (the mordant) for 4 minutes (**Figure 12.3c**).
5. Rinse thoroughly with distilled water (**Figure 12.3d**).
6. Place a piece of paper towel on the smear and soak it with solution B (the stain). Heat the slide in a boiling water bath for 5 minutes in an exhaust hood with the fan on. Add more stain to keep the slide from drying out (**Figure 12.3e**).
7. Remove the toweling and rinse off excess solution B with distilled water. Flood the slide with distilled water and allow it to sit for 1 minute while more silver nitrate residue floats to the surface (**Figure 12.3f**).
8. Then, rinse gently with water and carefully shake excess water off the slide (**Figure 12.3g**).
9. Allow the slide to air dry at room temperature (**Figure 12.3h**).
10. Examine the slide with the oil immersion objective. The best place to view your sample is at the edge of the smear where bacteria are less dense. Record your results in the report for Exercise 12.

Figure 12.3 Flagella Staining Procedure. Note the safety gloves on the hands.

(a) Place bacteria in 3 drops distilled water and spread out

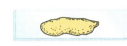

(b) Air dry for 15 minutes

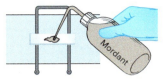

(c) Cover smear with mordant for 4 minutes

(d) Rinse thoroughly with distilled water

(e) Place paper towel over smear and soak with stain; heat for 5 minutes

(f) Flood slide with distilled water and allow to sit for 1 minute

(g) Shake excess water from slide

(h) Air dry at room temperature

Disposal

When you are finished, discard all items for sterilization and disposal.

> **HELPFUL HINTS**
>
> - Do not vortex the cultures, and be gentle when making smears to avoid detaching the flagella.
> - Perform all steps as gently as possible; rough handling of bacteria could cause flagella to be broken off and lost.
> - The use of scrupulously clean slides is essential for best results.
> - To prepare clean slides:
>   - [ ] soak slides for 1 week at room temperature in a 3% hydrochloric acid/95% ethanol solution.
>   - [ ] Wash slides several times with tap water followed by a final wash with distilled or deionized water.
>   - [ ] Pass the slide through the Bunsen burner flame and rest heated side up on a clean paper towel.

Laboratory Report 12

Name: _____

Date: _____

Lab Section: _____

Flagella Staining

1. Make a drawing of a representative microscope field of each preparation and fill in the table.

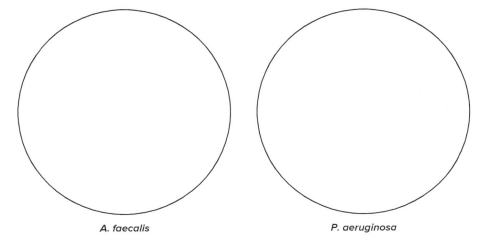

A. faecalis   *P. aeruginosa*

Magnification

Arrangement of flagella

Number of flagella

2. Are you satisfied with the results of your flagella stain? If not, how can you improve your results the next time you do a flagella stain?

ASSESSMENT
Critical Thinking and Learning Outcomes Review

1. Why are flagella so difficult to stain?

2. Why were you instructed to handle the bacterial cultures so carefully during this procedure?

3. Why must the glass slide be free of grease and oil before staining for flagella?

4. Name and define four types of flagella arrangements. Provide the name of a human pathogen for each case and describe the disease it causes.

    a.

    b.

    c.

    d.

5. What is a mordant? What purpose does the mordant serve during a flagella stain?

6. After preparing a flagella stain of *Proteus vulgaris,* your lab partner cannot detect any flagella. When you performed the experiment 10 minutes prior, flagella were easily visible. Speculate on what went wrong with your partner's slide prep.

7. What happens to the size of flagella when they are stained?

PART 3
Basic Culture Techniques

In Part 3, you will be introduced to the basic techniques needed to cultivate microbes in the lab. This includes learning how to prepare and sterilize culture media, to isolate bacteria in pure culture from various types of specimens, and to subculture bacteria and fungi in the laboratory. You will also learn how to determine the number of bacteria in a given culture.

Along with Parts 1 and 2 of this manual, Part 3 continues your acquisition of basic microbiological techniques. A thorough understanding of microscopic, slide, and culture techniques are the foundation on which the rest of this manual is built.

After completing the exercises in Part 3, you will be able to demonstrate the proper use of aseptic techniques and to estimate the number of microorganisms in a sample using serial dilution techniques. This will meet the following American Society for Microbiology Core Curriculum skills:

- **sterilizing and maintaining sterility of transfer instruments**
- **performing aseptic transfers**
- **obtaining microbial samples**
- **correctly choosing and using pipettes and pipetting devices**
- **correctly spreading diluted samples for counting**
- **estimating appropriate dilutions**
- **extrapolating plate counts to obtain correct CFU or PFU in the starting sample**

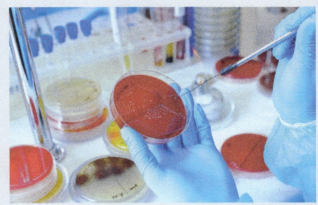

Salov Evgeniy/Shutterstock

Cultivating microbes in the laboratory is an essential part of microbiology. Whether for identifying unknown clinical isolates, or propagating well-studied laboratory strains, the basic techniques for manipulating microbes are the same. The wide variety of commercially available media formulations makes growing microbes easier than ever. However, extreme caution must be exercised when handling microbes, particularly bacteria, fungi, and viruses that can cause disease. Proper use of personal protective equipment and access to a carefully designed laboratory are critically important prior to conducting any microbiology research.

EXERCISE 13
Preparation of Microbiology Media and Equipment

SAFETY CONSIDERATIONS

- Several types of heat that can cause severe burns are used in this experiment.
- Do not operate the autoclave without approval from your instructor.
- Always wear heat-proof gloves when unloading the autoclave.
- Boiling agar can cause severe burns if spilled on your hands.
- Be especially careful with Bunsen burners and hot plates.
- If you are burned, seek immediate treatment.
- Wear personal protective equipment at all times.

MATERIALS

- suggested bacterial strain: *Escherichia coli*
- autoclave
- sterile plastic Petri plates
- sterile culture tubes
- test tube rack
- test tube caps
- defined culture medium ingredients as in Table 13.1
- complex culture medium ingredients as in Table 13.2
- 2-l Erlenmeyer flask
- 10-ml pipets
- weighing paper or boats
- balance
- agar
- heat-proof gloves
- water bath set at 50°C
- aluminum foil
- stir plate with stir bar
- Bunsen burner or hot plate
- container for biohazard waste
- safety glasses
- disposable gloves
- lab coat

LEARNING OUTCOMES

Upon completion of this exercise, students will demonstrate the ability to

1. Describe the concept of sterility
2. Compare and contrast how microbiological media and equipment can be sterilized
3. Describe the different types of culture media and their composition, and give several examples of what each is used for
4. Demonstrate the various ways culture tubes are capped
5. Describe how to prepare and transfer culture media
6. Explain how to safely use the autoclave
7. Prepare defined and undefined media, and prepare agar plates

SUGGESTED READING IN TEXTBOOK

1. Microbial Growth and Replication Pathways: Targets for Control, section 8.1; see also figure 8.1.
2. Microbes Can Be Controlled by Physical Means, section 8.2; see also figures 8.4–8.7.
3. Microorganisms Are Controlled with Chemical Agents, section 8.3; see also table 8.3 and figures 8.8 and 8.9.
4. Laboratory Culture of Microbes Requires Conditions That Mimic Their Normal Habitats, section 7.7; see also table 7.3 and Figure 7.28.

Pronunciation Guide

- *Escherichia coli* (esh-er-I-ke-a KOH-li)

Why Is the Following Bacterium Used in This Exercise?

One of the major objectives of this exercise is to prepare defined and complex media. After the media have been prepared, they can be inoculated with *Escherichia coli*.

E. coli is facultatively anaerobic and chemoorganoheterotrophic, having both a respiratory and a fermentative type of metabolism. As such, *E. coli* can grow under a variety of culture conditions. Given the ease with which *E. coli* can be manipulated, it has long been an indispensable tool in many research labs. It also is the perfect organism to use when training aspiring microbiologists!

PRINCIPLES

Methods of Sterilization

Sterilization is the process by which all living cells, spores, and acellular entities (e.g., viruses, viroids, and prions) are either destroyed or removed from an object or habitat. This differs from **antisepsis,** which is the removal or destruction of microorganisms on living tissues. In the microbiology lab, there are several methods of sterilization that rely on either chemical or physical means. The method used depends on the material you intend to sterilize, as well as the equipment that is available to you. In this exercise, we will explore two of the most common methods of sterilization: autoclaving and filtration.

Perhaps the most important tool in the microbiology lab is the **autoclave**. Here, items are sterilized by exposure to steam at high temperature and pressure (typically 121°C and 15 lb of pressure) an extended period of time (**Figure 13.1**). The exposure time depends on the item being sterilized; for example, culture media can be considered sterilized after 20 minutes, while infectious proteins called prions require comparatively a much longer time. Under these conditions, even bacterial endospores will not survive longer than about 12 to 13 minutes. This method is rapid and dependable, which is vitally important for microbiologists who study infectious diseases. Modern autoclaves are designed to ensure that all of the air has been expelled from the chamber so that only steam is present. They are carefully temperature controlled as well. Almost all culture media and anything else that will resist 121°C temperatures and steam can be sterilized in this way.

Often, dry plastic and glassware such as pipet tips and culture flasks must be sterilized. Another option for such items is **dry-heat sterilization.** The glassware is placed in an electric oven set to operate between 160° and 170°C. Since dry heat is not as effective as wet heat, the glassware must be kept at this temperature for about 2 hours or longer. Be careful! The temperature must not rise above 180°C or any cotton plugs or paper labels will char.

Two other sterilization techniques make use of radiation or gas. **Ultraviolet (UV) radiation** around 260 nm is quite lethal to many microorganisms because it causes thymine-thymine dimers in the DNA. However, UV radiation does not penetrate glass, dirt, water, and other substances very effectively. Because of this disadvantage, UV is used as a sterilizing agent only in a few particular situations. For example, UV lamps are sometimes placed in the back of biological safety cabinets to sterilize any exposed surfaces. **Ionizing radiation** can penetrate deep into objects, making it ideal for sterilizing things like surgical tools and various types of food. However, these devices are not commonly found in most microbiology labs.

Many heat-sensitive items such as disposable plastic Petri plates and syringes, sutures, and catheters can be sterilized with chemical gasses like **ethylene oxide**, **chlorine dioxide**, and vaporized hydrogen peroxide. Ethylene oxide and chlorine dioxide are both microbicidal and sporicidal and kill by damaging DNA and proteins. It is a particularly effective sterilizing agent for use in commercial and industrial settings because it rapidly penetrates packing materials, even plastic wraps.

Figure 13.1 **Autoclaves, the Workhorse of Microbiology. (a)** An example of a small autoclave and **(b)** the instruments about to be loaded into it. *(a: DenGuy/iStock/360/Getty Images; b: McGraw Hill)*

(a)

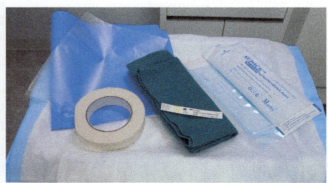
(b)

Vaporized hydrogen peroxide is typically used to decontaminate rooms like surgical suites or biocontainment labs.

Sometimes culture media must be made from components that will not withstand heating at 121°C. Such a medium can be sterilized by passing it through a **bacteriological filter**, which physically removes bacteria and larger microorganisms from the solution and thereby sterilizes them without heat. Sintered glass filters with ultrafine, fritted disks (0.9 to 1.4 μm pore size) and Seitz asbestos-pad filter funnels (3 mm thick with 0.1 μm pores) are both quite effective in sterilizing solutions. However, if pore sizes greater than 0.2 μm are used, there is an exceedingly high chance that the filtrate will not be sterile. There are now commercially available cellulose- and polycarbonate-based membranes in a variety of pore sizes. Generally, membranes with 0.2 μm pores are employed in sterilization of culture media. Conveniently, there are different devices available for filter sterilization of both large and small volumes. For example, a filter flask with a vacuum or syringe can be used to force liquid through a special membrane filter holder.

Types of Microbiological Culture Media

The growth of microorganisms depends on available nutrients and a favorable growth environment. In the laboratory, the nutrient preparations that are used for culturing microorganisms are called **media** (singular, **medium**). Three physical forms of media are used: **liquid (broth)**, **semisolid**, and **solid**. The major difference among these media is that solid and semisolid media contain a solidifying agent (usually **agar**), whereas a liquid medium does not. Liquid media, such as nutrient broth (**Figure 13.2a**), can be used to propagate large numbers of microorganisms in fermenters to create industrially useful compounds or for routine

Table 13.1	A Chemically Defined Medium
Ingredient	Quantity
Dipotassium phosphate K_2HPO_4	7 g
Potassium phosphate, monobasic, KH_2PO_4	2 g
Hydrated magnesium sulfate, $MgSO_4 \cdot 7H_2O$	0.2 g
Ammonium sulfate, $(NH_4)_2SO_4$	1 g
Glucose	5 g
Distilled water	1 l

molecular biology experiments. Semisolid media can also be used in fermentation studies, or for determining if an organism is motile. Solid media, are used extensively in microbiology labs. They allow for the surface growth of microorganisms in order to observe colony appearance, enable pure culture isolations, and depending on the formulation, can allow detection of biochemical reactions that bacteria can perform.

Remember that when culture media first comes out of the autoclave, it is extremely hot. This high temperature melts agar, thus transforming solid media into a liquefied state. While in the liquefied state, solid media can be poured into either a sterile tube or Petri plate. When the media cools to room temperatures, it will solidify; imagine a very firm version of Jell-O. If the medium in the tube is allowed to harden in a slanted position, the tube is called an **agar slant** (**Figure 13.2b,c**); if the tube is allowed to harden in an upright position, the tube is designated an **agar deep tube** (**Figure 13.2d**); and if the agar is poured into a Petri plate, the plate is simply referred to as an **agar plate** (**Figure 13.2e**).

Microbiological culture media are grouped into two major categories. **Chemically defined**, or **synthetic, media** are composed of known amounts of pure chemicals (**Table 13.1**). Such media are often used in

Figure 13.2 Culture Media. Different forms of culture media with the proper volume in each.

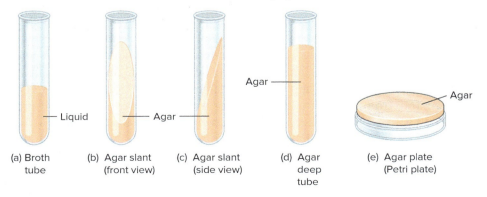

(a) Broth tube
(b) Agar slant (front view)
(c) Agar slant (side view)
(d) Agar deep tube
(e) Agar plate (Petri plate)

Table 13.2	A Complex Medium
Ingredient	Quantity
Casein peptone	17 g
Soybean peptone	3 g
NaCl	5 g
Dipotassium phosphate K_2HPO_4	2.5 g
Glucose	2.5 g
Distilled water	1 l

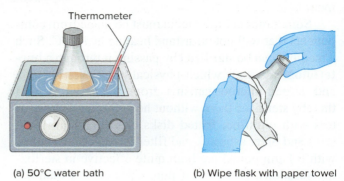

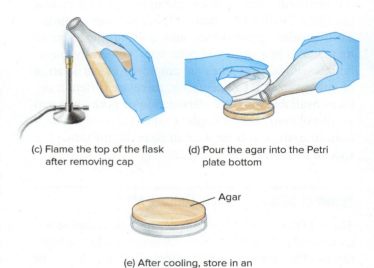

Figure 13.3 Pouring Agar Plates. Notice the safety gloves on the hands.

(a) 50°C water bath
(b) Wipe flask with paper towel
(c) Flame the top of the flask after removing cap
(d) Pour the agar into the Petri plate bottom
(e) After cooling, store in an inverted position

culturing autotrophic microorganisms such as algae or nonfastidious heterotrophs. In routine bacteriology laboratory exercises, **complex,** or **nonsynthetic, media** are employed (**Table 13.2**). These are composed of complex materials that are rich in vitamins and nutrients. Such formulations allow growth of a wide range of organisms. Three of the most commonly used components are beef extract, yeast extract, and peptones. Commercial sources of media and the composition of culture media used in this manual are found in the Appendix.

The preparation of media from commercial dehydrated products is simple and straightforward. Each bottle of dehydrated medium has instructions for preparation on its label. When reconstituting dehydrated media, always use a flask that holds twice the volume of media you are preparing. After the media is sterilized, it can be dispensed as needed.

If the medium lacks agar, the powder will usually dissolve without heating. If it contains agar, the powder will not dissolve at room temperature. Specific heating instructions are given for each type of medium.

Most of the exercises you will be doing in this manual will involve the use of sterile media in broth tubes and agar plates. Normally, 18 × 150 mm, 16 × 125 mm, or 13 × 100 mm culture tubes will be used. Whether they are glass or plastic is at the discretion of your instructor. Regardless, all culture tubes must be capped in order to maintain media sterility. This can be accomplished by using cotton or plastic foam plugs; plastic or metal caps are also common Capping the culture tubes keeps them free from contamination while allowing air into the culture tube and, at the same time, minimizing evaporation of the broth. It is sometimes desirable to use screw-cap culture tubes. This is especially true when the culture needs to be incubated or stored for long periods of time. Regardless of the type of cap used, make sure it is loose-fitting when it goes into the autoclave. Afterward when the sterilization cycle is complete, make sure it is secure.

When Petri plates are needed, the sterilized agar is removed from the autoclave and transferred to a 50°C water bath and kept there for at least 15 minutes before use (**Figure 13.3a**). The agar should be cooled to about 50°C before being poured to minimize the amount of steam condensation on the Petri plate lids after the agar has been poured. Agar does not solidify until its temperature drops to about 42°C. When the molten agar has cooled to 50°C, the flask is taken from the water bath and the outside is dried with a paper towel (**Figure 13.3b**). Its cap is removed and the top is briefly flamed using a Bunsen burner (**Figure 13.3c**). The agar is immediately poured into a sterile, dry Petri plate while holding the top carefully above the Petri plate bottom in order to avoid contamination (**Figure 13.3d**).

Replacing the top quickly prevents contamination. After allowing the agar to cool and harden, the plates should be stored in an inverted position (**Figure 13.3e**). As an alternative, agar media can be prepared in shatter-proof glass bottle with screw cap lids. After the media is allowed to cool, the caps should be tightened to prevent the media from drying out during storage. When needed, the agar can be melted in the microwave and poured into plates as described above.

Procedure

Preparing a Chemically Defined Medium

1. Prepare 500 ml of broth using the recipe outlined in Table 13.1. To a 1-l Erlenmeyer flask add 375 ml of distilled water.
2. Weigh out and add each ingredient to the water in the order listed. Using a magnetic stir plate, mix the broth well until the ingredients are completely dissolved.
3. Add the remaining 125 ml of water to the flask, allowing the water to drain down the wall of the flask to collect any undissolved ingredients.
4. Adjust the pH to 7.2 to 7.4 by adding just enough HCl or NaOH dropwise.
5. Dispense 3 to 5 ml of the glucose-mineral salts broth into each of 10 test tubes using a 10-ml pipette and then loosely cap the tubes. Other students can use the remaining 450 ml of broth for their tubes. Place your tubes in a test tube rack in preparation for autoclaving.

Preparing a Complex Medium

1. Prepare two separate 500 ml batches of tryptic soy broth according to the recipe outlined in Table 13.2.
2. Add 375 ml of distilled water to a 1-l Erlenmeyer flask and add the ingredients individually; mix after each addition.
3. Add the remaining 125 ml of water to rinse the sides of the flask.
4. Adjust the pH to 7.2 to 7.4 by adding just enough HCl or NaOH dropwise.
5. Using one batch of broth, dispense 3 to 5 ml of the broth into each of 10 tubes and loosely cap them. Place the tubes in a test tube rack in preparation for autoclaving.
6. With the other batch of broth, add 7.5 g of agar to give an agar concentration of 1.5%.
7. Cover with aluminum foil and place in the autoclave.

Procedure for Autoclaving

1. As the make and model of autoclaves vary widely, your instructor will demonstrate how to safely use the instrument in your laboratory.
2. Load the autoclave with the freshly prepared culture media.
3. Close and lock the autoclave door.
4. Set the autoclave time for 20 minutes or longer and select a slow rate of exhaust.
5. Make certain that the autoclave temperature is set to 121°C.
6. Start the autoclave according to the directions provided by your instructor.
7. When the period of sterilization is completed and the pressure in the chamber reads 0, carefully open the door and remove the containers, using heat-proof gloves.

Pouring Agar Plates

1. After autoclaving, cool the flask of sterile agar in a 50°C water bath.
2. Line up the desired number of sterile Petri plates on the benchtop. You will need about 20 plates.
3. Remove the aluminum foil cap from the flask and briefly flame the flask's neck. Lift the top of each plate, pour about 25 ml of agar (Note: The agar should be approximately 3 to 5 mm deep in the plate).
4. Pour the remaining plates without stopping.
5. When the plates are cool (i.e., the agar is solidified), invert them to prevent condensing moisture from accumulating on the agar surfaces.
6. Store plates upside-down in a bag to prevent the agar from drying out in the refrigerator.
7. Inoculate your media with the culture of *E. coli* provided by your instructor. Incubate your cultures at 37°C for 24 hours. Observe your cultures and then dispose of them in a biohazard waste container.

Disposal

When you are finished with all of the plates and tubes, discard them in the designated place for sterilization and disposal.

> **HELPFUL HINTS**
>
> - Don't overload the autoclave chamber; provide ample space between baskets of media to allow circulation of steam.
> - Before opening the door to the autoclave, you should always wear heat-proof gloves, stand at arm's length, and slowly open the door. This will prevent two problems from occurring:
>   - [ ] the trapped steam will dissipate toward the ceiling in a controlled fashion without burning your skin.
>   - [ ] the media will not boil out of the stoppered containers because of a too rapid change in internal pressure in the flask.

Laboratory Report 13

Name: _____

Date: _____

Lab Section: _____

Preparation of Microbiology Media and Equipment

1. After at least 24 hours of incubation, do your prepared plates and broth tubes appear to be sterile? How would you know if there was contamination present? Explain your answer.

2. List the steps you would go through to make tryptic soy agar slants.

3. LB is perhaps the most widely used culture medium in labs across the world. True aficionados know that LB is commonly sold in three different formulations: Luria, Lennox, and Miller. Do some research to determine the difference between these types of LB. Why might you choose one formulation over the other?

4. Provide the requested information using a *Difco Manual* or *BBL Manual*.
    a. Quantity of starch in Mueller-Hinton agar
    b. Quantity of lactose in eosin methylene blue agar
    c. Percentage of sodium chloride in mannitol salt agar
    d. Percentage of bile salts in MacConkey agar
    e. Quantity of beef extract in nutrient broth

5. Using a *Difco Manual* or *BBL Manual*, find the following:
    a. The purpose of malachite green in Lowenstein-Jensen agar
    b. The way in which bismuth sulfite agar selects for *Salmonella* and *Shigella* species
    c. The role of pH in Sabouraud dextrose agar
    d. The roles of thioglycollate and methylene blue in thioglycollate medium
    e. The purpose of sodium deoxycholate in XLD agar

ASSESSMENT
Critical Thinking and Learning Outcomes Review

1. What are the two major approaches for preparing sterile microbiological culture media? Give an example of each and describe when their use is most appropriate.

2. Explain the major differences between a defined (synthetic) medium and a complex (nonsynthetic) medium. Under what circumstances would you use one over the other?

3. Why must all culture media be sterilized before use?

4. Describe three ways for sterilizing instruments or equipment used in a microbiology lab.
    a.
    b.
    c.

5. Why are Petri plates inverted after they cool?

6. Why is culture medium cooled to about 50°C before it is poured into Petri plates?

7. How does an agar slant differ from an agar deep?

8. What is the source of carbon in the chemically defined medium in Table 13.1? The source of nitrogen?

EXERCISE 14 The Spread-Plate Technique

SAFETY CONSIDERATIONS
- Alcohol is extremely flammable.
- Keep the beaker of ethanol away from the Bunsen burner.
- Do not put a flaming glass rod back into the ethanol.
- Be certain you know the location of the fire extinguisher.

MATERIALS
- suggested bacterial strains:
  Escherichia coli, *Micrococcus luteus*, and a mixture of the two
- Bunsen burner
- inoculating loop
- 95% ethanol
- glass or plastic cell spreaders (also called a "hockey stick")
- wax pencil or lab marker
- 500-ml glass beaker
- pipettes with pipettor
- 3 tryptic soy agar plates
- rulers
- container for biohazard waste
- safety glasses
- disposable gloves
- lab coat

LEARNING OUTCOMES
Upon completion of this exercise, students will demonstrate the ability to
1. Explain the purpose of the spread-plate technique
2. Perform the spread-plate technique
3. Identify different types of bacterial colony morphology

SUGGESTED READING IN TEXTBOOK
1. Laboratory Culture of Microbes Requires Conditions That Mimic Their Normal Habitats, section 7.7; see also figure 7.31.

Medical Application
In the clinical laboratory, growth of a pure culture is absolutely necessary before any biochemical tests can be performed to identify a suspect microorganism.

Pronunciation Guide
- *Escherichia coli* (Esh-er-I-ke-a KOH-lie)
- *Micrococcus luteus* (my-kro-KOK-us LOO-tee-us)

Why Are the Following Bacteria Used in This Exercise?
After this exercise, you should be able to use the spread-plate technique to separate a mixture of two or more bacteria into well-isolated colonies. The bacteria to be used are *Escherichia coli* and *Micrococcus luteus*. *E. coli* is easy to culture since it grows on simple medium (e.g., tryptic soy agar) and produces dull white to cream colonies that are easy to see. *M. luteus* produces yellow colonies. By using color and colony morphology, you can see what a well-isolated colony of each of the above bacteria looks like. The isolated bacteria can then be picked and streaked onto fresh plates to obtain a pure culture.

PRINCIPLES
Isolating a microorganism in pure culture is essential to study it in the laboratory. In natural habitats, microbes usually grow together in populations containing a number of species; this can include bacteria, archaea, fungi, and viruses that prey upon these microbes. Certain techniques for isolating bacteria are based on the concept that if an individual bacterial cell is separated from other cells and provided adequate space on a nutrient surface, it will grow into a

Figure 14.1 Isolation Technique. Stages in the formation of an isolated colony, showing the microscopic events and the macroscopic result.

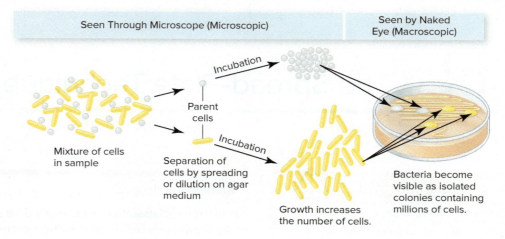

discrete mound of cells called a **colony** (**Figure 14.1**). In order to adequately study and characterize an individual bacterial species, one needs these well-isolated colonies in order to establish a **pure culture**. The **spread-plate technique** is an easy, direct way of achieving this result. In this technique, a small volume of dilute bacterial culture containing 100 to 300 cells or less is transferred to the center of an agar plate and is spread evenly over the surface with a sterile, L-shaped cell spreader (or "hockey stick"). If using a glass rod, it must first be sterilized by dipping it in ethanol and then flaming it to burn off the excess ethanol. After spreading and incubating the plate, some of the dispersed cells will develop into isolated colonies. A **colony** is a large number of bacterial cells on solid medium, which is visible to the naked eye as a discrete entity. The assumption is that each colony is derived from one parent cell and therefore represents a clone of a pure culture.

After incubation, the shape of the edge or margin can be determined by looking down at the colony from directly overhead. The nature of the colony elevation is apparent when viewed from the side as the plate is held at eye level. These variations in **colony morphology** are illustrated in **Figure 14.2**. Using a magnifying glass or a dissecting microscope can reveal even more detailed morphological features. After a well-isolated colony has been identified, it can then

Figure 14.2 Bacterial Colony Characteristics on Agar Media as Seen with the Naked Eye. The characteristics of bacterial colonies are described using the terms depicted below.

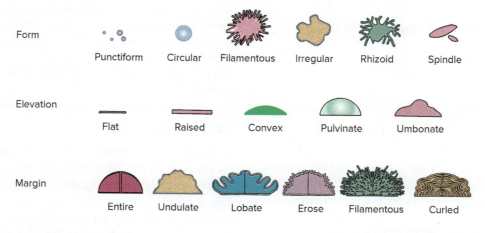

Appearance: Shiny or dull
Optical property: Opaque, translucent, or transparent
Pigmentation: Pigmented or Nonpigmented
Texture: Rough or smooth

Basic Culture Techniques

be picked and streaked onto a fresh plate to obtain a pure culture.

Procedure

1. With a wax pencil or lab marker, label the **bottom** of the agar plates with the name of the bacterium to be inoculated, your name, and date. Three plates are to be inoculated: (a) one with *E. coli,* (b) one with *M. luteus,* and (c) one with the mixture.
2. Using a pipettor, transfer 0.1 ml of the respective bacterial culture onto the center of an agar plate (**Figure 14.3a**). Note that you should not use a saturated, stationary phase culture. Your instructor will prepare an appropriate dilution of a saturated culture in advance.
3. If using a glass cell spreader, place it into a beaker of ethanol (**Figure 14.3b**) for 1 minute and then tap the rod on the side of the beaker to remove any excess ethanol. If using a sterile plastic cell spreader, skip to step 5.
4. Briefly pass the ethanol-soaked spreader through the flame to burn off the excess alcohol (**Figure 14.3c**), and allow it to cool by pressing it against the inside of the lid of the petri plate.
5. Spread the bacterial sample evenly over the agar surface with the sterilized spreader (**Figure 14.3d,e**), making sure the entire surface of the plate has been covered and all of the liquid has been absorbed. Rotate the plate with your other hand or use a turntable if available. Also make sure you do not touch the edge of the plate with your fingers.
6. If using a glass cell spreader, immerse it in ethanol, tap on the side of the beaker to remove any excess ethanol, and reflame. If using a plastic cell spreader, simply get a new one.
7. Repeat the procedure to inoculate the remaining two plates.
8. Invert the plates and incubate for 24 to 48 hours at room temperature or 37°C.

Figure 14.3 Technique for the Preparation of a Spread-plate. (**a**) Pipette a small sample onto the center of an agar plate. (**b**) Place the glass spreader into a beaker of ethanol for 1 minute. (**c**) Briefly flame the ethanol-soaked spreader by passing it through the flame several times and allow it to cool. (**d**) Spread the sample evenly over the agar surface with the sterilized spreader and incubate as desired. Note the safety gloves on the hands. (**e**) Example of a single-use sterile blue plastic cell spreader. *(e: unoL/Shutterstock)*

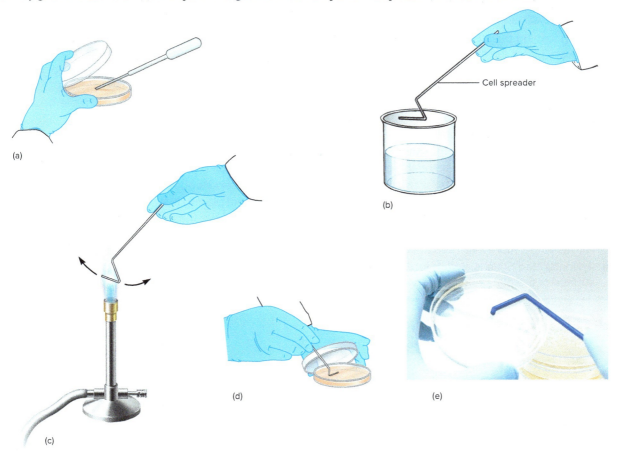

9. After incubation, measure some representative colonies and carefully observe their morphology (**Figure 14.4**). Record your results in the report for this exercise.

Disposal

When you are finished with all of the pipettes and plates, discard them in the designated place for sterilization and disposal.

Figure 14.4 **Spread Plate.** Macroscopic photomicrograph of two spread plates. Notice the well-isolated colonies and the wide variety of morphologies. *(a: Mateusz Kropiwnicki/Shutterstock; b: Lisa Burgess/McGraw Hill)*

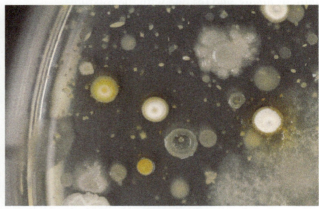

(a) (b)

HELPFUL HINTS

- When flaming the alcohol on the glass rod, touch it to the flame only long enough to ignite the alcohol, then remove it from the flame while the alcohol burns.
- Wait 5 to 10 seconds after flaming to allow the alcohol to burn off and to ensure that the glass is cool enough to spread the culture without sizzling. Hold the rod briefly on the surface of the agar to finish cooling.
- Do not return a flaming rod to the beaker. If you accidentally do this, remove the rod from the beaker and smother the flames with a book by quickly lowering the book on the beaker.
- Do not pour flaming alcohol into the sink.
- Do not pour water into the flaming alcohol.
- To prevent burns, avoid holding the glass rod so that alcohol runs onto your fingers.
- Keep all flammable objects, such as paper, out of reach of ignited alcohol.
- Avoid contamination of the Petri plate cover and the culture by not placing the cover on the table, desk, or any other object while spreading. Hold the cover, bottom side down, above the agar surface as much as possible.
- Turning the plate while carefully spreading the culture (but not hitting the sides of the plate with the glass rod) will result in a more even separation of the bacteria.
- An inoculated plate is always incubated in an inverted position to prevent condensation from falling onto the surface of the plate and interfering with discrete colony formation.

Laboratory Report 14

Name: _____

Date: _____

Lab Section: _____

The Spread-Plate Technique

1. Make drawings of several well-isolated colonies from each plate and fill in the table.

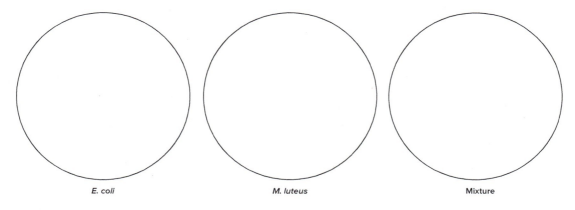

E. coli   M. luteus   Mixture

Form

Elevation

Margin

Appearance

Optical property

Pigmentation

Texture

2. With your ruler, measure the diameter of the average colony appearing on each plate by placing the ruler on the bottom of the plate. Hold the plate and ruler against the light to make your readings and be sure to measure a well-separated colony.
   a. Size of *E. coli* colony

   b. Size of *M. luteus* colony

3. In summary, what is the major purpose of this exercise?

107

ASSESSMENT
Critical Thinking and Learning Outcomes Review

1. What is a bacterial colony? What does it mean to say that a colony is a clonal population?

2. What is the purpose of the ethanol in the spread-plate technique?

3. Why is it necessary to use only diluted cultures that contain 100 to 300 cells for a successful spread plate?

4. Describe the form of some typical bacterial colonies.

5. What is the purpose of the spread-plate technique?

6. In all routine laboratory work, Petri plates are labeled on the bottom and not on the lids. Why?

EXERCISE 15 The Streak-Plate Technique

SAFETY CONSIDERATIONS

- Be careful with the Bunsen burner flame.
- Exercise caution when handling cultures that can cause human disease.

MATERIALS

- suggested bacterial strains: broth cultures of *Escherichia coli*, *Serratia marcescens*, *Staphylococcus aureus*, *Bacillus subtilis*, and a mixture of *Escherichia coli* and *Staphylococcus aureus*
- 5 tryptic soy agar plates
- Bunsen burner
- inoculating loop
- wax pencil or lab marker
- mannitol salt agar (MSA) plate
- eosin methylene blue (EMB) agar plate
- container for biohazard waste
- safety glasses
- disposable gloves
- lab coat

LEARNING OUTCOMES

Upon completion of this exercise, students will demonstrate the ability to

1. Explain the purpose of the streak-plate technique and differential media
2. Perform a streak-plate technique and isolate discrete colonies for subculturing

SUGGESTED READING IN TEXTBOOK

1. Laboratory Culture of Cellular Microbes Requires Media and Conditions That Mimic the Normal Habitat of a Microbe, section 7.7; see also figures 7.28 and 7.30.

Pronunciation Guide

Before reading this exercise, pronounce each microbe name out loud. This will help you avoid stumbling over the pronunciation as you read the exercise.

- *Escherichia coli* (esh-er-I-ke-a KOH-lie)
- *Bacillus subtilis* (bah-SIL-lus sub-til-lus)
- *Serratia marcescens* (se-RA-she-ah mar-SES-sens)
- *Staphylococcus aureus* (staf-il-oh-KOK-lus ORE-ee-us)

Why Are the Following Bacteria Used in This Exercise?

Another procedure that is used to obtain well-isolated, pure colonies is the streak-plate technique. Remember, *S. marcescens* produces red colonies; *B. subtilis*, white to cream colonies; and *E. coli*, off-white colonies. *S. aureus* colonies are typically a golden yellow color.

Medical Application

In the clinical laboratory, growth of a pure culture is absolutely necessary before any biochemical tests can be performed to identify a suspect microorganism.

PRINCIPLES

Isolated, pure colonies can be easily obtained by the **streak-plate technique.** In this technique, a bacterial mixture is transferred to the edge of an agar plate with an inoculating loop and then streaked out over the surface (**Figure 15.1a**). As the inoculating loop glides across the agar surface, individual cells will be removed from the loop, which will eventually give rise to separate, distinct colonies. Again, we assume that one colony comes from one cell. The key principle of this method is that by streaking across the entire surface of an agar plate, a dilution gradient is established allowing cells to be separated from one another. Because of this gradient, confluent growth occurs on part of the plate where the cells are not sufficiently separated, and individual, well-isolated colonies develop in other regions of the plate where few cells are deposited to form separate colonies that can be seen with the naked eye. Cells from the new colony can then be picked up with an inoculating loop and transferred to fresh media for maintenance of the pure culture.

Many kinds of media can be used to generate streak plates. For example, tryptic soy agar is a general-purpose or supportive medium. Often it is most advantageous to prepare streak plates with selective and/or differential media (**Figure 15.2**). **Selective media** favor the growth of particular microorganisms, while inhibiting the growth of others. For example, bile salts or dyes like basic fuchsin and crystal violet favor the growth of Gram-negative bacteria by inhibiting the

Figure 15.1 An Example of the Streak Plate Technique. Note the well-isolated colonies of *E. coli* (off-white), *M. luteus* (yellow), and *S. marcescens* (red). The goal of this technique is to thin the numbers of bacteria growing in each successive quadrant of the plate as it is rotated and streaked so that well-isolated colonies will appear in the fourth quadrant. (*Kathy Park Talaro*)

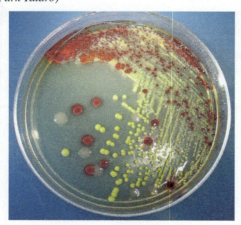

Figure 15.2 Comparison of Selective and Differential Media with General-Purpose Media. (**a**) A mixed sample of four differently pigmented bacteria are streaked onto two plates: one plate contains a general-purpose nonselective medium whereby all four species grow, and the other plate contains a selective medium, in which only one species grows. (**b**) Another mixed sample of three different bacteria is streaked onto two plates: one contains a general-purpose, nondifferential medium whereby all three species have a similar appearance, and on the other plate, all three species grow but show different reactions.

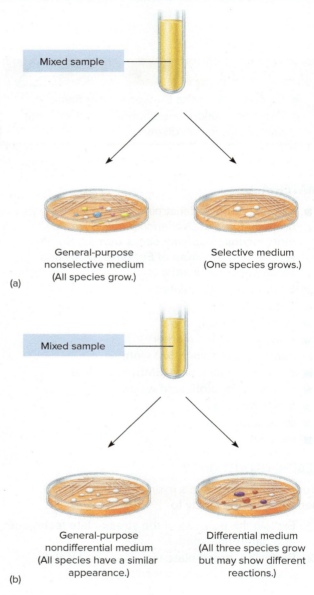

growth of Gram-positive bacteria without affecting Gram-negative organisms. **Differential media** are media that distinguish between different groups of bacteria and even permit tentative identification of microorganisms based on their biological characteristics and reactions. Blood agar is both a differential medium and an enriched one. It distinguishes between hemolytic and nonhemolytic bacteria. Hemolytic bacteria (e.g., many streptococci and staphylococci isolated from throats) produce clear zones around their colonies because of red blood cell destruction.

Two very important differential and selective media that are used to isolate and partially identify bacteria are mannitol salt agar and eosin methylene blue agar. Mannitol salt agar (MSA) is used to isolate staphylococci from clinical and nonclinical samples. It contains 7.5% sodium chloride, which inhibits the growth of most bacteria other than staphylococci. *Staphylococcus aureus* will ferment the mannitol and form yellow zones in the reddish agar (**Figure 15.3**) because phenol red becomes yellow in the presence of fermentation acids. Eosin methylene blue (EMB) agar is widely used for the detection of *E. coli* and related bacteria in water supplies and elsewhere. It contains the dyes eosin Y and methylene blue, which partially suppress the growth of Gram-positive bacteria. The dyes also help differentiate between Gram-negative bacteria. Lactose fermenters such as *Escherichia coli* will take up the dyes and form blue-black colonies with a metallic sheen (**Figure 15.4**). Lactose nonfermenters such as *Salmonella, Proteus,* and *Pseudomonas* form colorless to amber colonies.

In this exercise, we will combine the streak-plate technique with selective and differential media to isolate and partly identify *Staphylococcus aureus* and *Escherichia coli*.

Figure 15.3 **Cultures of *Staphylococcus* on Mannitol Salt Agar.** Notice that the medium has turned yellow around the growing bacteria on the left since the bacteria are able to ferment the mannitol and produce acid. The bacteria on the right half of the plate cannot ferment mannitol so the media remains pink. *(Heidi Smith)*

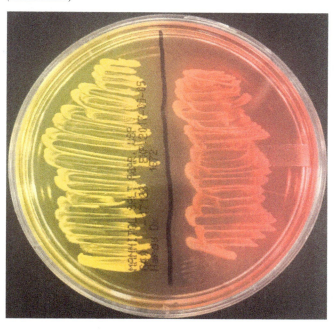

Figure 15.4 **Growth Patterns on Eosin Methylene Blue (EMB) Agar.** Eosin methylene blue agar is a medium with both selective and differential properties. On the left, bacteria that ferment lactose. On the top, bacteria that do not ferment lactose. On the right, bacteria that strongly ferment lactose. Note the metallic green sheen. On the bottom, bacteria that grow poorly (and are thus likely Gram-positive). *(Lisa Burgess/McGraw Hill)*

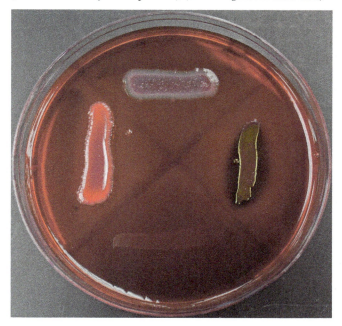

Procedure

1. Obtain 5 tryptic soy agar plates. Label the bottom of the plate with your name, date, and name of the five cultures provided by your instructor.
2. From each broth culture, aseptically remove a loopful of the bacterial mixture.
3. Streak out the loopful of bacteria on an agar plate as follows:

    a. Carefully lift the top of the Petri plate just enough to insert your inoculating loop easily (**Figure 15.5**). The top should cover the agar surface as completely as possible at all times in order to avoid contamination. Insert the inoculating loopful of bacteria and spread it over a small area at one edge of the plate as shown in **Figure 15.5** (plate 1) in order to

The Streak-Plate Technique 111

Figure 15.5 Procedure for Streak Plating. Steps in a quadrant streak plate. A typical streaking pattern is shown.

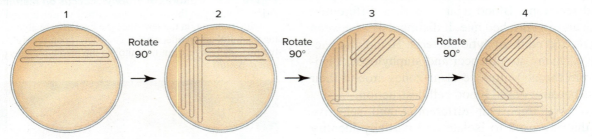

make effective use of the agar surface. This is accomplished by letting the loop rest gently on the surface of the agar and then moving it across the surface each time without digging into the agar.

b. Remove the inoculating loop and kill any remaining bacteria by flaming them. Then insert the loop under the lid and cool it at the edge of the agar.

c. Rotate the plate while carefully keeping in mind where the initial streaks ended (use the marked quadrants as a guide) and cross over the streaks you just made. Streak out the bacteria picked up as shown in plate 2 (Figure 15.5).

d. Remove the loop, flame it, cool in the agar as before, and repeat the streaking process as shown for plate 3.

e. Repeat this sequence once more to make a fourth set of streaks as shown for plate 4. Use fewer cross-streaks here than in the previous quadrant. You have now created a streak-plate!

f. Repeat the above procedure for each of the remaining bacterial cultures using fresh agar plates.

4. Using your best streak-plate technique, use the culture of *S. aureus* to inoculate an MSA plate. Inoculate the EMB agar plate using *E. coli*.

5. Incubate the plates at 37°C for 24 to 48 hours in an inverted position. Afterwards, examine each of the agar plates to determine the distribution and amount of growth and record your results in the report for this exercise. Make sure to note any color changes on the EMB and MSA plates.

Disposal

When you are finished with all of the plates and tubes, place them in the designated place for sterilization and disposal.

> **HELPFUL HINTS**
>
> - Each time the loop is flamed, allow it to cool at least 10 seconds before streaking the culture.
> - When you streak an agar plate, make sure the inoculating loop is bent at a slight angle; this minimizes the agar being gouged as you streak with the loop tip over the agar surface.
> - Use a gentle gliding stroke when making your streaks.
> - Use a loopful of culture when applying the first streak in quadrant 1 of the Petri plate. Do not return to the source tube for more culture when streaking quadrants 2 to 4.
> - An inoculated plate is always incubated in an inverted position to prevent condensation from falling onto the surface of the plate and interfering with discrete colony formation.

Laboratory Report 15

Name: _____

Date: _____

Lab Section: _____

The Streak-Plate Technique

1. Make a drawing of the distribution of the colonies on each Petri plate.

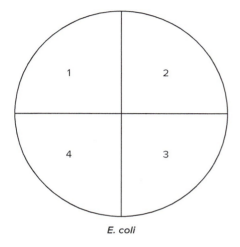

E. coli

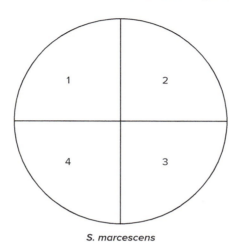

S. marcescens

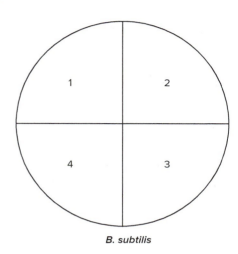

B. subtilis

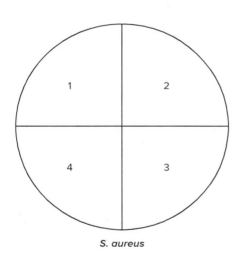

S. aureus

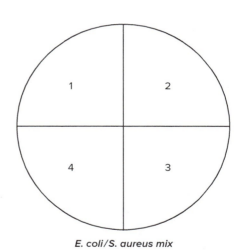

E. coli/S. aureus mix

2. From these plates, select three discrete colonies and describe them in the table below.

Bacterium	_____	_____	_____
Colony form	_____	_____	_____
Colony elevation	_____	_____	_____
Colony margin	_____	_____	_____
Colony size	_____	_____	_____
Colony color	_____	_____	_____

3. Draw your streaking patterns on mannitol salt agar and EMB agar.

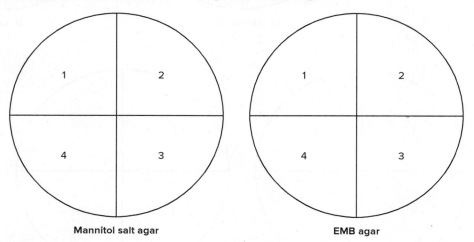

Mannitol salt agarEMB agar

114   Basic Culture Techniques

ASSESSMENT
Critical Thinking and Learning Outcomes Review

1. In the streak-plate technique, how are microorganisms diluted and spread out to form individual colonies?

2. Which area of a streak plate will contain the greatest amount of growth? The least amount of growth? Explain your answers.

3. Does each discrete colony represent the growth of one cell? Explain your answer. Why can a single colony on a plate be used to start a pure culture?

4. Why can MSA and EMB agar be described as both selective and differential media? What components of MSA and EMB agar make them selective and differential?

5. How can a streak plate become contaminated?

EXERCISE 16

The Pour-Plate Technique

SAFETY CONSIDERATIONS

- Be careful with the Bunsen burner flame and the hot water baths.

MATERIALS

- suggested bacterial strains:
  24- to 48-hour mixed tryptic soy broth culture of *Escherichia coli, Serratia marcescens,* and *Bacillus subtilis*
- 3 tryptic soy agar pour tubes
- 3 9-ml sterile 0.9% NaCl (saline) blanks
- 50°C water bath
- boiling water bath
- wax pencil or lab marker
- 3 Petri plates
- inoculating loop
- Bunsen burner
- 3 sterile 1-ml pipettes with pipettor
- container for biohazard waste
- safety glasses
- disposable gloves
- lab coat

LEARNING OUTCOMES

Upon completion of this exercise, students will demonstrate the ability to

1. Demonstrate the pour-plate technique
2. Perform a pour-plate technique to obtain isolated colonies

SUGGESTED READING IN TEXTBOOK

1. Laboratory Culture of Cellular Microbes Requires Conditions That Mimic Their Normal Habitats section 7.7; see also figure 7.31.

Pronunciation Guide

- *Escherichia coli* (esh-er-I-ke-a KOH-lie)
- *Bacillus subtilis* (bah-SIL-lus sub-til-lus)
- *Serratia marcescens* (se-RA-she-ah mar-SES-sens)

Why Are the Following Bacteria Used in This Exercise?

Another procedure that is used to obtain well-isolated, pure colonies is the pour-plate technique. Since *Serratia marcescens, Bacillus subtilis,* and *Escherichia coli* were used in the past two exercises, and the pure culture plates should have been saved, these same cultures are used in this exercise.

PRINCIPLES

The **pour-plate technique** is another technique that allows you to obtain isolated colonies and has been extensively used with bacteria, archaea, and fungi. Using this technique, the original sample is diluted

Figure 16.1 **The Pour-Plate Technique.** The original sample is diluted several times to decrease or dilute the population sufficiently. One milliliter of each dilution is then dispensed into the bottom of a Petri plate. Agar pours are then added to each plate. Isolated cells grow into colonies and can be used to establish pure cultures. The surface colonies are circular and large; subsurface colonies are lenticular (lens-shaped) and much smaller.

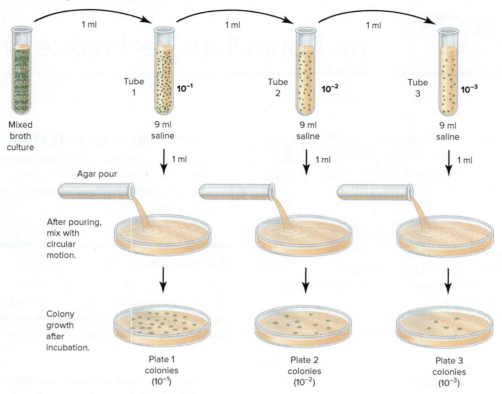

several times to reduce the microbial population sufficiently (down to 25 to 250 cells) to obtain separate colonies upon plating (**Figure 16.1**). The small volumes of serially diluted samples are added to sterile Petri plates and mixed with liquid tryptic soy agar that has been cooled to about 48°C to 50°C. Most bacteria and fungi will not be killed by the brief exposure to the warm agar. After the agar has hardened, each cell is fixed in place and will form an individual colony if the sample is dilute enough. The number of microorganisms that form colonies when cultured using spread plates or pour plates is called **colony-forming units** (**CFUs**). This value is used as a measure of the number of viable microorganisms in a sample. Assuming no chaining or cell clusters, the total number of colonies is equivalent to the number of viable microorganisms in the diluted sample. To prepare **pure cultures,** colonies growing on the surface or subsurface can be transfered into fresh medium.

Procedure

1. With a wax pencil or lab marker, label three sterile saline tubes 1 to 3.
2. Melt the tryptic soy agar deeps in a boiling water bath and cool in a 50°C bath for at least 15 minutes (*see Figure 13.3*).
3. With a wax pencil or lab marker, label the bottoms of three Petri plates 1 to 3, and add your name and date.
4. Inoculate saline tube 1 with 1 ml of the MIXED bacterial culture using aseptic technique (*see Figure 15.3*) and mix thoroughly. This represents a 10^{-1} dilution.
5. Using aseptic technique, immediately inoculate tube 2 with 1 ml from tube 1, which results in a 10^{-2} dilution.
6. Using aseptic technique, mix the contents of tube 2 and use 1 ml of this dilution to inoculate tube 3, resulting in a 10^{-3} dilution.

7. After tube 3 has been inoculated, mix its contents, remove the cap, flame the top, and aseptically transfer 1 ml into Petri plate 3. Then inoculate plates 1 and 2 in the same way, using 1 ml from tubes 1 and 2, respectively.
8. Add the contents of the melted tryptic soy agar pours to the Petri plates. Gently mix each agar plate with a circular motion while keeping the plate flat on the benchtop. Do not allow any agar to splash over the side of the plate! Set the plate aside to cool and harden.
9. Incubate the plates at 37°C for 24 to 48 hours in an inverted position.
10. Examine the pour plates and record your results in the report for this exercise.

Disposal

When you are finished with the pour plates, pipettes, and tubes, discard them in the designated place for sterilization or disposal.

> **HELPFUL HINTS**
>
> - Always allow sufficient time for the agar deeps to cool in the water bath after they have been boiled prior to the addition of bacteria.
> - When the poured agar has solidified in the Petri plates, it will become lighter in color and cloudy (opaque) in appearance. Wait until this occurs before attempting to move the plates.
> - An inoculated plate is always incubated in an inverted position to prevent condensation from falling onto the surface of the plate and interfering with discrete colony formation.

NOTES

Laboratory Report 16

Name: _____

Date: _____

Lab Section: _____

The Pour-Plate Technique

1. Examine each of the agar plates for colony distribution and amount of growth. Look for discrete surface colonies and record your results. Do the same for the subsurface colonies. Color each species of bacterium in a different color or label each. Fill in the table.

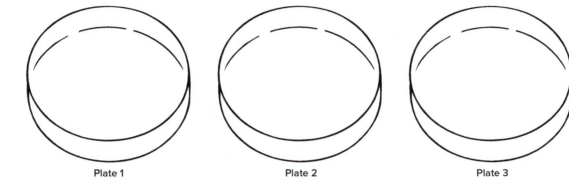

Plate 1   Plate 2   Plate 3

Surface Colonies

Form	_____	_____	_____
Elevation	_____	_____	_____
Margin	_____	_____	_____
Number	_____	_____	_____

Subsurface Colonies

Form	_____	_____	_____
Elevation	_____	_____	_____
Margin	_____	_____	_____
Number	_____	_____	_____

2. In summary, what is the major purpose of this exercise?

ASSESSMENT
Critical Thinking and Learning Outcomes Review

1. How do the results of the pour-plate method compare with those obtained using the streak-plate and spread-plate methods?

2. What is the main advantage of the pour-plate method over other methods of bacterial colony isolation? What are some problems?

3. Why are the surface colonies on a pour plate larger than those within the medium?

4. Why doesn't the 48°C to 50°C temperature of the melted agar kill most of the bacteria?

5. Explain how the pour-plate method can be used to isolate fungi.

6. Why is it important to invert the Petri plates during incubation?

EXERCISE 17
Cultivation of Anaerobic Bacteria

SAFETY CONSIDERATIONS

- Be careful with the Bunsen burner flame and water baths.
- The various gas-generating envelopes that liberate hydrogen react rapidly with water to produce flammable gas. Observe appropriate precautions when handling these products.
- The anaerobic jar used must not leak and the lid seal should be tight.
- Envelopes that have been activated with water but not used in anaerobic jars should be kept in a ventilating hood away from open flames or sparks for approximately 30 minutes until the reaction has subsided.

MATERIALS

- suggested bacterial strains:
  24- to 48-hour Eugon broth cultures of *Pseudomonas aeruginosa* and *Escherichia coli;* thioglycollate broth cultures of *Clostridium sporogenes*
- Eugon agar deeps
- boiling water bath
- 50°C water bath
- thioglycollate broth tubes
- inoculating loop
- tryptic soy agar plates
- Petri plates containing Brewer's anaerobic agar
- sterilized Brewer's anaerobic covers
- GasPak Jar, GasPak® Disposable Anaerobic System (BBL™ Microbiology Systems), GasPak Pouch (BBL™), Difco Gas Generating envelopes, BioBag (Marion Scientific), Anaero-Pack System (KEY ScientificProducts), Anaerobic Growth Bag (Ward's)
- Oxyrase for Broth, Oxyrase for Agar, OxyDish, OxyPlates (Oxyrase, Inc.)
- trypticase soy broth tubes containing 0.1 ml Oxyrase for Broth
- OxyDishes containing Oxyrase for Agar
- test tubes
- rubber stoppers
- test-tube rack
- wax pencil or lab marker
- container for biohazard waste
- safety glasses
- disposable gloves
- lab coat

LEARNING OUTCOMES

Upon completion of this exercise, students will demonstrate the ability to

1. Describe why some bacteria need an anaerobic environment to grow
2. Explain some of the different methods that are used to cultivate anaerobic bacteria
3. Successfully cultivate several anaerobic bacteria

SUGGESTED READING IN TEXTBOOK

1. Environmental Factors Affect Microbial Growth, section 7.5; see also figure 7.20.
2. Anaerobic Respiration Uses the Same Three Steps as Aerobic Respiration, section 11.5; see also table 11.2 and figure 11.16.
3. Fermentation Does Not Involve an Electron Transport Chain, section 11.6; see also figures 11.17 and 11.18, and table 11.3.

Pronunciation Guide

- *Clostridium sporogenes* (klos-STRID-ee-um spo-ROJ-ah-nees)
- *Escherichia coli* (esh-er-I-ke-a KOH-lie)
- *Pseudomonas aeruginosa* (soo-do-MO-nas a-ruh-jin-OH-sah)

Why Are the Following Bacteria Used in This Exercise?

This exercise is designed to give you expertise in the cultivation of anaerobic bacteria. Thus the authors have chosen an obligate anaerobe (*C. sporogenes*),

123

a facultative anaerobe (*E. coli*), and for comparison, a strict aerobe (*P. aeruginosa*). *Pseudomonas aeruginosa* is a straight or slightly curved motile rod (1.5 to 3.0 μm in length) that has a polar flagellum. This bacterium is aerobic, having a strictly respiratory type of metabolism with oxygen as the usual terminal electron acceptor. It is widely distributed in nature. *Clostridium sporogenes* is a straight rod 1.3 to 16.0 μm in length, motile with peritrichous flagella. Endospores are oval and subterminal and distend the cell. *C. sporogenes* is obligatorily anaerobic and it is widespread in the environment. *Escherichia coli* grows readily on nutrient agar. The colonies may be smooth, low convex, and moist. *E. coli* is facultatively anaerobic, having both a respiratory and a fermentative type of metabolism.

Medical Application

From a clinical laboratory perspective, many common pathogenic bacteria live under anaerobic conditions and vary depending on body site. Examples include: blood—*Bacteroides fragilis*; intestine—*Clostridium, Bacteroides*; genital area—*Actinomycetes, Bacteroides, Fusobacterium, Clostridium, Mobiluncus*; and skin and soft tissue—*Clostridium perfringens, Bacteroides fragilis, Peptostreptococcus*.

PRINCIPLES

One of the environmental factors to which bacteria and other microorganisms are quite sensitive is the presence of O_2. For example, some microorganisms will grow only in the presence of O_2 and are called **obligate aerobes**. **Facultative anaerobes** will grow either aerobically or in the absence of O_2, but better in its presence. Strict **obligate anaerobes** will grow only in the absence of O_2 and are actually harmed by its presence. **Aerotolerant anaerobes** are microorganisms that cannot use O_2 but are not harmed by it either. Finally, microorganisms that require a small amount of O_2 for normal growth but are inhibited by O_2 at normal atmospheric tension are called **microaerophiles**. These variations in O_2 requirements can be easily seen by inoculating a tube of molten agar with the bacterium in question, mixing the agar thoroughly without aerating it, and allowing it to solidify. The bacteria will grow in the part of the **agar deep culture** that contains the proper O_2 concentration (**Figure 17.1**).

The damaging effects of O_2 on anaerobic bacteria create difficult culturing problems. Ideally, one should provide not only an O_2-free environment, but also one that has an adequate amount of moisture for bacterial growth. It is also necessary to have CO_2 present for the growth of many anaerobic bacteria. There are a number of ways in which anaerobic bacteria may be cultured. Four of the most useful will be described.

One of the most convenient approaches is to employ a specially designed commercial anaerobic broth. Two of the most useful are cooked meat medium and thioglycollate broth. Thioglycollate medium can be purchased with methylene blue or resazurin as an oxidation-reduction indicator. When this medium begins to turn bluish or reddish, it is becoming too aerobic for the culture of anaerobic bacteria (**Figures 17.2a–e** and **17.3a–c**).

Anaerobic bacteria may also be grown in special Petri plates without the use of complex and expensive incubators. One of the most convenient plate methods

Figure 17.1 The Appearance of Various Agar Deep Cultures. Each dot represents an individual bacterial colony within the agar or on its surface. The surface, which is directly exposed to atmospheric oxygen, will be aerobic (oxic). The oxygen content of the medium decreases with depth until the medium becomes anaerobic (anoxic) toward the bottom of the tube. Notice how facultative anaerobes, although growing throughout the agar deep, grow in higher densities close to the surface of the tube.

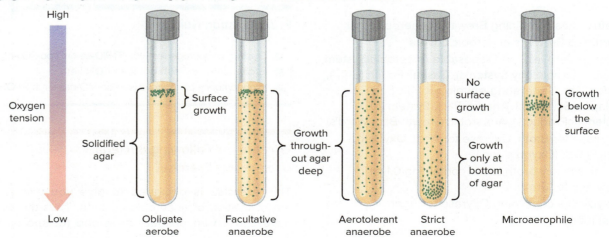

Figure 17.2 Thioglycollate Tubes. (**a**) Control, (**b**) aerotolerant, (**c**) microaerophile, (**d**) obligate anaerobe, and (**e**) facultative anaerobe. *(Lisa Burgess/McGraw Hill)*

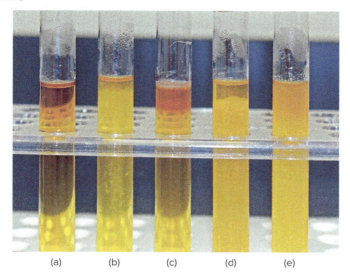

Figure 17.3 Effects of Oxygen on Growth. Close-up photos showing the relative positions of (**a**) facultative anaerobe, (**b**) aerobe, and (**c**) anaerobe microorganisms in fluid thioglycollate medium. The red dye reazurin indicates the presence of oxygen. *(Lisa Burgess/McGraw Hill)*

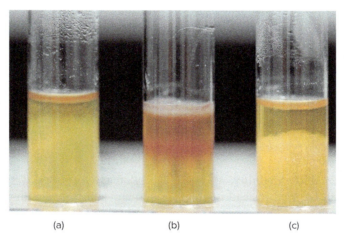

Figure 17.4 Brewer's Petri Plate. An anaerobic space is created between the Brewer cover and the thioglycollate agar.

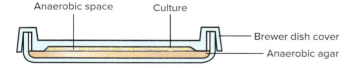

uses the **Brewer's** (named after John H. Brewer, an industrial bacteriologist, in 1942) **anaerobic Petri plate,** together with special anaerobic agar (**Figure 17.4**). Brewer's special cover fits on a normal Petri plate bottom in such a way that its circular ridge rests on the agar, thereby protecting most of the surface from the exposure to O_2. Brewer's anaerobic agar contains a high concentration of thioglycollic acid. The free sulfhydryl groups of thioglycollate reduce any O_2 present and create an anaerobic environment under the Brewer cover.

Another way of culturing bacteria anaerobically on plates is the **GasPak Anaerobic System.** In the GasPak System (**Figure 17.5**), hydrogen and CO_2 are generated by a GasPak envelope after the addition of water. A palladium catalyst (pellets) in the chamber lid catalyzes the formation of water from hydrogen and O_2, thereby removing O_2 from the sealed chamber.

Cultivation of Anaerobic Bacteria

Figure 17.5 The GasPak System. (a) Schematic diagram of the GasPak System. Hydrogen and carbon dioxide are generated in the GasPak envelope. The catalyst (palladium) in the chamber lid catalyzes the formation of water from hydrogen and oxygen, thereby removing oxygen from the sealed chamber. (b) A GasPak System in use. *(b: Lisa Burgess/McGraw Hill)*

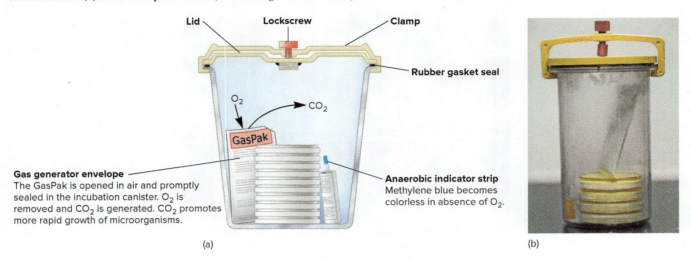

Figure 17.6 Procedure for GasPak Pouch. (a) Dispense GasPak liquid activating reagent into channel of pouch. (b) Place plates inside the pouch. (c) Lock in anaerobic environment with sealing bar and incubate. Note the safety gloves on the hands.

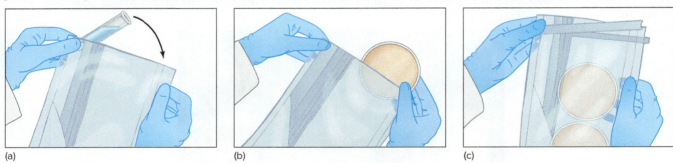

For greater convenience and visibility, **GasPak pouches** (**Figure 17.6**) can be used instead of the regular GasPak incubation chamber. In this procedure, a special activating reagent is dispensed into the reagent channel (**Figure 17.6a**). Inoculated plates are then put into the pouch (**Figure 17.6b**). The anaerobic environment is locked in with a sealing bar (**Figure 17.6c**) and the pouch incubated. Growth can be observed at one's convenience.

One of the simplest methods for growing anaerobes has recently been introduced by Oxyrase, Inc. One simply adds 0.1 ml of **Oxyrase for Broth** to 5.0 ml of broth medium (Mueller-Hinton, Eugon, Trypticase Soy, Nutrient, Schaedler, Columbia, or Brain Heart Infusion). Most anaerobes can be inoculated immediately after this addition. Also from Oxyrase, Inc. are the **OxyDish** and **Oxyrase for Agar.** Oxyrase for Agar is mixed with your agar medium and poured into an OxyDish. The OxyDish contains an inner ring that forms a tight seal with the agar surface, which can be easily broken and reformed. Within minutes, the enzyme system and substrates in the Oxyrase for Agar reduce oxygen in the agar medium and the trapped headspace in the dish. The dish can be opened and closed several times while still maintaining an anaerobic environment. This system eliminates the complications and expense of bags, jars, anaerobic incubators, and chambers.

Finally, a more up-to-date method for working with anaerobes is to use an **anaerobic chamber** (**Figure 17.7**). An anaerobic chamber is an enclosed chamber that can maintain an anaerobic environment. Notice the special airlock port, which can be filled with an inert gas such as nitrogen or carbon dioxide, that is used to add or remove items. The airtight gloves enable a person to handle items within the chamber. As in the GasPak systems, a palladium catalyst stored in a fan box and a hydrogen gas mix of 5% ensure that any oxygen that makes it into the chamber is removed and converted to water.

Figure 17.7 An Anaerobic Chamber. (a) Components of an anaerobic chamber. An airlock is used for transferring materials into and out of the chamber. Work gloves are used to reach into the anoxic chamber. (b) An anaerobic chamber in use by an instructor and students. *(a: Coy Laboratory Products, Inc.; b: Nathan Rigel/McGraw Hill)*

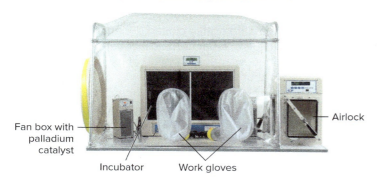
(a)

(b)

Procedure

The Relationship of O_2 to Bacterial Growth

1. Melt three Eugon agar deeps and heat them in a boiling water bath for a few minutes in order to drive off any O_2.
2. Cool the deeps in a water bath 48°C to 50°C.
3. With a wax pencil or lab marker, label each tube with the name of the bacterium to be inoculated, your name, and date.
4. Using aseptic technique (*see Figure 5.3*), inoculate each cooled deep with 1 or 2 loopfuls of one of each of the three different bacteria (*P. aeruginosa, C. sporogenes,* and *E. coli*).
5. Mix the bacteria throughout the agar without aerating it by rolling each tube between the palms of your hands.
6. Allow the agar to harden and incubate the three tubes for 24 to 48 hours at 37°C.
7. Observe each tube for growth and record your results in the report for this exercise.

Broth Culture of Anaerobic Bacteria

1. With a wax pencil or lab marker, label three freshly steamed thioglycollate broth tubes with *P. aeruginosa, C. sporogenes,* and *E. coli,* as well as your name and date.
2. Using aseptic procedures, inoculate the three broth tubes. Do not shake these tubes to avoid oxidizing the medium! Methylene blue or resazurin is present in the medium as an oxidation-reduction indicator. If more than one-third of the broth is bluish or reddish in color, the tube should be reheated in a water bath in order to drive off the O_2 before use.
3. Incubate the tubes at 37°C for 24 to 48 hours.
4. Observe each tube for growth and record your results in the report for this exercise.

Plate Culture of Anaerobic Bacteria

1. With a wax pencil or lab marker, divide the bottom of each of the culture plates (one regular tryptic soy agar Petri plate, one Petri plate containing Brewer's anaerobic agar, and one Brewer's plate with cover) in thirds and label each section with *P. aeruginosa, C. sporogenes,* and *E. coli*.
2. With an inoculating loop, streak each third with the proper bacterium.
3. Incubate the tryptic soy agar plate inverted at 37°C in a regular aerobic incubator.
4. Carefully cover the Brewer's plate with a sterile Brewer anaerobic cover. The circular rim of the cover should press against the agar surface, but not sink into it. Incubate inverted at 37°C.
5. Place the regular anaerobic agar plate in a GasPak Anaerobic Jar or a GasPak Disposable Anaerobic Pouch.
6. Follow the instructions supplied with these products.
7. Incubate at 37°C for 24 to 48 hours.
8. Record your results in the report for this exercise.

Simplified Method for Growing Anaerobes

1. Using aseptic technique, inoculate each trypticase soy broth tube containing Oxyrase for Broth with *P. aeruginosa, C. sporogenes,* and *E. coli*.
2. Repeat the inoculation on the OxyDish containing tryptic soy agar and Oxyrase for Agar by streaking each bacterium into a third of this plate.

Cultivation of Anaerobic Bacteria

3. Incubate for 24 to 48 hours at 37°C.
4. Record your growth results in the report for this exercise.

Disposal

When you are finished with all of the anaerobic tubes and plates, discard them in the designated place for sterilization and disposal.

> **HELPFUL HINTS**
>
> - If screw-cap tubes are used with any tubed media (such as thioglycollate broth), the cap should be tightly closed during incubation outside of an anaerobic incubator to avoid unnecessary penetration of oxygen from the atmosphere.
> - In an anaerobic GasPak jar, the caps should be loosened slightly to allow neutralization of oxygen within the tubes by the hydrogen gas generated within the system.
> - During incubation, if the indicator strip turns blue, you must remove the GasPak envelope and replace it with a new one.

Name: _____

Date: _____

Laboratory Report 17

Lab Section: _____

Cultivation of Anaerobic Bacteria

1. Describe the relationship of O₂ to bacterial growth using Eugon culture tubes.

Drawing of growth in Eugon culture tubes

P. aeruginosa   C. sporogenes   E. coli

Growth pattern   _____   _____   _____

Cracking of agar due to gas (+ or −)   _____   _____   _____

Classification according to growth pattern   _____   _____   _____
 (Figure 17.1)

2. Draw and describe growth on the anaerobic broth cultures of bacteria using thioglycollate tubes.

Drawing of culture in anaerobic broth culture tubes

P. aeruginosa   C. sporogenes   E. coli

Comments   _____   _____   _____

129

3. Drawings of plate cultures of aerobic and anaerobic bacteria.

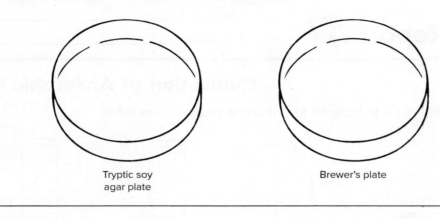

Tryptic soy agar plate   Brewer's plate

Comments _____   _____

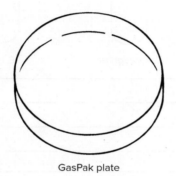

GasPak plate   OxyDish

Comments _____   _____

4. Describe the growth on the Oxyrase broth tubes.

Drawing of growth in Oxyrase broth tubes

	P. aeruginosa	C. sporogenes	E. coli
Growth pattern	_____	_____	_____
Classification according to growth pattern (Figure 17.1)	_____	_____	_____

ASSESSMENT
Critical Thinking and Learning Outcomes Review

1. Explain how an anaerobic atmosphere can be created in a jar.

2. Differentiate between the following:

    a. an obligate anaerobe

    b. an obligate aerobe

    c. a facultative anaerobe

    d. an aerotolerant anaerobe

    e. a microaerophile

3. What are the ingredients in Brewer's anaerobic agar that remove O_2 from the medium? Briefly explain how an Oxyrase plate works.

4. Of the methods used in this exercise to create an anaerobic environment, which works the best, and why?

5. What is the role of the palladium catalyst in an anaerobic jar (the GasPak system) and the anaerobic chamber?

EXERCISE 18 Determination of Bacterial Numbers

SAFETY CONSIDERATIONS
- Be careful with the Bunsen burner flame and water baths.

MATERIALS
- suggested bacterial strains:
  24-hour tryptic soy broth culture of *Escherichia coli*
- 4 sterile 99-ml saline or phosphate buffer blanks
- 1-ml or 1.1-ml pipettes with pipettor
- 6 Petri plates
- 6 agar pour tubes of tryptone glucose yeast agar (plate count agar)
- 50°C water bath
- boiling water bath
- Bunsen burner
- cuvettes
- spectrophotometer
- 4 tubes of tryptic soy broth
- wax pencil or lab marker
- container for biohazard waste
- safety glasses
- disposable gloves
- lab coat

LEARNING OUTCOMES
Upon completion of this exercise, students will demonstrate the ability to

1. Describe several different ways to quantify the number of bacteria in a given sample
2. Determine quantitatively the number of viable cells in a bacterial culture by the standard plate count technique
3. Measure the turbidity of a culture with a spectrophotometer and relate this to the number (biomass) of bacteria

SUGGESTED READING IN TEXTBOOK
1. Microbial Population Size Can Be Measured Directly or Indirectly, section 7.8; see also figures 7.33-7.35.

Pronunciation Guide
- *Escherichia coli* (esh-er-I-ke-a KOH-lie)

Why Is the Following Bacterium Used in This Exercise?

All of the learning objectives of this exercise are related to determining bacterial numbers. When working with large numbers and a short time frame, one of the most reliable microorganisms is one that has been used in previous experiments, namely, *Escherichia coli*. *E. coli* has a generation time at 37°C of 0.35 hours. Thus it reproduces very rapidly and is easy to quantify (i.e., the number [biomass] of *E. coli* cells in a bacterial culture can be easily determined by spectrophotometry).

Medical Application

In the clinical laboratory and in research laboratories, it is frequently necessary to have an accurate count of living (viable) cells in a given culture. If done properly, counting procedures can produce very accurate results.

133

PRINCIPLES

Many studies require the quantitative determination of bacterial populations. The two most widely used methods for determining bacterial numbers are the **standard, or viable, plate count method** and **spectrophotometric (turbidimetric) analysis.** Although the two methods are somewhat similar in the results they yield, there are distinct differences. For example, the standard plate count method reveals quantitative information related only to live bacteria. The spectrophotometric analysis is based on turbidity and indirectly measures all bacteria (cell biomass), dead or alive.

The standard plate count method consists of diluting a sample with sterile saline or phosphate buffer diluent until the bacteria are dilute enough to count accurately. That is, the final plates in the series should have between 25 and 250 colonies. Fewer than 25 colonies are not acceptable for statistical reasons, and more than 250 colonies on a plate are likely to produce colonies too close to each other to be distinguished as distinct **colony-forming units (CFUs).** The assumption is that each viable bacterial cell is separate from all others and will develop into a single discrete colony. Thus the number of colonies should give the number of live bacteria that can grow under the incubation conditions employed. A wide series of dilutions (e.g., 10^{-4} to 10^{-10}) is normally plated because the exact number of live bacteria in the sample is usually unknown. Greater accuracy is achieved by plating duplicates or triplicates of each dilution.

A more rapid and sensitive method for measuring bacterial growth and cell numbers (biomass) is **spectrophotometry** (**Figure 18.1**). Spectrophotometry depends on the fact that bacterial cells scatter light that strikes them. Because microbial cells in a population are of roughly constant size, the amount of light scattering is directly proportional to the biomass of cells present and indirectly related to cell number. The extent of light scattering (i.e., decrease in transmitted light) can be measured by a **spectrophotometer** and is called the **absorbance** (optical density) of the medium. This method is faster than the standard plate count method but is limited because sensitivity is restricted to bacterial suspensions of 10^7 cells or greater.

DILUTION RATIOS

According to the *American Society for Microbiology Style Manual,* dilution ratios may be reported with either colons (:) or shills (/), but note that there is a difference between them. A shill indicates the ratio of a part to a whole; for example, 1/2 means 1 of 2 parts,

Figure 18.1 Turbidity and Microbial Mass Measurement. Determination of cell numbers (microbial mass) is by light absorption. As the bacterial population and turbidity increase, more light is scattered and the absorbance reading (absorbance) given by the spectrophotometer increases.

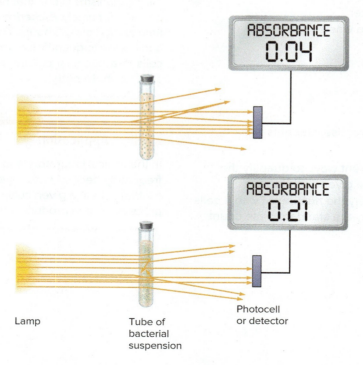

134   Basic Culture Techniques

with a total of 2 parts. A colon indicates the ratio of 1 part to 2 parts, with a total of 3 parts. Thus 1/2 equals 1:1, but 1:2 equals 1/3.

Procedure

Standard Plate Count

1. With a wax pencil or lab marker, label the bottom of six Petri plates with the following dilutions: 10^{-4}, 10^{-5}, 10^{-6}, 10^{-7}, 10^{-8}, and 10^{-9}. Label four bottles of saline or phosphate buffer 10^{-2}, 10^{-4}, 10^{-6}, and 10^{-8}.
2. Using aseptic technique, the initial dilution is made by transferring 1.0 ml of liquid sample or 1 g of solid material to a 99-ml sterile saline blank (**Figure 18.2**). This is a 1/100 or 10^{-2} dilution (*see appendix A*). Cap the bottle.
3. The 10^{-2} blank is then shaken vigorously 25 times by placing one's elbow on the bench and moving the forearm rapidly in an arc from the bench surface and back (**Figure 18.3**). This serves to distribute the bacteria and break up any clumps of bacteria that may be present.
4. Immediately after the 10^{-2} blank has been shaken, uncap it and using a new pipette, aseptically transfer 1.0 ml to a second 99-ml saline blank. Since this is a 10^{-2} dilution, this second blank represents a 10^{-4} dilution of the original sample. Cap the bottle.
5. Shake the 10^{-4} blank vigorously 25 times and using a new pipette, transfer 1.0 ml to the third 99-ml blank. This third blank represents a 10^{-6} dilution of the original sample. Cap the bottle. Repeat the process once more to produce a 10^{-8} dilution.
6. Shake the 10^{-4} blank again and using a new pipette, aseptically transfer 1.0 ml to one Petri plate and 0.1 ml to another Petri plate. Do the same for the 10^{-6} and the 10^{-8} blanks (Figure 18.2).

Figure 18.2 Quantitative Plating Procedure. After dilutions are transferred to empty Petri plates, molten agar is added. Note the safety gloves on the hands.

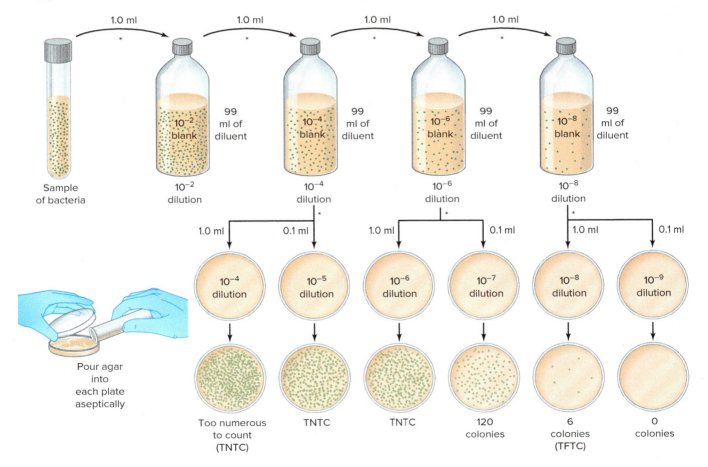

*Discard pipette after each transfer.

Determination of Bacterial Numbers 135

Figure 18.3 Standard Procedure for Shaking Water Blanks.

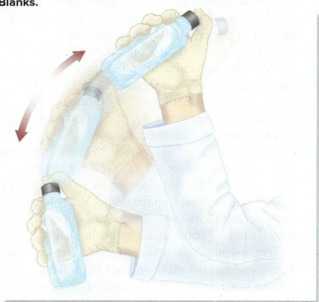

Figure 18.4 Serial Dilution Plates of *E. coli.* (Lisa Burgess/McGraw Hill)

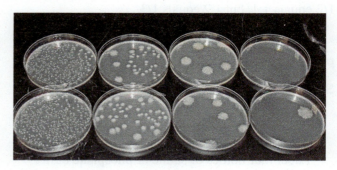

Then, the number of bacteria in 1 ml of the original sample can be calculated as follows:

Bacteria/ml = $(130) \div (10^{-6}) = 1.3 \times 10^8$

or

130,000,000.

11. Record your results in the report for this exercise.

Turbidimetry Determination of Bacterial Numbers

1. Put one empty tube and four tubes of the sterile tryptic soy broth in a test-tube rack. With the exception of the empty tube, each tube contains 3 ml of sterile broth. Use four of these tubes (tubes 2 to 5) of broth to make four serial dilutions of the culture (**Figure 18.5**). Make sure to mix by gently pipetting up and down, not by vortexing.
2. Standardize and use the spectrophotometer (**Figure 18.6**) as follows:

   a. Turn on the spectrophotometer.
   b. Set the correct wavelength in nanometers (550 to 600 nm). Your instructor will inform you which wavelength to use.
   c. Place the cuvette that contains just sterile broth in the cuvette holder. This tube is called the **blank** because it has a sample concentration equal to zero. It should therefore have an absorbance of zero (or a transmittance of 100%). This is at the right end of the scale.
   d. Place the other cuvettes, which contain the diluted bacterial suspension, in the cuvette chamber one at a time. Repeat step c between experimental readings to confirm settings.

7. Remove one agar pour tube from the 48°C to 50°C water bath. Carefully remove the cover from the 10^{-4} Petri plate and aseptically pour the agar into it. The agar and sample are immediately mixed by gently moving the plate in a figure-eight motion while it rests on the tabletop. Repeat this process for the remaining five plates.
8. After the pour plates have cooled and the agar has hardened, they are inverted and incubated at 37°C for 24 hours or 20°C for 48 hours.
9. At the end of the incubation period, select all of the Petri plates containing between 25 and 250 colonies. Plates with more than 250 colonies cannot be counted and are designated **too numerous to count (TNTC).** Plates with fewer than 25 colonies are designated **too few to count (TFTC).** Count the colonies on each plate (**Figure 18.4**) using a colony counter, a handheld counter, or a smartphone app, as indicated by your instructor.
10. Calculate the number of bacteria (CFU) per milliliter or gram of sample by dividing the number of colonies by the dilution factor. The number of colonies per milliliter reported should reflect the precision of the method and should not include more than two significant figures. For example, suppose the plate of the 10^{-6} dilution yielded a count of 130 colonies.

Figure 18.5 Two-fold Serial Dilution for Standard Curve.

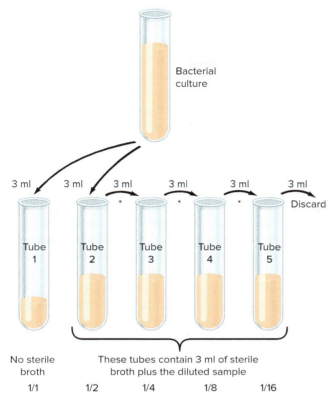

*Discard pipette after each transfer.

e. Close the hatch and read the absorbance values of each bacterial dilution, and record your values. Remember to mix the bacterial suspension just before reading its absorbance.

f. Record your values in the report for this exercise. Using the plate count data, calculate the colony-forming units per milliliter for each dilution.

Disposal

When you are finished with all of the pipettes, tubes, and plates, discard them in the designated place for sterilization and disposal.

> **HELPFUL HINTS**
>
> - When mixing dilution tubes, do not use a vortex mixer.
> - Always clean the cuvette by wiping with a Kimwipe or other lint-free paper towel before it is inserted into the cuvette holder. Do not use a paper towel, as it may scratch the glass.
> - Always mix the original culture tubes so the bacteria are uniformly dispersed.
> - Read the absorbance immediately after you insert the cuvette; otherwise, the bacteria may settle to the bottom of the cuvette and the absorbance reading will be incorrect.
> - Always align the reference line on the cuvette to the mark at the front of the holder.

Figure 18.6 Typical Spectrophotometers. (**a**) An analog model. (**b**) A digital model that features direct concentration readout. (**c**) A digital model with a compartment for multiple samples and programmable capabilities. *(Javier Izquierdo/McGraw Hill)*

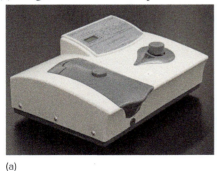
(a)

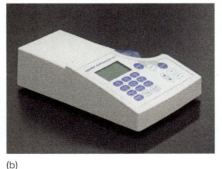
(b)

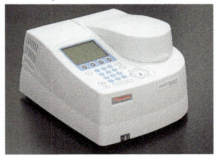
(c)

Determination of Bacterial Numbers 137

NOTES

Laboratory Report 18

Name: _____

Date: _____

Lab Section: _____

Determination of Bacterial Numbers

1. Record your observations and calculated bacterial counts per milliliter in the following table.

Petri Plate	Dilution	ml of Dilution Plated	Number of Colonies	Bacterial Count per ml of Sample[a]
1	_____	_____	_____	_____
2	_____	_____	_____	_____
3	_____	_____	_____	_____
4	_____	_____	_____	_____

[a]This value is also expressed as colony-forming units per milliliter (CFU/ml).

2. Record your data from the turbidimetry experiment in the following table.

Turbidity-Absorbance Standard Curve		
Dilution	**CFU/ml**	**Absorbance**
Undiluted	_____	_____
1/2 dilution	_____	_____
1/4 dilution	_____	_____
1/8 dilution	_____	_____
1/16 dilution	_____	_____

3. Construct a cell biomass standard curve by plotting the absorbance on the *y*-axis and the dilution ratios on the *x*-axis.

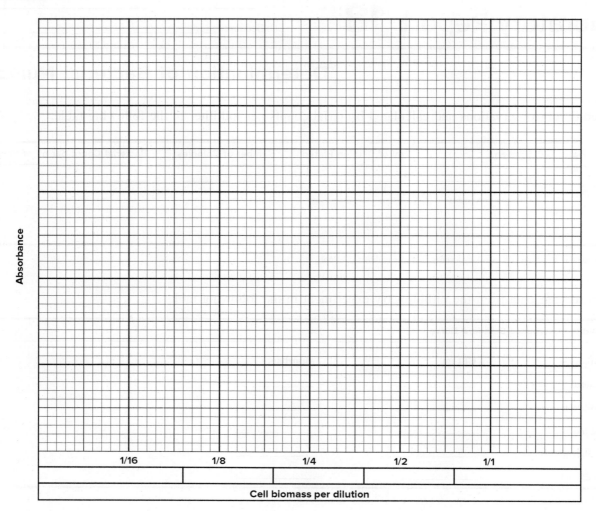

ASSESSMENT
Critical Thinking and Learning Outcomes Review

1. Why is the viable plate count technique considered to be an indirect measurement of cell density, whereas the turbidimetry method is not a "count" at all?

2. Why is absorbance used in constructing a calibration curve instead of percent transmittance?

3. What is the purpose of constructing a calibration curve?

4. Why is it necessary to perform a plate count in conjunction with the turbidimetry procedure?

Determination of Bacterial Numbers

5. Give several reasons why it is necessary to shake the water blanks 25 times.

6. What is a CFU?

7. How would you prepare a series of dilutions to get a final dilution of 10^{-10}? Outline each step.

8. Why was 550 to 600 nm used in the spectroscopy portion of this experiment?

9. How would you define cell biomass?

10. What are some advantages to spectrophotometric determination of bacterial numbers? Some disadvantages?

PART 4
Microbial Biochemistry

Bacteria accomplish growth and multiplication using nutrients obtained from the environment. The biochemical transformations that occur both inside and outside of bacteria are governed by biological catalysts called **enzymes.**

This part of the laboratory manual presents exercises (**differential** or **biochemical tests**) that have been designed to experimentally investigate for some of the biochemical activities of bacteria. This will be accomplished by observing the ability of bacteria to use enzymes and degrade carbohydrates, lipids, proteins, and amino acids. The metabolism of these organic molecules often produces by-products that can be used in the identification and characterization of bacteria.

After completing the exercises in part four, you will be able to demonstrate the ability to use appropriate microbiological media and test systems. This will meet the following American Society for Microbiology Core Curriculum skills:

- **using biochemical test media**
- **accurately recording macroscopic observations**

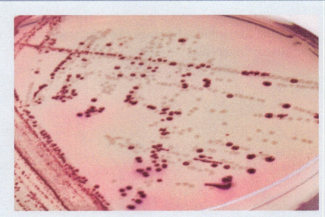

Lisa Burgess/McGraw Hill

The diversity of microbial metabolic processes is truly astonishing. The ability to break down and modify so many different compounds in order to obtain carbon, energy, and electrons varies across even the closest of species. Biochemical tests like the ones used in this section enable us not only to understand important details about the metabolic capabilities of a microbe, but also to determine what makes it different from closely related microorganisms. These are skills that are extremely useful in every subdiscipline of microbiology, including medical, applied, food, and environmental microbiology.

EXERCISE 19

Carbohydrates I: Fermentation and β-Galactosidase Activity

SAFETY CONSIDERATIONS

- Be careful with the Bunsen burner flame.
- Seventy percent ethanol is flammable—keep away from open flames.
- *Salmonella* spp. are potential pathogens (BSL2). Use aseptic technique throughout this experiment.
- Keep all culture tubes upright in a test-tube rack or in empty cans.

MATERIALS

- suggested bacterial strains:
  Escherichia coli, *Alcaligenes faecalis*, *Salmonella enterica* ser. Typhimurium, and a 7- to 10-day Sabouraud dextrose plate culture of *Saccharomyces cerevisiae*
- phenol red dextrose, lactose, and sucrose peptone broths with Durham tubes
- Bunsen burner
- inoculating loop
- test-tube rack
- incubator set at 37°C
- ONPG disks, ONPG Hardy Disks, or ONPG tablets for β-galactosidase
- inoculating needle
- 1-ml pipette and pipettor
- sterile 0.85% NaCl (for β-galactosidase test)
- 3 sterile test tubes
- wax pencil or lab marker
- container for biohazard waste
- safety glasses
- disposable gloves
- lab coat

LEARNING OUTCOMES

Upon completion of this exercise, students will demonstrate the ability to

1. Explain the biochemical process of fermentation
2. Describe how the carbohydrate fermentation patterns of some bacteria result in the production of an acid, or an acid and a gas
3. Summarize how a Durham tube or a tryptic agar base tube and a differentiation disk can be used to detect acid and gas production
4. Perform a carbohydrate and a β-galactosidase fermentation test
5. Explain the function of enzymes, using β-galactosidase as an example

SUGGESTED READING IN TEXTBOOK

1. Fermentation Does Not Involve an Electron Transport Chain, section 11.6; see also figures 11.17 and 11.18.

Pronunciation Guide

- *Alcaligenes faecalis* (al-kah-LIJ-e-neez fee-KAL-iss)
- *Escherichia coli* (esh-er-I-ke-a KOH-lie)
- *Salmonella enterica* (sal-mon-EL-ah en-TEH-ree-kah)
- *Saccharomyces cerevisiae* (sak-a-row-MY-sees seri-VISS-ee-eye)
- *Staphylococcus aureus* (staf-il-oh-KOK-kus ORE-ee-us)

Why Are the Following Microorganisms Used in This Exercise?

In this exercise, you will observe how microbial fermentation can yield acid, gas, or acid and a gas. You also will test microorganisms for the presence of the enzyme β-galactosidase, indicating that the organism can use lactose as the sole carbon source. To accomplish these objectives, the authors have chosen three

145

Lactose (β-form) —β-galactosidase/H_2O→ **β-galactose** + **β-glucose**

bacteria and one yeast, each with a different fermentation profile. *Escherichia coli* is a facultatively anaerobic Gram-negative rod that produces acid and gas by catabolizing D-glucose (dextrose) and other carbohydrates. *E. coli* is β-galactosidase positive since it can also use lactose as its sole source of carbon. *Alcaligenes faecalis* is an obligately aerobic rod that uses acetate, propionate, butyrate, and some other organic acids as a sole carbon and energy source; because carbohydrates are not used, it does not produce acid or gas. *A. faecalis* is β-galactosidase negative. *Salmonella enterica* serovar Typhimurium is a facultatively anaerobic Gram-negative rod having both a respiratory and fermentative type of metabolism; thus D-glucose and other types of carbohydrates are metabolized with the production of acid and sometimes gas. *S. enterica* is β-galactosidase positive. The yeast, *Saccharomyces cerevisiae*, will ferment glucose (but not sucrose) to produce gas, but no acid. *S. cerevisiae* is β-galactosidase negative.

Phenol red broth is a differential medium used to detect the fermentation of specific carbohydrates. This information can be used as an aid in the clinical microbiology laboratory for the detection of many species of bacteria.

PRINCIPLES

Fermentations are energy-producing biochemical reactions under anaerobic conditions in which organic molecules serve as electron donors and organic intermediates serve as electron acceptors. The ability of microorganisms to ferment carbohydrates and the types of products formed are very useful in identification. A given carbohydrate may be fermented to a number of different end products depending on the microorganism involved (**Figure 19.1**). These end products (alcohols, acids, gases, or other organic molecules) are characteristic of the particular microorganisms. For example, if fermenting bacteria are grown in a liquid culture medium containing the carbohydrate glucose, they may produce organic acids as by-products of the fermentation. These acids are released into the medium and lower its pH. If a pH indicator such as phenol red is included in the medium, acid production will change the medium from its original color to yellow below pH 6.8, purple-pink above pH 7.4, and red between these two points (**Figure 19.2**).

Gases produced during the fermentation process can be detected by using a small, inverted tube, called a **Durham tube** (named after Herbert Edward Durham, English bacteriologist, 1866–1945), within the liquid culture medium. After the proper amount of broth is added, Durham tubes are inserted into each culture tube. During autoclaving, the air is expelled from the Durham tubes, and they become filled with the medium. If gas is produced, the liquid medium inside the Durham tube will be displaced, entrapping the gas in the form of a bubble (Figure 19.2 and **Table 19.1**).

Some microorganisms, such as *E. coli*, can use lactose as their sole source of carbon. An essential enzyme in the metabolism of this sugar is **β-galactosidase.** β-galactosidase hydrolyzes lactose to galactose and glucose as shown in the diagram at the top of this page.

Instead of lactose, an artificial substrate, ONPG (*o*-nitro-phenyl-β-D-galactopyranoside), can be used in a laboratory assay. β-galactosidase catalyzes the hydrolysis of ONPG as follows:

$$ONPG + H_2O \xrightarrow{\beta\text{-galactosidase}} galactose + o\text{-nitrophenol}$$

ONPG is colorless but upon hydrolysis yields *o*-nitrophenol, which is yellow in an alkaline solution. If an ONPG disk or KEY tablet is incubated with a bacterial culture and the culture turns yellow, this is the positive test for β-galactosidase activity.

Each of the above tests is important in the identification of certain bacteria.

Procedure

Durham Tube
First Period

1. Label five of the specified culture tubes with your name, date, and type of culture medium.

146   Microbial Biochemistry

Figure 19.1 Major Fermentation Pathways. Microorganisms produce various waste products when they ferment sugars. The end products released (green boxes) are often characteristic of the microorganisms and can be used as identification tools.

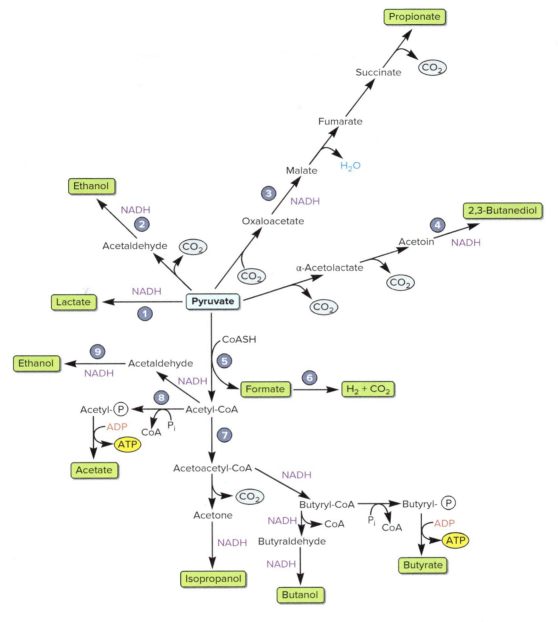

1. Lactic acid bacteria e.g., (*Streptococcus, Lactobacillus* species), *Bacillus* spp., enteric bacteria
2. Yeast, *Zymomonas* spp.
3. Propionic acid bacteria (*Cutibacterium* spp.)
4. *Enterobacter, Serratia, Bacillus* species
5. Enteric bacteria
6. Enteric bacteria
7. *Clostridium* spp.
8. Enteric bacteria
9. Enteric bacteria

2. Label the first tube *E. coli;* the second, *S. enterica;* the third, *A. faecalis;* the fourth, *S. cerevisiae;* and the fifth, "control."
3. Using aseptic technique (*see Figure 5.3*), inoculate each tube with the corresponding microbial culture. Leave the fifth tube uninoculated. Care should be taken during this step not to tip or shake the fermentation tube, as this may accidentally force a bubble of air into the Durham tube and give a false-positive result.
4. Place the five tubes in a test-tube rack and incubate at 37°C for 24 to 48 hours.

Carbohydrates I: Fermentation and β-Galactosidase Activity

Figure 19.2 Carbohydrate Fermentation. (a) Possible carbohydrate fermentation patterns of microorganisms, with phenol red as the pH indicator. (b) The tube on the left shows acid and gas production. The next tube is the control. The third tube from the left shows growth but no carbohydrate fermentation. *(b: Javier Izquierdo/McGraw Hill)*

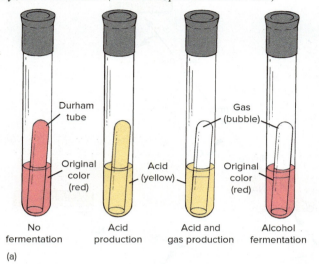
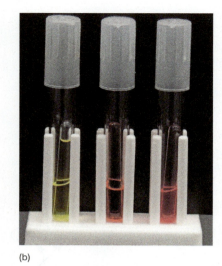

Table 19.1	Phenol Red Broth Interpretations
Interpretation (Symbol)	**Appearance**
Fermentation with acid and gas end products (A/G)	Yellow and bubble in Durham tube
Fermentation with acid end products but no gas (A/–)	Yellow and no bubble in Durham tube
No fermentation (–/–)	Red and no bubble in Durham tube
Degradation of protein and alkaline end products (–/–)	Pink and no bubble in Durham tube

Second Period

1. Ideally, the tubes should be examined carefully between 2 and 4 hours, at 8 hours, and at 18 hours in order to avoid false negatives due to reversal of the fermentation reactions that may occur with long incubations.
2. Examine all carbohydrate broth cultures for evidence of acid (A), or acid and gas (A/G) production. Use Figure 19.2 as a reference. Acid production is detected by the medium turning yellow, and gas production by a gas bubble in the Durham tube.
3. The control tube should be negative for acid and gas production, and should have no turbidity.
4. Based on your observations, determine and record in the report for this exercise whether or not each microorganism was capable of fermenting the carbohydrate substrate with the production of acid, or acid and gas. Compare your results with those of other students who used other sugars.

Procedure

ONPG Disks or Tablets

1. Dispense 1 ml of sterile 0.85% NaCl into four sterile test tubes.
2. Label each tube with the microorganism to be inoculated. Add your name and the date.
3. Suspend a thick loopful of concentrated microorganisms into each tube.
4. Place an ONPG disk or tablet into each tube and incubate at 37°C.
5. Check at 20 minutes and 4 hours (follow the instructions from the manufacturer).
6. A positive β-galactosidase test is indicated by a yellow color; no color change indicates a negative test.
7. Record your results in the report for this exercise.

Disposal

When you are finished with all of the tubes and pipettes, discard them in the designated place for sterilization and disposal.

> **HELPFUL HINTS**
>
> - The amount of inoculum placed in each tube should be small. Too much inoculum may lead to overgrowth and neutralization of acidic fermentation products by the bacteria, thus resulting in false negatives.
> - Do not vortex the fermentation tubes when inoculating because this can force air into the Durham tubes and result in false-positive recordings for gas production.

Laboratory Report 19

Name: _____

Date: _____

Lab Section: _____

Carbohydrates I: Fermentation and β-Galactosidase Activity

1. Complete the following table on carbohydrate fermentation.

Biochemical Results	Bacteria			
	E. coli	*S. enterica*	*A. faecalis*	*S. cerevisiae*
Lactose (A), (A/G), (G), (−)	_____	_____	_____	_____
Dextrose (A), (A/G), (G), (−)	_____	_____	_____	_____
Sucrose (A), (A/G), (G), (−)	_____	_____	_____	_____

   A = Acid production.
   A/G = Acid and gas production.
   G = Gas production.
   − = No growth or no change (alkaline).

2. Indicate whether the following bacteria have β-galactosidase activity.

   a. *A. faecalis* _____

   b. *E. coli* _____

   c. *S. enterica* _____

   d. *S. cerevisiae* _____

3. In summary, what is the major purpose of this exercise?

ASSESSMENT
Critical Thinking and Learning Outcomes Review

1. Define fermentation.

2. Do all microorganisms produce the same end product from pyruvate? Explain your answer.

3. What is the purpose of the phenol red or bromcresol purple in the fermentation tube?

4. What is the function of the Durham tube in the fermentation tube?

5. What are some of the fermentation end products produced by the different microorganisms used in this experiment?

6. What reaction is catalyzed by β-galactosidase?

EXERCISE 20

Carbohydrates II: Triple Sugar Iron Agar Test

SAFETY CONSIDERATIONS

- Be careful with the Bunsen burner flame.
- ***Shigella flexneri* is a potential pathogen (BSL2).** Use aseptic technique throughout this experiment.
- Keep all culture tubes upright in a test-tube rack or in empty cans.

MATERIALS

- suggested bacterial strains:
  Alcaligenes faecalis, Escherichia coli, Proteus vulgaris, Pseudomonas aeruginosa, and *Shigella flexneri* (ATCC 29903)
- triple sugar iron (TSI) agar slants
- Bunsen burner
- inoculating needle
- incubator set at 37°C
- test-tube rack
- wax pencil or lab marker
- container for biohazard waste
- safety glasses
- disposable gloves
- lab coat

LEARNING OUTCOMES

Upon completion of this exercise, students will demonstrate the ability to

1. Explain the biochemical reactions involved in the triple sugar iron agar test
2. Differentiate among members of the family *Enterobacteriaceae*
3. Distinguish between the *Enterobacteriaceae* and other intestinal bacteria
4. Perform a TSI test

SUGGESTED READING IN TEXTBOOK

1. Catabolism of Other Organic Molecules Other Than Glucose, section 11.7; see also figure 11.19.

2. *Gammaproteobacteria* Is the Largest Bacterial Class, section 21.2; see also table 21.3.

Pronunciation Guide

- *Alcaligenes faecalis* (al-kah-LIJ-e-neez fee-KAL-iss)
- *Escherichia coli* (esh-er-I-ke-a KOH-lie)
- *Proteus vulgaris* (PRO-tee-us vul-GA-ris)
- *Pseudomonas aeruginosa* (soo-do-MO-nas a-ruh-jin-OH-sah)
- *Shigella flexneri* (shi-GEL-la flex-NER-i)

Why Are the Following Bacteria Used in This Exercise?

This exercise will provide you with experience in using the triple sugar iron (TSI) agar test to differentiate among the members of the family *Enterobacteriaceae* and other members of the phylum Proteobacteria. The authors have chosen three common bacteria in the family *Enterobacteriaceae*: *Escherichia coli, Proteus vulgaris,* and *Shigella flexneri*. All three are facultatively anaerobic Gram-negative rods. In a TSI tube, *E. coli* produces an acid butt, an acid or alkaline slant, and no H_2S, but does produce gas. *P. vulgaris* produces an acid butt, an acid or alkaline slant, H_2S, and gas. *S. flexneri* produces an acid butt, an alkaline slant, no H_2S, and no gas. For the other Proteobacteria, the authors have chosen *Alcaligenes faecalis* and *Pseudomonas aeruginosa*. These are Gram-negative, aerobic, nonfermentative rods belonging to the family *Pseudomonadaceae*. In a TSI tube, *A. faecalis* produces an alkaline butt, alkaline slant, no H_2S, and no gas; *P. aeruginosa*, an acid butt, alkaline slant, no H_2S, and no gas.

Medical Application

Triple sugar iron (TSI) agar is a differential media used to evaluate the fermentation of sucrose, glucose, and lactose as well as the production of hydrogen sulfide and/or gas.

TSI agar is most often used in the clinical laboratory to help identify Gram-negative rods, such as members of the *Enterobacteriaceae*.

PRINCIPLES

As originally described in 1911 by F. F. Russell, the **triple sugar iron (TSI) agar test** is generally used for the identification of bacteria belonging to the family *Enterobacteriaceae*. It is also used to distinguish the *Enterobacteriaceae* from other Gram-negative intestinal bacilli by their ability to catabolize glucose, lactose, and/or sucrose, and to liberate sulfides from ferrous ammonium sulfate or sodium thiosulfate. (*See Exercise 23 for the biochemistry of H_2S production from the reduction of inorganic sulfur-containing compounds such as thiosulfate.*) TSI agar slants contain a 1% concentration of lactose and sucrose, and a 0.1% glucose concentration. The term **saccharolytic** can be used to describe microorganisms that are capable of breaking the glycosidic bonds in carbohydrates. The pH indicator, phenol red, is also incorporated into the medium to detect acid production from carbohydrate fermentation (*see Exercise 19*).

TSI slants are inoculated by either streaking the slant surface using a zigzag streak pattern and then stabbing the agar deep with a straight inoculating needle or stabbing the butt first and then streaking the slant (*see Figure 5.5*). Because the medium consists of a slant and a butt, it is possible to observe the formation of two reaction areas in the same tube. Since the slant is exposed to atmospheric oxygen, it is aerobic. The butt is not exposed to atmospheric oxygen and is anaerobic. A small amount of acid produced by the fermentation of the 0.1% glucose is sufficient to turn the butt yellow. However, fermentation of the 1% lactose and/or sucrose results in enough acid to overcome the alkalinization due to oxidative carboxylation of proteins near the surface of the slant. Incubation is for 18 to 24 hours in order to detect the presence of sugar fermentation, gas production, and H_2S production. The following reactions may occur in the TSI tube (**Figure 20.1**):

1. **Yellow butt (A) and red slant (A)** due only to the fermentation of glucose (phenol red indicator turns yellow due to the persisting acid formation in the butt). The **slant remains red (alkaline) (K)** because of the limited glucose in the medium and, therefore, limited acid formation, which does not persist since the oxidative decarboxylation forms alkaline amines that increase the pH.
2. A **yellow butt (A) and acid slant (A)** due to the fermentation of glucose and lactose and/or sucrose (yellow slant and butt due to the high concentration of these sugars) leads to excessive acid formation in the entire medium.
3. **Gas formation** noted by splitting or lifting of the agar.

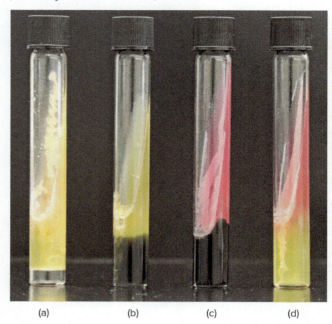

Figure 20.1 Triple Sugar Iron Reactions (TSI-3) and Their Interpretation. (a) The tube on the left has a yellow slant (acid), yellow butt (acid), gas production at the bottom of the tube, and no H_2S production. (b) The second tube from the left has a yellow slant (acid), yellow butt (acid), and the black precipitate indicates H_2S production, but no gas. (c) The third tube from the left has a red slant (alkaline), yellow butt (acid), and the black precipitate indicates H_2S production, but no gas. (d) The tube on the right has a red slant (alkaline), yellow butt (acid), no H_2S production, and no gas production. This would indicate a nonlactose fermenter. *(Javier Izquierdo/McGraw Hill)*

	Tube a	Tube b	Tube c	Tube d
Slant	A	A	K	K
Butt	A	A	A	A
Gas	+	−	−	−
H_2S	−	+	+	−

4. **Gas formation (H₂S)** seen by blackening of the agar.
5. **Red butt (K)** and **slant (K)** indicate that none of the sugars were fermented and neither gas nor H$_2$S were produced. The bacteria are thus not from the *Enterobacteriaceae*.

Table 20.1 gives reactions usually expected from some of the more frequently encountered genera of the *Enterobacteriaceae*. **Figure 20.2** summarizes the possible reactions and results in TSI for the various bacteria used in this experiment.

Table 20.1	Results of TSI Reaction			
Bacterium	**Butt**	**Slant**	**H$_2$S**	**Gas**
Enterobacter	A	A	–	+
Escherichia	A	A	–	+
Klebsiella	A	A	–	+
Citrobacter	A	K or A	V	+
Proteus vulgaris	A	A or K	+	+
Edwardsiella	A	K	+	+
Morganella	A	K	–	+
Serratia	A	K or A	–	V
Shigella	A	K	–	–
Salmonella typhi	A	K	+	–

A = acid, K = alkaline, V = varies between species.

Procedure

First Period

1. Label each of the TSI agar slants with the name of the bacterium to be inoculated. Use one of the tubes as a control. Place your name and date on each tube.
2. Using aseptic technique (*see Figure 5.3*), streak the slant with the appropriate bacterium and then stab the butt. Screw the caps on the tubes but do not tighten. Caps must be loose to allow free gas exchange for the anaerobic reactions.
3. Incubate for only 18 to 24 hours at 37°C for changes in the butt and on the slant. Tubes should be incubated and checked daily for up to 7 days in order to observe blackening.

Figure 20.2 The Possible Reactions and Results in TSI Agar for the Various Bacteria Used in This Experiment. *(a: Source: Dr. W.A. Clark/CDC; b: Source: CDC; c: Source: Dr. W.A. Clark/CDC; d: Source: Janice Haney Carr/CDC; e: Source: Dr. William A. Clark/CDC)*

	No carbohydrate fermentation, H$_2$S or gas production	**Glucose fermentation only,** no H$_2$S or gas production	**Glucose fermentation with H$_2$S production,** but no gas production	**Lactose and/or sucrose and glucose fermentation,** with gas production	**Lactose and/or sucrose and glucose fermentation with H$_2$S and gas production**
Glucose	K, red butt	A, yellow butt	A, yellow butt	A, yellow butt	A, yellow butt
Lactose, sucrose	K, red slant	K, red slant	K, red slant	A, yellow slant	A, yellow slant
Cysteine	–	–	Black precipitate	–	Black precipitate
Gas	–	–	–	Gas formation	Gas formation
Example organism	*Alcaligenes faecalis*	*Shigella flexneri*	*Pseudomonas aeruginosa*	*Escherichia coli*	*Proteus vulgaris*
	(a)	(b)	(c)	(d)	(e)

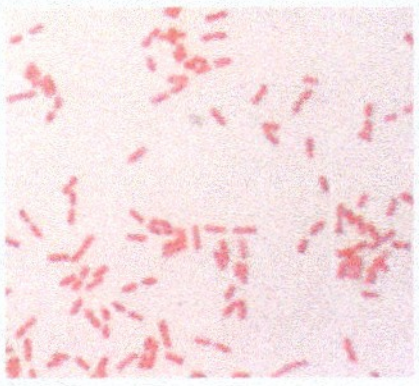

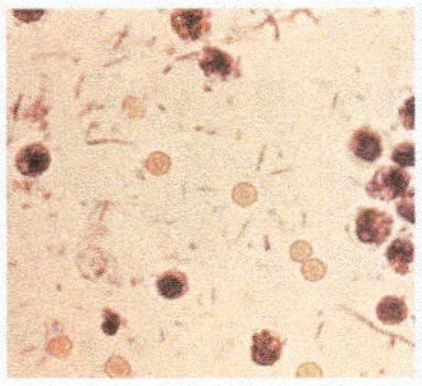

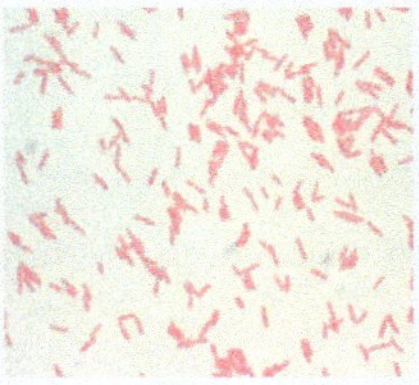

Carbohydrates II: Triple Sugar Iron Agar Test

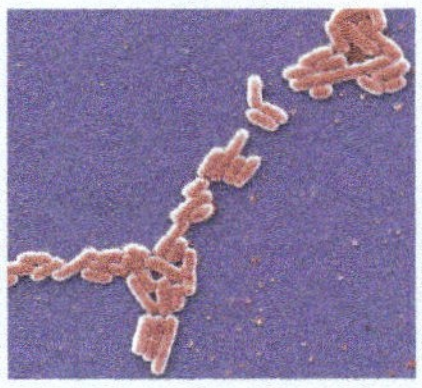

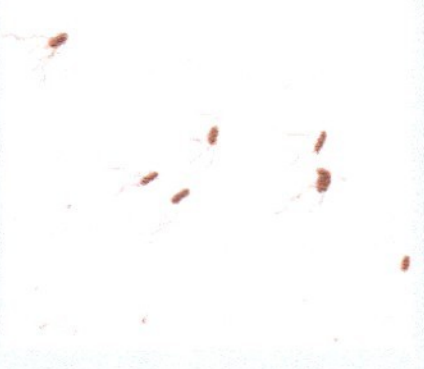

Second Period

1. Examine all slant cultures for the color of the slant and butt, and for the presence or absence of blackening within the medium.
2. Record your results in the report for this exercise.

Disposal

When you are finished with all of the tubes, discard them in the designated place for sterilization and disposal.

HELPFUL HINTS

- If screw-cap tubes are used, leave the caps loose about ¼ turn after inoculating the tubes to prevent excessive disruption of the agar should large amounts of gas be produced during incubation.
- Record the butt as acid production if the black color of FeS masks the color in the butt.
- If improperly stabbed (e.g., if your needle goes down the side of the tube), often the butt will not change color.
- Examine the tubes no earlier or later than 18 to 24 hours as incorrect patterns of carbohydrate fermentation may occur.

Laboratory Report 20

Name: _____

Date: _____

Lab Section: _____

Carbohydrates II: Triple Sugar Iron Agar Test

1. Complete the following table on the TSI test.

	Carbohydrate Fermentation		Gas Production	
Bacterium	Butt Color	Slant Color	H_2S	Fissure or Lifting of Agar
A. faecalis	_____	_____	_____	_____
E. coli	_____	_____	_____	_____
P. vulgaris	_____	_____	_____	_____
P. aeruginosa	_____	_____	_____	_____
S. flexneri	_____	_____	_____	_____

2. In summary, what is the major purpose of this exercise?

ASSESSMENT
Critical Thinking and Learning Outcomes Review

1. For what bacteria would you use the TSI test?

2. Why must TSI test observations be made between 18 and 24 hours after inoculation?

3. Distinguish between an acid and an alkaline slant.

4. What is the purpose of thiosulfate in the TSI agar?

5. What is meant by a saccharolytic bacterium? What reaction would it give in a TSI tube?

6. Why is there more lactose and sucrose than glucose in TSI agar?

7. What is the pH indicator in TSI agar?

ASSESSMENT
Critical Thinking and Learning Outcomes Review

1. For what bacteria would you use the TSI test?

2. Why must TSI test observations be made between 18 and 24 hours after inoculation?

3. Distinguish between an acid and an alkaline slant.

4. What is the purpose of thiosulfate in the TSI agar?

5. What is meant by a saccharolytic bacterium? What reaction would it give in a TSI tube?

6. Why is there more lactose and sucrose than glucose in TSI agar?

7. What is the pH indicator in TSI agar?

Carbohydrates III:

Figure 21.1 Test for Starch Hydrolysis after Adding Gram's Iodine. (a) Example of negative (above) and positive (below) starch hydrolysis. In negative hydrolysis, no color change occurs as indicated by the brown color around the streak. In positive hydrolysis, the complete breakdown of all starch is shown by the clear halo. (b) The biochemistry of starch hydrolysis. *(Javier Izquierdo/McGraw Hill)*

Starch Test Negative (–)

starch $\xrightarrow{\text{no α-amylase}}$ starch + added Gram's iodine = purple to brown color around growth

α-amylase

- Be careful with the Bunsen burner flame.
- Use caution to avoid dripping bacteria-laden iodine solution out of the plates while making observations.

MATERIALS

- suggested bacterial strains:

- Gram's iodine
- wax pencil or lab marker

- disposable gloves
- lab coat

LEARNING OUTCOMES

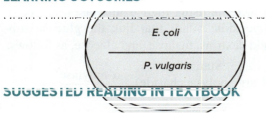

SUGGESTED READING IN TEXTBOOK

1. Catabolism of Organic Molecules Other Than Glucose, section 11.7, see also figure 11.19.

Pronunciation Guide

- *Bacillus subtilis* (bah-SIL-lus SUB-til-is)
- *Escherichia coli* (esh-er-I-ke-a KOH-lie)

The major objective of this exercise is for you to gain expertise in performing a starch hydrolysis test. If a bacterium produces α-amylase, it can hydrolyze starch; if α-amylase is not produced, the bacterium will not hydrolyze starch. The three bacteria the authors have chosen vary in their ability to produce α-amylase. *Bacillus subtilis* is amylase positive; *Escherichia coli* is amylase

starch has not been hydrolyzed, and the test is negative.

2. If the results are difficult to read, an alternative

300 units), and **amylopectin,** a large branched polymer. Both amylopectin and amylose are rapidly hydrolyzed

HELPFUL HINTS

- Carefully adding iodine to only a small part of the growth at one end of the streak does not contaminate the plate, and it may be

(Disaccharide) + (Monosaccharide)

A starch agar plate contains beef extract, peptone, soluble starch, and agar. **Starch agar** is used to test for the breakdown of starch by alpha-amylase.

Gram's iodine (also used for the Gram stain) can be used to indicate the presence of starch. When

a brownish-orange, discard it and replace with fresh iodine

158 Microbial Biochemistry

Laboratory Report 21

Name: _____
Date: _____
Lab Section: _____

Carbohydrates III: Starch Hydrolysis

1. In the following plate, sketch the presence or absence of starch hydrolysis.

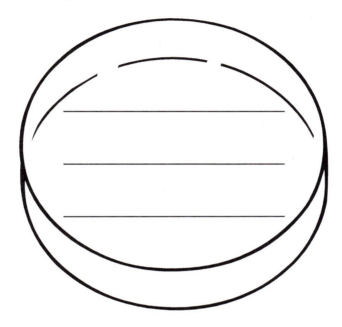

2. In summary, what is the major purpose of this exercise?

ASSESSMENT
Critical Thinking and Learning Outcomes Review

1. Describe the function of hydrolases.

2. Describe the chemistry of starch hydrolysis.

3. The chemical used to detect microbial starch hydrolysis on starch plates is _____.

4. What does starch hydrolysis by a bacterium indicate?

5. Amylase is an enzyme that attacks starch. The smallest product of this hydrolysis is called _____.

6. How is it possible that bacteria may grow heavily on starch agar but not necessarily produce α-amylase?

EXERCISE 22 Lipid Hydrolysis

SAFETY CONSIDERATIONS

- Be careful with the Bunsen burner flame.
- Exercise caution when handling cultures that can cause human disease.

MATERIALS

- suggested bacterial strains:
  Escherichia coli and *Staphylococcus aureus*
- Spirit Blue agar supplemented with 3% Bacto lipase reagent
- tributyrin agar
- inoculating loop
- wax pencil or lab marker
- Bunsen burner
- container for biohazard waste
- safety glasses
- disposable gloves
- lab coat

LEARNING OUTCOMES

Upon completion of this exercise, students will demonstrate the ability to

1. Outline the biochemistry of lipid hydrolysis
2. Determine the ability of bacteria to hydrolyze lipids by producing specific lipases
3. Explain how it is possible to detect the hydrolysis of lipids by a color change reaction

SUGGESTED READING IN TEXTBOOK

1. Catabolism of Organic Molecules Other Than Glucose, section 11.7; see also figures 11.20 and 11.21.

Pronunciation Guide

- *Escherichia coli* (esh-er-I-ke-a KOH-lie)
- *Staphylococcus aureus* (staf-il-oh-KOK-kus ORE-ee-us)

Why Are the Following Bacteria Used in This Exercise?

After this exercise, you will be able to differentiate between bacteria that can produce lipases and those that cannot. A lipase-positive and a lipase-negative bacterium have been selected to demonstrate this difference. *Escherichia coli* is a facultatively anaerobic Gram-negative rod that does not produce lipase. *E. coli* occurs in the intestines of humans and can cause urinary tract infections. *Staphylococcus aureus* is a Gram-positive coccus that produces lipase. It is a normal inhabitant of the skin and occasionally the mucous membranes of warm-blooded vertebrates. *S. aureus* can cause a variety of skin and wound infections.

Medical Application

The lipase test is used to detect and enumerate lipolytic bacteria (e.g., some species of *Staphylococcus, Flavobacterium, Clostridium,* and *Pseudomonas*), many of which are human pathogens. Lipase activity is also monitored in food microbiology to detect lipolytic microbes, especially in dairy products.

PRINCIPLES

Lipids are high-molecular-weight compounds possessing large amounts of stored energy. The two common lipids catabolized by bacteria are triglycerides (triacylglycerols) and phospholipids. Triglycerides are hydrolyzed by enzymes called **lipases** into glycerol and free fatty acid molecules (**Figure 22.1a**). Glycerol and free fatty acid molecules can then be taken up by the bacterial cell and further metabolized through reactions of glycolysis, the β-oxidation pathway, and the citric acid cycle. These lipids can also enter other metabolic pathways where they are used for the synthesis of phospholipids that are needed for biogenesis of the cell envelope. Since phospholipids are components of all cells, the ability to

161

Figure 22.1 Lipid Hydrolysis. (a) The biochemistry of lipase activity. (b) Procedure for inoculating Spirit Blue agar and tributyrin plates. (c) Examples of positive (1–3) and negative (4) reactions on a Spirit Blue plate. Note the disappearance of the oily sheen which indicates lipid hydrolysis. *(c: Nathan Rigel/McGraw Hill).*

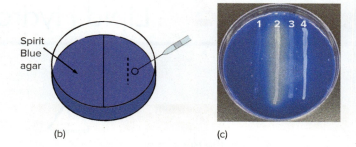

hydrolyze host-cell phospholipids is an important factor in the spread of pathogenic bacteria. In addition, lipase-producing bacteria that contaminate food products, especially dairy products, cause spoilage by hydrolyzing the lipids.

When lipids are added to an agar-solidified culture medium and are cultured with lipolytic bacteria, the surrounding medium becomes acidic due to the release of fatty acids. By adding a pH indicator to the culture medium, it is possible to enhance detection of lipid hydrolysis (**lipolysis**). In addition to the local change in pH, the opacity in the agar will turn clear as lipids are degraded, resulting in a halo surrounding the bacterial growth. The Spirit Blue agar used in this exercise is blue in color. The agar will turn clear around lipolytic bacterial colonies. This is a positive test. Bacterial growth that is not surrounded by a halo indicates a negative reaction.

Tributyrin agar does not contain a pH indicator. Here, a clear zone (halo) will appear around the bacterial growth indicating lipase activity. If no clear zone appears, the bacterium is lipase negative.

Procedure

First Period

1. With a wax pencil or lab marker, divide the bottom of a Spirit Blue agar and tributyrin agar plate in half and label half of each plate *E. coli* and the other half *S. aureus*. Place your name and date on the plate.

2. Using aseptic technique, inoculate the plates with the respective bacteria with a single streak of culture (**Figure 22.1b**).
3. Incubate the plate in an inverted position for 24 to 48 hours at 37°C.

Second Period

1. Examine the plates for evidence of lipid hydrolysis (**Figure 22.1c**). Clearing, not lightening, of the medium is considered a positive test for lipid hydrolysis. If no lipid hydrolysis has taken place, the agar around the colony will remain opaque.
2. Record your results in the report for this exercise.

Disposal

When you are finished with all of the plates, discard them in the designated place for sterilization and disposal.

> **HELPFUL HINTS**
> - When trying to observe color changes in agar media, it helps to put the plates against different-colored backgrounds.
> - Try different light angles by holding the plates in various positions with respect to the light source and looking at the cultures from both above and below.

Laboratory Report 22

Name: _____
Date: _____
Lab Section: _____

Lipid Hydrolysis

1. Based on your observations, did you detect lipid hydrolysis? Explain your answer.

2. Sketch and describe what is happening on each plate with respect to lipid hydrolysis.

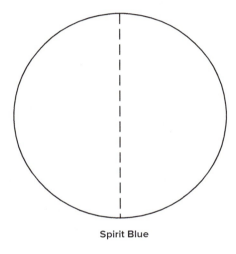
Spirit Blue

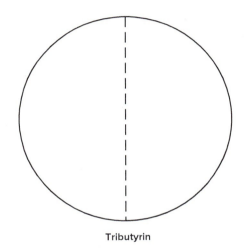
Tributyrin

3. In summary, what is the major purpose of this exercise?

ASSESSMENT
Critical Thinking and Learning Outcomes Review

1. What are the major functions of lipases exported by bacteria?

2. Look up the composition of Spirit Blue agar. For bacteria that are not lipolytic, what do they use for a carbon source?

3. What are two functions of lipids in bacterial cells?

4. Give some examples of foods that might be spoiled by lipolytic bacteria. Conversely, what foods might not be spoiled as rapidly by such bacteria?

5. How is the ability of certain bacteria to attack phospholipids related to pathogenesis?

6. What is the difference between a triglyceride and a phospholipid?

7. What are several pathways that bacteria use to metabolize lipids?

EXERCISE 23 Proteins I: The IMViC Tests

SAFETY CONSIDERATIONS

- Be careful with the Bunsen burner flame.
- Barritt's reagent contains naphthol, which is toxic and may cause peeling of the skin; thus wear gloves when using this reagent.
- Kovacs' reagent is also caustic to the skin and mucous membranes due to the concentrated HCl and *p*-dimethylaminobenzaldehyde.
- In case of contact with either Barritt's or Kovacs' reagent, immediately flush with plenty of water for at least 15 minutes.

MATERIALS

- suggested bacterial strains: *Enterobacter aerogenes, Escherichia coli Klebsiella pneumoniae,* and *Proteus vulgaris*
- SIM (sulfur indole motility) agar deep tubes
- Kovacs' reagent
- Bunsen burner
- inoculating loop and needle
- MR-VP broth tubes
- methyl red indicator
- Barritt's reagent (solutions A and B)
- Simmons citrate agar slants
- empty test tubes
- pipettes with pipettor
- wax pencil or lab marker
- container for biohazard waste
- safety glasses
- disposable gloves
- lab coat

LEARNING OUTCOMES

Upon completion of this exercise, students will demonstrate the ability to

1. Describe how some bacteria degrade the amino acid tryptophan
2. Determine the ability of some bacteria to oxidize glucose with the production of acid end products
3. Differentiate between glucose-fermenting enteric bacteria
4. Explain the purpose of the Voges-Proskauer test
5. Differentiate among enteric bacteria on the basis of their ability to ferment citrate
6. Perform the IMViC series of tests
7. Understand the biochemical process of hydrogen sulfide production by bacteria
8. Describe how motility can be detected

SUGGESTED READING IN TEXTBOOK

1. Catabolism of Organic Molecules Other Than Glucose, section 11.7; see also figure 11.22.
2. Class *Gammaproteobacteria* Is the Largest Bacterial Class, section 21.2; see also table 21.3.

Pronunciation Guide

- *Enterobacter aerogenes* (en-ter-oh-BAK-ter a-RAH-jen-eez)
- *Escherichia coli* (esh-er-I-ke-a KOH-lie)
- *Klebsiella pneumoniae* (kleb-se-EL-lah new-MOH-nee-eye)
- *Proteus vulgaris* (PRO-te-us vul-GA-ris)
- *Salmonella* (sal-mon-EL-ah)
- *Shigella* (shi-GEL-la)
- Enterobacteriaceae (en-ter-oh-BAK-ter-ac-e-ee)

Why Are the Following Bacteria Used in This Exercise?

In this exercise, you will learn how to perform the IMViC series of tests that distinguish between different enteric (pertaining to the small intestine) bacteria. To illustrate the various IMViC reactions, the authors have chosen four enteric bacteria. *Enterobacter aerogenes* is a facultatively anaerobic Gram-negative rod that has peritrichous flagella. It is a motile lactose fermenter. *E. aerogenes* is widely distributed in nature, occurring in fresh water, soil, sewage, plants, vegetables, and animal and human feces. It is indole negative, MR negative, VP positive, and citrate positive. *Escherichia coli* is a facultatively anaerobic Gram-negative rod that is motile with peritrichous flagella or nonmotile. It is a lactose fermenter. *E. coli* occurs as normal flora in the lower part of the intestine of warm-blooded animals. It is indole positive, MR positive, VP negative, and citrate negative. *Klebsiella pneumoniae* is a facultatively anaerobic Gram-negative rod. It is nonmotile and a lactose fermenter. *K. pneumoniae* is a human pathogen, causing respiratory tract infections. *K. pneumoniae* is one of several bacterial pathogens that is increasingly difficult to treat due to widespread antibiotic resistance. It is indole positive, often MR negative, VP positive, and citrate positive. *Proteus vulgaris* is a Gram-negative facultatively anaerobic rod that occurs in the intestines of humans and a wide variety of animals, in manure, and in polluted waters. It has peritrichous flagella, is motile, and does not ferment lactose. *P. vulgaris* is indole positive, MR positive, VP negative, and sometimes citrate positive.

Medical Application

The biochemical tests described in this exercise are used in the clinical laboratory for the identification of Gram-negative enteric bacteria. Among them are the causative agents of many different gastrointestinal and urinary tract infections. While running these tests using large volumes of media is appropriate for teaching purposes, commercial vendors have adapted these tests to a small-scale format allowing rapid processing of multiple clinical specimens. This enhances efficiency and speed dramatically in the clinical laboratory, where positive treatment outcomes depend on correct initial diagnoses.

PRINCIPLES

The identification of enteric (intestinal) bacteria is of prime importance in determining certain food-borne and water-borne diseases. Many of the bacteria that are found in the intestines of humans and other mammals belong to the family *Enterobacteriaceae*. These bacteria are short, Gram-negative, nonsporing bacilli. They can be subdivided into lactose fermenters and nonfermenters.

The differentiation and identification of these enteric bacteria can be accomplished by using the **IMViC test** (*i*ndole, *m*ethyl red, *V*oges-Proskauer, and *c*itrate; the "i" is for ease of pronunciation).

In addition to these tests, you will also monitor the production of hydrogen sulfide and motility using differential media in this exercise.

Indole Production

The amino acid **tryptophan** is found in nearly all proteins. Bacteria that contain the enzyme **tryptophanase** can hydrolyze tryptophan to its metabolic products, namely, indole, pyruvic acid, and ammonia. While bacteria use the pyruvic acid and ammonia to satisfy nutritional needs, the indole is not used and accumulates in the medium. The presence of indole can be detected by the addition of **Kovacs' reagent.** Kovacs' reagent reacts with the indole, producing a bright red compound on the surface of the medium (**Figure 23.1**). Such bacteria are **indole positive;** in contrast, the absence of a red color indicates tryptophan was not hydrolyzed, and the bacteria are thus **indole negative.**

Methyl Red Test

All enteric bacteria catabolize glucose for their energy needs; however, the end products vary depending on the enzyme pathways present in the bacteria. The pH indicator **methyl red** detects a pH change to the acid range as a result of acidic end products such as lactic, acetic, and formic acids. This test is of value in distinguishing between *E. coli* (a mixed acid fermenter) and *E. aerogenes* (a butanediol fermenter). **Mixed acid fermenters** produce a mixture of fermentation acids and thus acidify the medium. In contrast, **butanediol fermenters** form butanediol, acetoin, and fewer organic acids. In this case, the pH of the medium does not fall as low as during mixed acid fermentation. As illustrated in **Figure 23.2**, at a pH of 4, the methyl red indicator turns red—a **positive methyl red test.** At a pH of 6, the indicator turns yellow—a **negative methyl red test.**

Voges-Proskauer Test

The **Voges-Proskauer test** identifies bacteria that ferment glucose, leading to **2,3-butanediol** accumulation in the medium. The addition of 40% KOH and a 5% solution of alpha-naphthol in absolute ethanol (Barritt's reagent) allows detection of **acetoin,** a precursor in the synthesis of 2,3-butanediol. A positive reaction is indicated by development of a cherry-red color. Absence of a red color is a **negative VP test** (**Figure 23.3**).

Citrate Utilization Test

The **citrate utilization test** determines the ability of bacteria to use **citrate** as a sole carbon source for their energy needs. This depends on the presence of a **citrate permease** that facilitates transport of citrate into the bacterium.

Figure 23.1 **Indole Test.** The SIM tube on the left is indole negative and the tube on the right is indole positive. *(Nathan W. Rigel/McGraw Hill)*

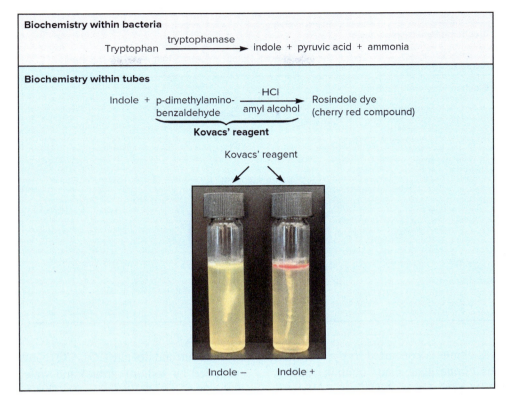

Figure 23.2 **Methyl Red Test.** The tube on the left is MR negative and the tube on the right is MR positive. *(Lisa Burgess/McGraw Hill)*

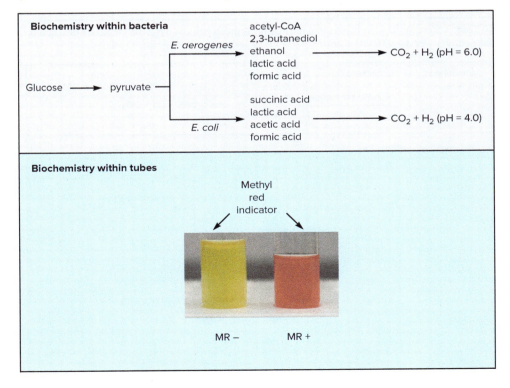

Figure 23.3 Voges-Proskauer Test. The tube on the left is VP negative and the tube on the right is VP positive. The copper color at the top of the VP− tube is due to the reaction of the KOH with the alpha-naphthol and should not be considered a positive result. *(Lisa Burgess/McGraw Hill)*

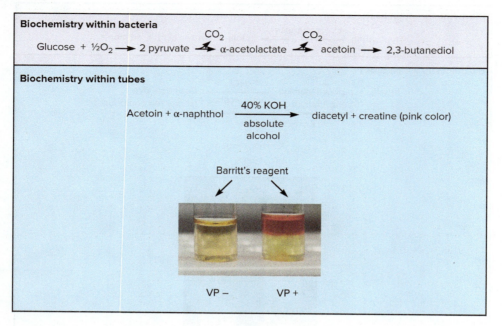

Once inside the cell, citrate is converted to pyruvic acid and CO_2. Simmons citrate agar slants contain sodium citrate as the only carbon source, NH_4^+ as a nitrogen source, and the pH indicator bromothymol blue. This test is done on slants since O_2 is necessary for citrate utilization. When bacteria oxidize citrate, they remove it from the medium and liberate CO_2. CO_2 combines with sodium (supplied by sodium citrate) and water to form sodium carbonate—an alkaline product. This raises the pH, turning the pH indicator from green to blue. The absence of a color change (i.e. the medium stays green) is a **negative citrate test** (**Figure 23.4**).

Figure 23.4 Citrate Test. Citrate utilization results in a blue color. A green color is negative for citrate utilization. *(Auburn University Photographic Services/McGraw Hill)*

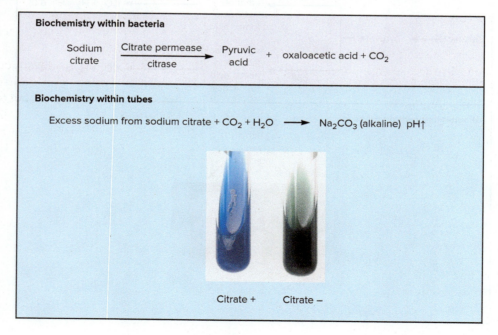

168   Microbial Biochemistry

Figure 23.5 **Hydrogen Sulfide Test.** Degradation of cysteine (**a**) or thiosulfate (**b**) to yield hydrogen sulfide gas. (**c**) Reaction of hydrogen sulfide gas with ferrous ammonium sulfate yields ferrous sulfide, an insoluble black precipitate.

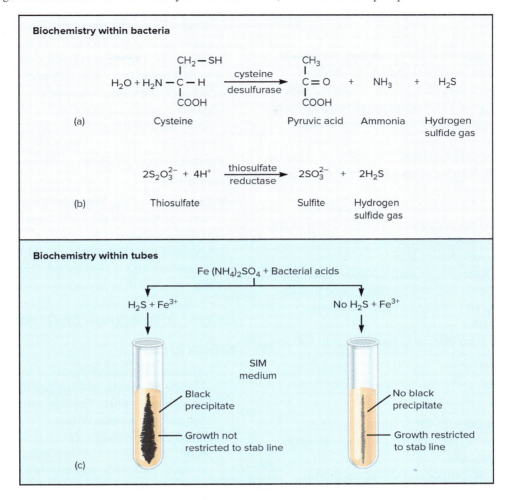

Hydrogen Sulfide Test

When proteins are hydrolyzed by some bacterial enzymes, the constituent amino acids are released and recycled as nutrients. In the presence of the enzyme cysteine desulfurase, the amino acid cysteine loses its sulfur atom through the addition of hydrogen from water to form hydrogen sulfide gas (**Figure 23.5a**). Bacteria can also produce hydrogen sulfide (H_2S) via other enzymes that drive reduction of inorganic sulfur-containing compounds such as thiosulfate ($S_2O_3^{2-}$), sulfate (SO_4^{2-}), or sulfite (SO_3^{2-}) (**Figure 23.5b**).

The SIM medium used in this exercise to detect indole can also be used to monitor hydrogen sulfide production. Peptones in the media serve as a source of cysteine, while ferrous ammonium sulfate ($Fe(NH_4)SO_4$) is the hydrogen sulfide indicator. Once hydrogen sulfide is produced, it combines with the ferrous ammonium sulfate forming ferrous sulfide, an insoluble, black precipitate. Presence of this black precipitate indicates the presence of hydrogen sulfide gas; absence of a black precipitate indicates a negative reaction (**Figure 23.5c**).

SIM agar can also be used to detect bacterial motility. In this assay, motility is indicated when the growth of the culture is not restricted to the stab line of the inoculation; since the SIM formulation is semisolid, flagellated bacteria swim away from the stab. In contrast, nonmotile bacterial growth is restricted to the stab line.

Procedure

Indole/Hydrogen Sulfide Production Test

First Period

1. Label each of the SIM deep tubes with the name of the bacterium to be inoculated, your name, and date.

2. Using aseptic technique, inoculate each tube by a stab inoculation.
3. Incubate the tubes for 24 hours at 37°C.

Second Period

1. Remove the tubes from the incubator. While wearing disposable gloves, add 5–10 drops of Kovacs' reagent to each tube and shake the tube gently. A deep red color develops in the presence of indole. Negative reactions remain colorless or light yellow.
2. Based on your observations, determine whether or not each bacterium was capable of hydrolyzing tryptophan.
3. Examine the SIM cultures for the presence or absence of a black precipitate along the line of the stab inoculation. A black precipitate of FeS indicates the presence of H_2S.
4. Examine the tubes for growth away from the stab line, which would indicate motility. Note that this may be obscured if the organism is a strong producer of H_2S.
5. Record your observations in the lab report for this exercise.

Methyl Red Test

First Period

1. Label each of the MR-VP broth media tubes with the name of the bacterium to be inoculated, your name, and date.
2. Using aseptic technique, inoculate each tube with the appropriate bacterium.
3. Incubate all tubes at 37°C for 24 hours. Note that for some slow fermenters, it may take 4 to 5 days.

Second Period

1. Transfer half of each culture into an empty test tube and set these aside for the Voges-Proskauer test.
2. To the culture remaining in each tube, add 4 to 5 drops of methyl red indicator. Carefully note any color change. Development of a red color indicates positive reaction.
3. Based on your observations, determine and record in the report for this exercise whether or not each bacterium was capable of fermenting glucose.

Voges-Proskauer Test

Second Period

1. Use the remaining aliquot from the methyl red test. While wearing disposable gloves and eye protection, add 10 drops of Barritt's solution A and 5 drops of solution B to each culture, mixing vigorously to aerate. Positive reactions occur within 20 minutes and are indicated by the presence of a red color.
2. Based on your observations, determine and record in the report for this exercise whether or not each bacterium was capable of fermenting glucose, resulting in the production of acetylmethylcarbinol.

Citrate Utilization Test

First Period

1. Label each of the Simmons citrate agar slants with the name of the bacterium to be inoculated, your name, and date.
2. Using aseptic technique, inoculate each bacterium into its proper tube by stabbing into the agar and then streaking the surface of the slant.
3. Incubate these cultures for 24 hours at 37°C.

Second Period

1. Examine the slant cultures for the presence or absence of growth and for any change in color from green to blue. The development of a deep blue color is a positive test.
2. Based on your observations, determine and record in the report for this exercise whether or not each bacterium was capable of using citrate as an energy source.

Disposal

When you are finished with all of the tubes and pipettes, discard them in the designated place for sterilization and disposal.

> **HELPFUL HINTS**
>
> - Incubate the SIM agar deeps for only 24 hours prior to adding Kovacs' reagent because the indole may be further metabolized resulting in false negatives.
> - Use no more than 5 drops of methyl red, as it may impart a red color to the medium that is unrelated to specific metabolic end products.

Laboratory Report 23

Name: _____

Date: _____

Lab Section: _____

Proteins I: The IMViC Tests

1. Based on your observations of the SIM tubes, record how each bacterium reacted with respect to hydrogen sulfide production, indole production, and motility.

Bacterium	H_2S Production?	Indole Production?	Motility?
E. coli			
E. aerogenes			
P. vulgaris			
K. pneumoniae			

2. Based on your observations, record whether or not each bacterium was capable of fermenting glucose, with the production of either acids or acetylmethylcarbinol.

	Methyl Red Test		Voges-Proskauer Test	
Bacterium	Medium Color	Result?	Medium Color	Result?
E. coli				
E. aerogenes				
P. vulgaris				
K. pneumoniae				

3. Based on your observations, record whether or not each bacterium was capable of using citrate as an energy source.

Bacterium	Presence or Absence of Growth	Color of Medium	Citrate Utilization?
E. coli			
E. aerogenes			
P. vulgaris			
K. pneumoniae			

4. In summary, what is the major purpose of this exercise?

ASSESSMENT
Critical Thinking and Learning Outcomes Review

1. What is the component in the SIM deep tubes that makes this medium suitable to detect the production of indole by bacteria?

2. What organic molecule is necessary to detect mixed acid fermentation by bacteria?

3. Why did you mix the MR-VP culture?

4. Can a bacterium that ferments using the 2,3-butanediol pathway also use the mixed acid route? Explain your answer.

5. Why is a chemically defined medium necessary for the detection of citrate utilization by bacteria?

6. What does formation of a black precipitate in a SIM tube indicate? How is motility visualized?

EXERCISE 24

Proteins II: Gelatin and Casein Hydrolysis

SAFETY CONSIDERATIONS

- Be careful with the Bunsen burner flame.
- Keep culture tubes upright in a test-tube rack.

MATERIALS

- suggested bacterial strains: *Escherichia coli*, *Bacillus subtilis*, *Enterobacter aerogenes*, *Proteus vulgaris*, *Klebsiella pneumoniae*, and *Pseudomonas aeruginosa*
- skim milk agar
- inoculating loop
- gelatin deep tubes
- Litmus milk broth tubes
- Bunsen burner
- test-tube rack
- wax pencil or lab marker
- metric ruler
- container for biohazard waste
- safety glasses
- disposable gloves
- lab coat

LEARNING OUTCOMES

Upon completion of this exercise, students will demonstrate the ability to

1. Understand the biochemical process of deamination
2. Determine the ability of some bacteria to secrete proteolytic enzymes capable of hydrolyzing the protein casein or gelatin by performing different biochemical tests
3. Explain what a zone of proteolysis indicates
4. Understand how the changes in Litmus milk broth indicate degradation of protein or fermentation of lactose

SUGGESTED READING IN TEXTBOOK

1. Catabolism of Organic Molecules Other Than Glucose, section 11.7.
2. Microbiology of Fermented Foods: Beer, Cheese, and Much More, section 40.5; see also figures 40.6 and 40.7.

Pronunciation Guide

- *Bacillus subtilis* (bah-SIL-lus SUB-til-us)
- *Escherichia coli* (esh-er-I-ke-a KOH-lie)
- *Pseudomonas aeruginosa* (soo-do-MO-nas a-ruh-jin-OH-sa)
- *Enterobacter aerogenes* (en-ter-oh-BAK-ter a-RAH-jen-eez)
- *Proteus vulgaris* (PRO-te-us vul-GA-ris)
- *Klebsiella pneumoniae* (kleb-se-EL-lah nu-MO-ne-EYE)

Why Are the Following Bacteria Used in This Exercise?

In this exercise, you will learn how to perform gelatin and casein hydrolysis tests to detect the presence of proteolytic enzymes. To demonstrate casein hydrolysis, the authors have chosen three bacteria that have been used in prior exercises: *Escherichia coli* will produce a negative reaction; *Bacillus subtilis* and *Pseudomonas aeruginosa* will produce positive reactions. *Enterobacter aerogenes* is gelatinase positive, but gelatin is very slowly liquefied by most strains. In contrast, *Proteus vulgaris* also is gelatinase positive and liquefies gelatin very rapidly. *Escherichia coli* is gelatinase negative.

PRINCIPLES

Gelatin Hydrolysis

When boiled in water, the vertebrate connective tissue collagen changes into gelatin, a soluble mixture of polypeptides and amino acids. Certain bacteria are able to hydrolyze (digest) gelatin by secreting a proteolytic enzyme called **gelatinase**. The resulting amino acids can then be used as nutrients by the bacteria. Since hydrolyzed gelatin is no longer able to solidify, it is easy to detect gelatinase activity by observing formation of liquid. The ability of some bacteria to digest gelatin is an important characteristic in their differentiation. For example, when grown on a gelatin-containing medium like Thiogel, *Clostridium perfringens* causes liquefaction, whereas *Bacteroides fragilis* does not. Gelatin hydrolysis can also be used to assess the pathogenicity of certain bacteria. The production of gelatinase can sometimes be correlated with the ability of a bacterium to break down tissue collagen and spread throughout the body of a host. Gelatin liquefaction (the formation of a liquid) can be tested for by stabbing gelatin deep tubes. The media formulation contains gelatin, peptone, and beef extract, with the gelatin serving both as the solidifying agent and as a substrate for the enzyme gelatinase. Importantly, gelatin liquefies at 37°C. Following inoculation and incubation at 37°C, the cultures are placed in a refrigerator or ice bath at 4°C until the bottom resolidifies. If gelatin has been hydrolyzed, the medium will remain liquid after refrigeration. If gelatin has not been hydrolyzed, the medium will resolidify during the time it is in the refrigerator. Depending on the bacterium, gelatin deeps may require up to a 14-day incubation period for results to become apparent.

Casein Hydrolysis

Casein is a large milk protein incapable of permeating the plasma membrane of bacteria. Therefore, before casein can be used by some bacteria as their source of carbon and energy, it must be degraded into amino acids. Bacteria accomplish this by secreting **proteolytic enzymes (proteases)** that catalyze the hydrolysis (**proteolysis**) of casein to yield amino acids, which are then transported into the cell and catabolized.

The casein present in skim milk agar makes the agar cloudy. Following inoculation of the skim milk agar, bacteria that secrete proteases will produce a **zone of proteolysis** (a clear area surrounding the colony). Clearing of the cloudy agar (a positive reaction) is the result of a hydrolytic reaction that yields soluble amino acids (**Figure 24.1**). In a negative reaction, there is no protease activity, and thus the medium surrounding the bacterial colony remains opaque.

Figure 24.1 Procedure for Determining Casein Hydrolysis. (a) Streak inoculation of a skim milk agar. (b) Schematic of a plate exhibiting zones of proteolysis. The presence of a clear halo indicates casein hydrolysis. (c) An example of a single bacterial streak on skim milk agar. Note the clearing around the dense bacterial growth. *(Lisa Burgess/McGraw Hill)*

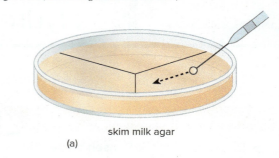
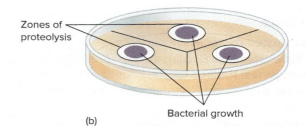
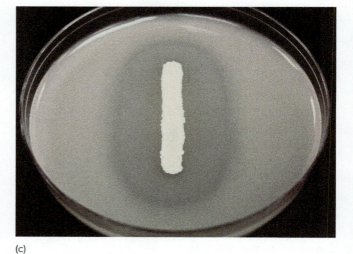

Litmus Milk

Another method to assess casein hydrolysis is by using Litmus milk broth. Despite its relatively simple composition, Litmus milk can be used to characterize the metabolic properties of bacterial strains. The two key components of this media are skim milk and the pH indicator litmus. The protein (casein) and sugar (lactose) present in skim milk can be consumed in different ways, depending on the metabolic capabilities of a given bacterium. The products of this metabolism are easily monitored by

174 Microbial Biochemistry

examining the media color and consistency after inoculation and incubation and comparing them to an uninoculated control tube (**Figure 24.2a**).

- **Alkaline** (**Figure 24.2b**): media turns blue/purple. Proteolysis releases ammonia from peptides, driving pH up.
- **Reduction** (**Figure 24.2c**): media turns white. Using litmus as an electron acceptor makes it colorless.
- **Peptonization** (**Figure 24.2d**): complete digestion of milk proteins resulting in clearing of the media. The media may also exhibit a slightly pink or brown color if peptonization is incomplete.
- **Acid** (**Figure 24.2e**): media turns pink. Fermentation of lactose generates lactic acid, which lowers the pH. The acid can then react with milk proteins (i.e. casein) forming a solid clot.
- **Coagulation** (**Figure 24.3**): media turns solid as the milk proteins coagulate into curd. In cases where gas is produced as a by-product of fermentation, the coagulated milk protein clots can break apart.

Procedure

Gelatin Hydrolysis

First Period

1. Label three gelatin deeps with your name, date, and the bacterium to be inoculated (*E. coli*, *E. aerogenes*, and *P. vulgaris*). Label a fourth tube "control."

Figure 24.2 Examples of Litmus milk Reactions. Shown here are Litmus milk tubes that have been inoculated with different bacteria and incubated overnight. The difference in color and consistency of media in each tube (a–d) should be compared to the uninoculated control (tube e). *(Lisa Burgess/McGraw Hill)*

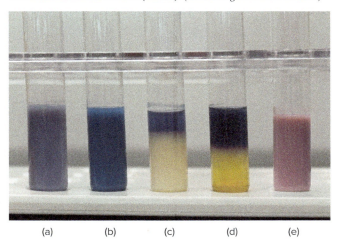

(a)　(b)　(c)　(d)　(e)

Figure 24.3 Coagulation in a Litmus milk tube. Note the clumping of protein in the bottom of the tube and how the clot is broken up due to gas production. *(Lisa Burgess/McGraw Hill)*

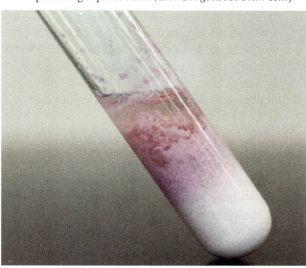

2. Using aseptic technique and an inoculating needle, inoculate three of the deeps with the appropriate bacterium by stabbing the medium three-fourths of the way to the bottom of the tube.
3. Incubate the four tubes for at least 24 hours at 37°C. The incubation time depends on the species of bacteria; some may require incubation for up to 2 weeks. If the latter is the case, observe on days 7 and 14. Check with your instructor for details.

Second Period

1. Remove the gelatin deep tubes from the incubator and place them in the refrigerator (at 4°C) for 30 minutes. Alternatively, you may place them in an ice water bath for 5 minutes.
2. When the bottom resolidifies, remove the tubes and gently slant them. Check to see if the surface of the medium is fluid or liquid. If the gelatin is liquid, this indicates that gelatin has been hydrolyzed by the bacterium. If no hydrolysis has occurred, the medium will remain semisolid and thus appear like the uninoculated control. Record your observations in the lab report section for this exercise.

Casein Hydrolysis

First Period

1. Obtain a skim milk agar plate and label it with your name and date using a wax pencil or lab marker.

Proteins II: Gelatin and Casein Hydrolysis

2. Divide the plate into three sections: label one *E. coli,* the second *B. subtilis,* and the third *P. aeruginosa.*
3. As shown in **Figure 24.1** inoculate each third of the skim milk plate with the appropriate bacterium as per the label. Using your inoculating loop, carefully draw a line of bacteria in the appropriate section.
4. Incubate the plate in an inverted position at 37°C for at least 24 hours.

Second Period

1. Examine the skim milk agar for the presence or absence of a clear zone surrounding the growth of each of the bacterial test organisms. You can see the clear zones best against a black background.
2. Based on your observations, determine which of the bacteria were capable of hydrolyzing the casein. Record this information in the lab report for this exercise

Litmus Milk

First Period

1. Label six Litmus milk broth tubes with your name, date, and the bacterium to be inoculated (*E. coli*, *P. vulgaris*, *B. subtilis*, *P. aeruginosa*, and *K. pneumoniae*). Label a sixth tube "control."
2. Using aseptic technique and an inoculating loop, inoculate the broth tubes with the appropriate bacterium using your best aseptic technique. Do not inoculate the control tube.

3. Incubate the tubes for at least 24 hours at 37°C. Some bacteria may require incubation for up to 48 hours.

Second Period

1. Remove the Litmus milk tubes from the incubator.
2. Observe the tubes for any color change as compared to the uninoculated control. Also take note of the consistency of the media by tilting the tubes at a slight angle. Record your observations.

Disposal

When you are finished with all of the tubes, pipettes, and plates, discard them in the designated place for sterilization and disposal.

> **HELPFUL HINTS**
>
> - Aseptic technique must be followed because contaminating microorganisms might be capable of hydrolyzing casein and thus lead to erroneous results.
> - Do not shake the gelatin tubes when moving them to a refrigerator. Note that gelatin digestion may have occurred only at the surface.
> - Be aware the bacteria that produce pigment (e.g. *Pseudomonas* sp.) may mask the color reactions in the Litmus milk tubes.

Laboratory Report 24

Name: _____

Date: _____

Lab Section: _____

Proteins II: Gelatin and Casein Hydrolysis

1. Complete the following table for the gelatin deep tubes.

Bacterium	Observed Reactions
E. aerogenes	
E. coli	
P. vulgaris	

2. Complete the following table for the skim milk agar plate.

Bacterium	Observed Reactions
E. coli	
B. subtilis	
P. aeruginosa	

3. Complete the following table for the Litmus milk tubes.

Bacterium	Observed Reactions
E. coli	
P. vulgaris	
B. subtilis	
P. aeruginosa	
K. pneumoniae	

4. In summary, what is the major purpose of this exercise?

ASSESSMENT
Critical Thinking and Learning Outcomes Review

1. Define the following terms:

    a. protein

    b. hydrolysis

    c. casein

    d. protease

    e. gelatin

    f. peptide bond

    g. litmus

2. How can skim milk agar, gelatin deeps, and Litmus milk broth be used to demonstrate proteolysis?

3. Why are some bacteria able to grow on skim milk agar even though they do not produce any proteases?

4. Draw the chemical reaction for the hydrolysis reaction catalyzed by a protease.

5. How can gelatin hydrolysis be beneficial to certain bacteria?

6. What is unique about gelatin at 37°C versus 4°C?

7. In a Litmus milk tube, what does an alkaline reaction reveal about fermentation and proteolysis? What does an acid reaction reveal about fermentation and proteolysis?

8. How does peptonization differ from coagulation?

NOTES

EXERCISE 25 — Proteins III: Catalase Activity

SAFETY CONSIDERATIONS
- Be careful with the Bunsen burner flame.
- Hydrogen peroxide is caustic to the skin and mucous membranes.

MATERIALS
- suggested bacterial strains:
  slant or plate cultures of *Staphylococcus aureus*, *Streptococcus pyogenes*, *Enterococcus faecalis*, and *Micrococcus luteus*
- 3% hydrogen peroxide (H_2O_2)
- Bunsen burner
- Pasteur pipette with pipettor
- test-tube rack
- wax pencil or lab marker
- clean glass slides
- wooden applicator stick
- container for biohazard waste
- safety glasses
- disposable gloves
- lab coat

LEARNING OUTCOMES
Upon completion of this exercise, students will demonstrate the ability to
1. Outline the biochemical process of hydrogen peroxide detoxification through the production of the enzyme catalase
2. Describe how catalase production can be determined
3. Perform a catalase test

SUGGESTED READING IN TEXTBOOK
1. Environmental Factors Affect Microbial Growth, section 7.5.

Pronunciation Guide
- *Enterococcus faecalis* (en-te-ro-KOK-kus fee-KAL-iss)
- *Micrococcus luteus* (my-kro-KOK-us LOO-tee-us)
- *Staphylococcus aureus* (staf-il-oh-KOK-kus ORE-ee-us)
- *Streptococcus pyogenes* (strep-to-KOK-us pi-AH-gen-eez)

Why Are the Following Bacteria Used in This Exercise?

In this exercise, you will learn to perform the catalase test. The catalase test is very useful in differentiating between groups of Gram-positive cocci. The authors have chosen the following four bacteria to accomplish the above objective. *Staphylococcus aureus* is catalase positive when grown in an aerobic environment. *S. aureus* is often part of the normal flora of human skin, but it is also capable of causing a variety of diseases. *Enterococcus faecalis* is catalase-negative, and occurs widely in the environment. *Micrococcus luteus* is catalase positive. *M. luteus* occurs primarily on mammalian skin and in soil, but commonly can be isolated from food products and air. *Streptococcus pyogenes* is catalase negative, and can cause several diseases such as pharyngitis (i.e., strep throat) and impetigo.

PRINCIPLES

Some bacteria contain flavoproteins that reduce O_2, resulting in the production of **reactive oxygen species (ROS)** including the superoxide radical ($O_2\cdot^-$), hydrogen peroxide (H_2O_2), or the hydroxyl radical ($OH\cdot$). These are extremely toxic because they are powerful oxidizing agents and destroy cellular constituents very rapidly. A bacterium must be able to protect itself against such O_2 products or it will be killed. Indeed, ROS are one of the most potent weapons available to the human immune system. The rapid generation of ROS during a bacterial infection is called the **respiratory burst**.

$$O_2 + e^- \rightarrow O_2\cdot^- \text{ (superoxide radical)}$$

$$O_2\cdot^- + e^- + 2H^+ \rightarrow H_2O_2 \text{ (hydrogen peroxide)}$$

$$H_2O_2 + e^- + H^+ \rightarrow H_2O + OH\cdot \text{ (hydroxyl radical)}$$

Many bacteria possess enzymes that afford protection against toxic O_2 products. Obligate aerobes and facultative anaerobes usually contain the enzyme **superoxide dismutase,** which catalyzes the destruction of superoxide, and either **catalase** or **peroxidase,** which catalyzes the destruction of hydrogen peroxide as follows:

$$2O_2\cdot^- + 2H^+ \xrightarrow{\text{superoxide dismutase}} O_2 + H_2O_2$$

$$2H_2O_2 \xrightarrow{\text{catalase}} 2H_2O + O_2$$

$$H_2O_2 + NADH + H^+ \xrightarrow{\text{peroxidase}} 2H_2O + NAD^+$$

Most strict anaerobes lack the enzymes needed to detoxify ROS and therefore cannot tolerate growth in an oxygen-rich environment.

Catalase production can be detected using a simple test. Bacteria are added to a drop of 3% hydrogen peroxide on a clean glass slide. If catalase is present, bubbles appear instantly as a result of the release of O_2 from the breakdown of the hydrogen peroxide by the enzyme; the absence of vigorous bubbling indicates a negative catalase test (**Figure 25.1**).

Catalase activity is very useful in differentiating between Gram-positive cocci with similar cellular arrangements. For example, the morphologically similar *Enterococcus* (catalase negative) and *Staphylococcus* (catalase positive) can be differentiated using the catalase test, as well as *Streptococcus* (catalase negative) from *Staphylococcus* (catalase positive). The catalase test can also be used to distinguish between Gram-positive rods like *Bacillus* (catalase positive) and *Clostridium* (catalase negative).

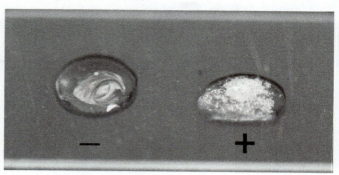

Figure 25.1 Interpretation of the Catalase Test. The absence of bubbles is a negative catalase test. The production of bubbles of oxygen gas after the addition of a small amount of bacterial cells to hydrogen peroxide is a positive catalase test. *(Lisa Burgess/McGraw Hill)*

Procedure

1. Label two clean glass slides with your name. Using a wax pencil or lab marker, draw two circles on the slide. Label each circle with the name of the bacterium to be tested (*S. aureus*, *S. pyogenes*, *E. faecalis*, and *M. luteus*).
2. Using a Pasteur pipette, place two drops of 3% hydrogen peroxide into the center of each circle.
3. With a sterile plastic loop or wooden applicator, transfer a small amount of colony material to the spot of hydrogen peroxide. Immediately record your observations in the lab report for this exercise

Disposal

When you are finished with all of the tubes, pipettes, and glass slides, discard them in the designated place for sterilization and disposal.

> **HELPFUL HINTS**
>
> - Always use fresh hydrogen peroxide, since it is unstable.
> - The metal in a regular inoculating loop that is used to transfer bacteria to a glass slide may non-specifically degrade the hydrogen peroxide and give a false-positive reaction.
> - Rapid, vigorous bubbling indicates a positive catalase reaction.

Laboratory Report 25

Name: _____

Date: _____

Lab Section: _____

Proteins III: Catalase Activity

1. Complete the following table on catalase activity.

Bacterium	Presence or Absence of Bubbling?	Catalase Activity?
E. faecalis		
S. aureus		
S. pyogenes		
M. luteus		

2. In summary, what is the major purpose of this exercise?

ASSESSMENT
Critical Thinking and Learning Outcomes Review

1. What is the importance of catalase to bacteria? Do you expect catalase to be produced by all aerobes? Explain your answer.

2. Do anaerobic bacteria require catalase? Explain your answer.

3. Why can using metal inoculating loops lead to false catalase test results?

4. What groups of bacteria can be differentiated with the catalase test?

5. What are three products that result when flavoproteins reduce O_2?
    a.

    b.

    c.

6. Not all bacterial pathogens possess catalase to defend against the respiratory burst of the human immune system. How are such pathogens able to combat this attack?

7. What is the substrate and what are the products of the catalase reaction?

EXERCISE 26 Proteins IV: Oxidase Test

SAFETY CONSIDERATIONS

- Be careful with the Bunsen burner flame.
- The oxidase reagent is caustic. Avoid contact with eyes and skin.
- In case of contact, immediately flush eyes or skin with plenty of water for at least 15 minutes.

MATERIALS

- suggested bacterial strains:
  Alcaligenes faecalis, *Escherichia coli*, and *Pseudomonas aeruginosa*
- tryptic soy agar plates
- tetramethyl-*p*-phenylenediamine dihydrochloride (oxidase reagent), Difco ampules (1%); OxiDrops®
- Bunsen burner
- platinum or plastic loops (instead of standard inoculating loops)
- wax pencil or lab marker
- Pasteur pipette with pipettor
- Oxidase Disks, Dry Slides or BBL™ Dry Slide™; Oxidase Test Strips; SpotTest Oxidase Reagent (Difco); OxyStrips® Oxy-Swab®
- wooden applicator sticks
- Whatman No. 2 filter paper
- sterile cotton swabs
- container for biohazard waste
- safety glasses
- disposable gloves
- lab coat

LEARNING OUTCOMES

Upon completion of this exercise, students will demonstrate the ability to

1. Explain the biochemistry underlying oxidase enzymes
2. Describe the experimental procedure that enables one to distinguish between groups of bacteria based on cytochrome oxidase activity
3. List examples of oxidase-positive and oxidase-negative bacteria
4. Perform an oxidase test

SUGGESTED READING IN TEXTBOOK

1. Electron Transport Chains: Sets of Sequential Redox Reactions, section 10.4; see also figures 10.6–10.11.
2. *Gammaproteobacteria* Is the Largest Bacterial Class, section 21.2; see also table 21.3.

Pronunciation Guide

- *Alcaligenes faecalis* (al-kah-LIJ-e-neez fee-KAL-iss)
- *Escherichia coli* (esh-er-I-ke-a KOH-lie)
- *Pseudomonas aeruginosa* (soo-do-MO-nas a-ruh-jin-OH-sah)

Why Are the Following Bacteria Used in This Exercise?

This exercise gives you experience in performing the oxidase test. The oxidase test distinguishes between groups of bacteria based on cytochrome *c* oxidase activity. Three bacteria will be used. *Alcaligenes faecalis* is a Gram-negative, aerobic rod (coccal rod or coccus) that possesses a strictly respiratory type of metabolism with oxygen as the terminal electron acceptor. It is thus oxidase positive. *Escherichia coli* is a facultatively anaerobic Gram-negative rod that has both respiratory and fermentative types of metabolism and is oxidase negative. *Pseudomonas aeruginosa* is a Gram-negative, aerobic rod having a strictly respiratory type of metabolism with oxygen as the terminal electron acceptor and thus is oxidase positive.

Medical Application

The oxidase test is a useful procedure in the clinical laboratory because some Gram-negative pathogenic species of bacteria (such as *Neisseria gonorrhoeae, P. aeruginosa, Moraxella,* and *Vibrio* species) are oxidase positive, in contrast to species in the family *Enterobacteriaceae,* which are oxidase negative.

PRINCIPLES

Oxidase enzymes play an important role in the operation of the electron transport system during aerobic respiration. **Cytochrome c oxidase** uses O_2 as an electron acceptor during the oxidation of reduced cytochrome c to form water and oxidized cytochrome c.

The ability of bacteria to produce cytochrome oxidase can be determined by the addition of the oxidase test reagent or test strip (tetramethyl-*p*-phenylenediamine dihydrochloride or an Oxidase Disk, *p*-minodimethylaniline) to colonies that have grown on a plate medium. Alternatively, a bacterial sample can be collected using a wooden application stick and either be rubbed on a Dry Slide Oxidase reaction area, on a KEY test strip, or filter paper moistened with the oxidase reagent. The light pink oxidase test reagent (disk, strip, or slide) serves as an artificial substrate, donating electrons to cytochrome oxidase and in the process becoming oxidized to dark blue and then dark purple (**Figures 26.1** and **26.2**) compound in the presence of free O_2 and the oxidase. The presence of this dark purple coloration represents a

Figure 26.1 Oxidase Test. Note the blue to dark purple color after the colony has been added to the strip with oxidase reagent (right) next to a negative reaction (left). *(Javier Izquierdo/McGraw Hill)*

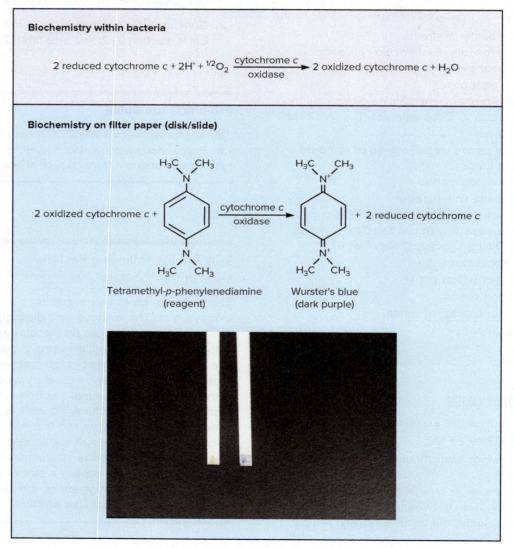

Figure 26.2 Oxidase Test. The swab on the left shows a blue to purple reaction due to oxidase production and is a positive oxidase test. The swab on the right shows a yellow-orange color from a culture that is oxidase negative and thus a negative oxidase test. *(Auburn University Photographic Services/McGraw Hill)*

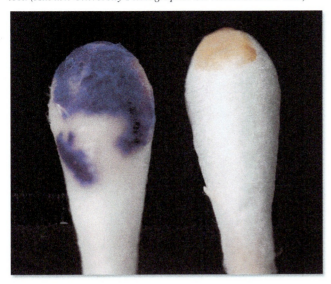

positive test. No color change or a light pink coloration on the colonies indicates the absence of oxidase and is a negative test.

Procedure for Oxidase Test

First Period

2. With a wax pencil or lab marker, divide the bottom of a tryptic soy agar plate into three sections and label each with the name of the bacterium to be inoculated, your name, and date.
3. Using aseptic technique (*see Figure 5.3*), make a single streak-line inoculation on the agar surface with the appropriate bacterium.
4. Incubate the plate in an inverted position for 24 to 48 hours at 37 °C.

Second Period

1. Add 2 to 3 drops of the oxidase reagent to the surface of the growth of several isolated colonies of each test bacterium or to some paste that has been transferred to a piece of filter paper. Using another colony, place an Oxidase Disk on it. Add a drop of sterile water. If Dry Slides or test strips are available, use a wooden applicator stick to transfer a sample to the slide, test strip, or filter paper moistened with oxidase reagent. Alternatively, drop a KEY Oxidase Test Strip onto the surface of a slant culture and moisten it with water if necessary. If using the sterile cotton swabs, touch a swab to a colony on the plate. Remove the cap from an oxidase ampule and deliver several drops of reagent to the culture on the swab.
2. Observe the colony or sample for the presence or absence of a color change from pink to blue, and finally to dark purple. This color change will occur within 20 to 30 seconds. Color changes after 20 to 30 seconds are usually disregarded since the reagent begins to change color with time due to auto-oxidation. Oxidase-negative bacteria will not produce a color change or will produce a light pink color.
3. Based on your observations, determine and record in the report for this exercise whether or not each bacterium was capable of producing oxidase.

Disposal

When you are finished with all of the plates, sticks, filter paper, pipettes, and cotton swabs, discard them in the designated place for sterilization and disposal.

> **HELPFUL HINTS**
>
> - Students should note the color change immediately following the addition of oxidase reagent. Color changes after 20 seconds are not valid.
> - Using nichrome or other iron-containing inoculating devices may cause false-positive reactions. This is due to the fact that the metal will catalyze an unwanted oxidation of the reagent.

NOTES

Laboratory Report 26

Name: _____

Date: _____

Lab Section: _____

Proteins IV: Oxidase Test

1. Complete the following table on the oxidase test.

Bacterium	Color of Colonies after Adding		Oxidase Production (+ or −)	
	Reagent	Disk or Slide	Reagent	Disk or Slide
A. faecalis	_____	_____	_____	_____
E. coli	_____	_____	_____	_____
P. aeruginosa	_____	_____	_____	_____

2. In summary, what is the major purpose of this exercise?

ASSESSMENT
Critical Thinking and Learning Outcomes Review

1. What metabolic property characterizes bacteria that possess oxidase activity?

2. What is the importance of cytochrome oxidase to bacteria that possess it?

3. Do anaerobic bacteria require cytochrome oxidase? Explain your answer.

4. What is the function of the test reagent in the oxidase test?

5. The oxidase test is used to differentiate among which groups of bacteria?

6. Why should nichrome or other iron-containing inoculating devices not be used in the oxidase test?

EXERCISE 27 Proteins V: Urease Activity

SAFETY CONSIDERATIONS

- Be careful with the Bunsen burner flame.
- **Salmonella spp. are potential pathogens (BSL2).** Use aseptic technique throughout this experiment.
- Keep all culture tubes upright in a test-tube rack or in a can.

MATERIALS

- suggested bacterial strains:
  Escherichia coli, *Klebsiella pneumoniae*, *Proteus vulgaris*, and *Salmonella enterica* serovar Typhimurium
- urea broth tubes or urea agar slants
- Bunsen burner
- test-tube rack
- inoculating loop
- incubator set at 37°C
- urea disks or urease test tablets
- 4 sterile test tubes
- wax pencil or lab marker
- sterile forceps
- container for biohazard waste
- safety glasses
- disposable gloves
- lab coat

LEARNING OUTCOMES

Upon completion of this exercise, students will demonstrate the ability to

1. Explain the biochemical process of urea hydrolysis
2. Determine the ability of bacteria to degrade urea by means of the enzyme urease
3. Explain when the urease test is used
4. Perform a urease test

SUGGESTED READING IN TEXTBOOK

1. *Gammaproteobacteria* Is the Largest Bacterial Class, section 21.2.
2. Direct Contact Diseases Can Be Caused by Bacteria, section 38.3.

Pronunciation Guide

- *Escherichia coli* (esh-er-I-ke-a KOH-lie)
- *Klebsiella pneumoniae* (kleb-se-EL-lah nu-mo-ne-ah)
- *Proteus vulgaris* (PRO-tee-us vul-GA-ris)
- *Salmonella enterica* (sal-mon-EL-ah en-TE-ree-kah)

Why Are the Following Bacteria Used in This Exercise?

In this exercise, you will perform a urease test to determine the ability of bacteria to degrade urea by means of the enzyme urease. The authors have chosen two urease-positive bacteria (*Klebsiella pneumoniae* and *Proteus vulgaris*) and two urease-negative bacteria (*Escherichia coli* and *Salmonella enterica*).

Medical Application

In the clinical laboratory, members of the genera *Proteus, Providencia, Klebsiella,* and *Morganella* can be distinguished from other enteric nonlactose-fermenting bacteria (*Salmonella, Shigella, Escherichia*) by their fast urease activity. *Proteus mirabilis* is a major cause of human urinary tract infections.

From a bacterium's perspective, urease can also be a beneficial enzyme. One of the causes of peptic ulcer disease and gastritis is the bacterium *Helicobacter pylori*. Movement into the mucous layer of the stomach

Figure 27.1 Urea Hydrolysis on/in Urea Agar Slants. From left to right: **(a)** Urease negative. **(b)** Urease positive. *(Javier Izquierdo/McGraw Hill)*

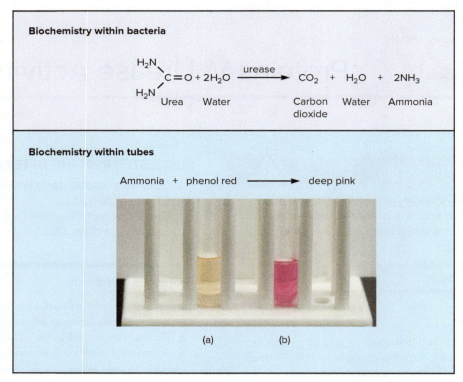

is aided by the fact that *H. pylori* is a strong producer of urease. Urease activity is thought to create a localized alkaline environment when hydrolysis of urea produces ammonia, protecting the bacterium from the gastric acid in the stomach.

PRINCIPLES

Some bacteria can use urea as a source of nitrogen. They can do this since they are able to produce an enzyme called **urease** that attacks the nitrogen and carbon bond in amide compounds such as urea, forming the end products ammonia, CO_2, and water (**Figure 27.1**).

Urease activity (the **urease test**) is detected by growing bacteria in a medium containing urea and a pH indicator such as phenol red (initial pH 6.5) or adding tablets that contain the pH indicator to culture tubes. When urea is hydrolyzed, ammonia accumulates in the medium and makes it alkaline. This increase in pH (to 8.4) causes the indicator to change from orange-red to deep pink or purplish red (cerise), indicating a positive test for urea hydrolysis. Failure to develop a deep pink color is a negative test.

Procedure

First Period

1. Label each of the urea broth tubes or slants with the name of the bacterium to be inoculated, your name, and date. Label one of the broth tubes and/or slants as a control.
2. Using aseptic technique (*see Figure 5.3*), inoculate each bacterial tube or slant with the appropriate bacterium by means of a loop inoculation.
3. Incubate the tubes for 24 to 48 hours at 37°C.

Urea Disks or Tablets

1. Add between 0.5 ml and 1 ml of sterile distilled water to four sterile test tubes according to the manufacturer's instructions.
2. Transfer one or two loopfuls of bacteria to each tube. Label with your name and date.
3. Using sterile forceps, add one disk or tablet to each tube.
4. Incubate up to 4 hours at 37°C. Check for a color change each hour. Depending on the

manufacturer, tests may be incubated up to 24 hours if necessary.

Second Period

1. Examine all of the urea broth cultures and disks or tablet tubes and/or slants to determine their color (**Figure 27.1**). As the pH becomes more alkaline, the phenol red changes from yellow (pH 6.8) to a bright pink (cerise) color (pH > 8.0). If your urea slant is negative, continue to incubate for an additional 7 days to check for slow urease production.
2. Based on your observations, determine and record in the report for this exercise whether each bacterium was capable of hydrolyzing urea.

Disposal

When you are finished with all of the tubes, discard them in the designated place for sterilization and disposal.

> **HELPFUL HINTS**
>
> ■ Some bacteria have a delayed urease reaction that may require an incubation period longer than 48 hours (e.g., up to 7 days).

NOTES

Laboratory Report 27

Name: _____

Date: _____

Lab Section: _____

Proteins V: Urease Activity

1. Complete the following table on urease activity.

Bacterium	Color of Urea Broth or Slants	Color of Disks or Tablets	Urea Hydrolysis (+ or −)
E. coli	_____	_____	_____
K. pneumoniae	_____	_____	_____
P. vulgaris	_____	_____	_____
S. enterica	_____	_____	_____

2. In summary, what is the major purpose of this exercise?

ASSESSMENT
Critical Thinking and Learning Outcomes Review

1. Explain the biochemistry of the urease reaction.

2. What is the purpose of the phenol red in the urea broth or slant medium?

3. When would you use the urease test?

4. In a positive urease test, why does the urea broth or slant change color?

5. What is the main advantage of the urea disk over the broth tubes or slants with respect to the detection of urease?

6. What is the media composition of urea broth?

EXERCISE 28

Proteins VI: Lysine and Ornithine Decarboxylase Tests

SAFETY CONSIDERATIONS

- Be careful with the Bunsen burner flame.
- Keep all culture tubes upright in a test-tube rack or in a can.

MATERIALS

- suggested bacterial strains: *Enterobacter aerogenes*, *Citrobacter freundii*, *Klebsiella pneumoniae*, and *Proteus vulgaris*
- Moeller's lysine decarboxylase broth with lysine (LDC)
- lysine iron agar slants (LIA)
- Moeller's ornithine decarboxylase broth with ornithine (ODC)
- Moeller's lysine decarboxylase broth without lysine (DC), which will serve as the control
- Moeller's ornithine decarboxylase broth without ornithine (OD), which will serve as the control
- Pasteur pipettes with pipettor
- inoculating loop
- test-tube rack
- sterile distilled water
- sterile mineral oil
- incubator set at 37°C
- sterile test tubes
- ornithine, lysine, and decarboxylase KEY Rapid Substrate Tablets and strips
- Bunsen burner
- ninhydrin in chloroform (Dissolve 50 mg ninhydrin in 0.4 ml of dimethylsulfoxide [DMSO], then add 25 ml of chloroform to the DMSO solution.)
- 10% KOH
- wax pencil or lab marker
- container for biohazard waste
- safety glasses
- disposable gloves
- lab coat

LEARNING OUTCOMES

Upon completion of this exercise, students will demonstrate the ability to

1. Explain the biochemical process of decarboxylation
2. Outline why decarboxylases are important to some bacteria
3. Explain how the decarboxylation of lysine can be detected in culture
4. Perform lysine and ornithine decarboxylase tests

SUGGESTED READING IN TEXTBOOK

1. Catabolism of Organic Molecules Other Than Glucose, section 11.7; see also figure 11.4.

Pronunciation Guide

- *Citrobacter freundii* (SIT-ro-bac-ter FRUN-dee)
- *Enterobacter aerogenes* (en-ter-oh-BAK-ter a-RAH-jen-eez)
- *Klebsiella pneumoniae* (kleb-se-EL-lah nu-MO-ne-ah)
- *Proteus vulgaris* (PRO-te-us vul-GA-ris)

Why Are the Following Bacteria Used in This Exercise?

This exercise provides you with the experience using the lysine and ornithine decarboxylase tests to differentiate between bacteria. Two lysine decarboxylase-positive (*Enterobacter aerogenes* and *Klebsiella pneumoniae*) and two lysine decarboxylase-negative (*Proteus vulgaris* and *Citrobacter freundii*) bacteria, and two ornithine decarboxylase-positive (*E. aerogenes* and *Citrobacter freundii*) and two ornithine decarboxylase-negative (*K. pneumoniae* and *P. vulgaris*) bacteria were chosen to demonstrate the lysine and ornithine decarboxylase tests.

Medical Application

In the clinical laboratory, decarboxylase differential tests (lysine, ornithine, and arginine) are used to differentiate between organisms in the family *Enterobacteriacea*.

Cadaverine is a foul-smelling nitrogenous base produced by decarboxylation of lysine. In the environment, it is produced in decaying protein material by the action of bacteria, particularly species of *Vibrio*.

Putrescine is a diamine first found in decaying animal tissue but now known to occur in almost all tissues and in cultures of certain bacteria. It is formed by decarboxylation of ornithine.

PRINCIPLES

Decarboxylation is the removal of a carboxyl group from an organic molecule. Bacteria growing in liquid media decarboxylate amino acids most rapidly when conditions are anaerobic and slightly acidic. Decarboxylation of amino acids, such as lysine and ornithine, results in the production of an amine (or diamine) and CO_2 as illustrated below.

$$R-\underset{NH_2}{CH}-COOH \xrightarrow{\text{decarboxylase}} R-CH_2-NH_2 + CO_2$$

An amino acid → An amine + Carbon dioxide gas

Bacteria that are able to produce the enzymes **lysine decarboxylase** and **ornithine decarboxylase** can decarboxylate lysine and ornithine and use the resulting amines as precursors for the synthesis of other needed molecules. In addition, when certain bacteria carry out fermentation, acidic waste products are produced, making the medium acidic and inhospitable. Many decarboxylases are activated by a low pH. They remove the acid groups from amino acids, producing alkaline amines, which raise the pH of the medium, making it more hospitable.

Decarboxylation of lysine or ornithine can be detected by culturing bacteria in a medium containing the desired amino acid, glucose, and a pH indicator (bromcresol purple). The pH indicator turns purple at a pH of 6.8 and yellow at a pH below 5.2. Before incubation, sterile mineral oil is layered onto the broth to prevent oxygen from reaching the bacteria and inhibiting the reaction. The acids produced by the bacteria from the fermentation of glucose will initially lower the pH of the medium and cause the pH indicator to change from purple to yellow. The acid pH activates the enzyme that causes decarboxylation of lysine or ornithine to amines and the subsequent neutralization of the medium. This

Figure 28.1 **Amino Acid Decarboxylase Test.** A yellow to orange color (left tube) indicates that decarboxylation did not occur after several days of incubation and is a negative test for decarboxylase. The appearance of a purple color (right tube) throughout the broth after a yellow color and several days of incubation indicates decarboxylation did occur and is a positive test for decarboxylase. *(Barry Chess)*

results in another color change from yellow back to purple (**Figures 28.1** and **28.2**) and is a positive test. Negative tubes will be yellow.

Lysine iron agar (LIA) is also used for the cultivation and differentiation of members of the *Enterobacteriaceae* based on their ability to decarboxylate lysine and to form H_2S. Bacteria that decarboxylate lysine turn the medium purple. Bacteria that produce H_2S appear as black colonies.

The lysine decarboxylase test is useful in differentiating *Pseudomonas* (L. −), *Klebsiella* (L. +), *Enterobacter* (L. +), and *Citrobacter* (L. −) species. The ornithine decarboxylase test is helpful in distinguishing between *Klebsiella* (O. −) and *Proteus* (O. −), and *Enterobacter* (O. +) bacteria.

A quick test for ornithine or lysine decarboxylase is to use the KEY Rapid Substrate Tablets and strips. These tablets contain the respective amino acids in a mixture of salts correctly buffered for each test.

Figure 28.2 Lysine and Ornithine Decarboxylase Test. The tubes labeled 1, 3, and 4 (yellow) are negative for decarboxylase activity; weak acid production (pH less than 5.2) from glucose fermentation has turned them yellow. If the bacterium is only capable of glucose fermentation, the medium will remain yellow. The tube labeled 2 (purple) is positive for decarboxylase activity due to the accumulation of alkaline end products. *(Source: R. E. Weaver, MD, PhD/CDC)*

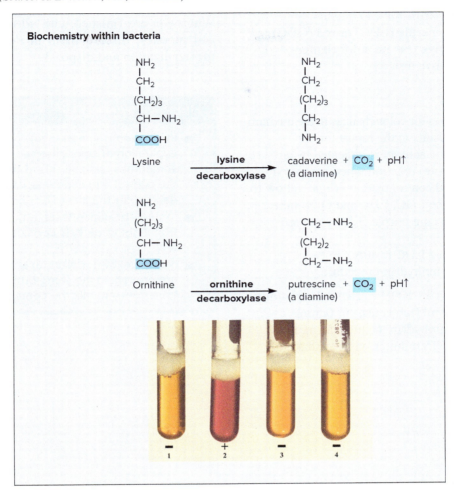

In addition, a pH indicator is present in the tablet, which changes color as the decarboxylation reaction progresses. In the lysine decarboxylase test tablet, the indicator is bromcresol purple, which turns purple as the test becomes positive. The indicator in the ornithine decarboxylase test tablet is phenol red, which turns red in a positive test.

Procedure

First Period (Standard Method)

1. Label four LDC tubes and/or LIA slants with the names of the respective bacteria (*K. pneumoniae, E. aerogenes, P. vulgaris,* and *C. freundii*) to be inoculated. Do the same for one control DC tube. Add your name and date to the tubes. Note that the medium is purple.

2. Do the same with the four ODC tubes and one OD control tube.
3. As shown in *Figure 5.3*, aseptically inoculate the tubes with the proper bacteria.
4. With a sterile Pasteur pipette, layer about 1 ml of sterile mineral oil on top of the inoculated media. LIA slants do not need mineral oil.
5. Incubate the cultures for 24 to 48 hours at 37°C.

Tablet/Strip Method

1. Label eight sterile test tubes with the respective bacteria, your name, and date.
2. Pipette 1 ml of sterile distilled water in each tube for regular tablets and 0.5 ml for ODC test strips.
3. Add a loopful of bacteria or 0.1 ml of thick bacterial culture to each tube.

4. Add four ornithine test strips to the first four tubes and four lysine tablets to the other four tubes.
5. Incubate the LDC tubes at 37°C for 24 to 48 hours and the ODC test strips for 4 to 6 hours.
6. A color change to purple (LDC) or red (ODC) constitutes a positive test; no color change (yellow) is a negative test.

Second Period

1. Examine the cultures for color changes in the medium and record your results in the report for this exercise. Enzymatic activity is indicated by an alkaline (dark purple) reaction when compared with the inoculated control medium (light yellow to orange color) in the LDC, LIA, and ODC tubes. Positive KEY tests are purple (LDC) and red (ODC).
2. The KEY ODC and LDC results can be confirmed by the ninhydrin procedure.
   a. Add 1 drop of 10% KOH to each tube and mix.
   b. Add either 1.0 ml (tablet test) or 0.5 ml (strip test) of ninhydrin in chloroform. Let stand for 10 to 15 minutes without shaking.
   c. Purple color in the bottom chloroform layer is positive for decarboxylation.

Disposal

When you are finished with all of the tubes, slants, and pipettes, discard them in the designated place for sterilization and disposal.

> **HELPFUL HINTS**
>
> - In biochemical tests involving visual evaluation of color changes that are sometimes minimal, it is always useful to hold the control and experimental tubes next to each other to discern any color differences.
> - In decarboxylase tests, any trace of purple, from light to dark, is considered a positive test.
> - If both yellow (or orange) and purple colors are present in the same broth tube, shake the tube gently to mix the contents.

Laboratory Report 28

Name: _____

Date: _____

Lab Section: _____

Proteins VI: Lysine and Ornithine Decarboxylase Tests

1. Results from the decarboxylase tests.

Bacterium	Color of LIA	Color of LDC	Color of ODC	LD Tablets	OD Tablets
C. freundii	_____	_____	_____	_____	_____
E. aerogenes	_____	_____	_____	_____	_____
K. pneumoniae	_____	_____	_____	_____	_____
P. vulgaris	_____	_____	_____	_____	_____

2. In summary, what is the major purpose of this exercise?

ASSESSMENT
Critical Thinking and Learning Outcomes Review

1. Explain what occurs during decarboxylation.

2. Why does the LDC broth or lysine iron agar turn purple when lysine is decarboxylated?

3. Why does the LDC medium always turn yellow regardless of the ability of the bacteria to produce lysine decarboxylase?

4. Why is the lysine decarboxylase test negative if both LDC and DC broths turn purple?

5. Why is sterile mineral oil added to LDC test media?

6. How does the pH indicator bromcresol purple indicate a change in pH?

EXERCISE 29

Proteins VII: Phenylalanine Deamination

SAFETY CONSIDERATIONS

- Be careful with the Bunsen burner flame.
- The ferric chloride solution is an irritant; do not breathe its vapors or get it on your skin.
- Keep all culture tubes upright in a test-tube rack.

MATERIALS

- suggested bacterial strains: *Escherichia coli* and *Proteus vulgaris*
- phenylalanine deaminase agar slants or phenylalanine deaminase test tablets
- 10% aqueous ferric chloride solution (or 10% $FeCl_3$ in 50% HCl)
- inoculating loop
- Pasteur pipette with pipettor
- test-tube rack
- incubator set at 37°C
- wax pencil or lab marker
- container for biohazard waste
- safety glasses
- disposable gloves
- lab coat

LEARNING OUTCOMES

Upon completion of this exercise, students will demonstrate the ability to

1. Explain the biochemical process of phenylalanine deamination
2. Describe how to perform the phenylalanine deamination test
3. Perform a phenylalanine test

SUGGESTED READING IN TEXTBOOK

1. Catabolism of Organic Molecules Other Than Glucose, section 11.7; see also figures 11.2 and 11.22.

Pronunciation Guide

- *Escherichia coli* (esh-er-I-ke-a KOH-lie)
- *Proteus vulgaris* (PRO-tee-us vul-GA-ris)

Why Are the Following Bacteria Used in This Exercise?

In this exercise, you will learn how to perform the phenylalanine deaminase test to differentiate between various enteric bacteria. The ability of certain bacteria to oxidatively degrade phenylalanine is of taxonomic importance. The two enteric bacteria chosen to show this differentiation are *Escherichia coli* and *Proteus vulgaris*. *P. vulgaris* produces the enzyme phenylalanine deaminase whereas *E. coli* does not.

Medical Application

In the clinical laboratory, phenylalanine deamination can be used to differentiate the genera *Morganella*, *Proteus*, and *Providencia* (phenylalanine deaminase +) from the *Enterobacteriaceae* (phenylalanine deaminase −). Bacteria in these genera can cause urinary tract infections and are capable of causing opportunistic infections elsewhere in the body.

PRINCIPLES

The enzyme **phenylalanine deaminase** (a flavoprotein oxidase) catalyzes the removal of the amino group (NH_3^+) from phenylalanine (**Figure 29.1**). The resulting products include organic acids, water, and the ammonium ion. Certain enteric bacteria (e.g., *Proteus, Morganella,* and *Providencia*) can use the organic acids in biosynthesis reactions.

Phenylalinine agar is used to detect the deamination of the amino acid phenylalanine. This agar

Figure 29.1 Phenylalanine Deamination. (a) Uninoculated control. (b) Phenylalanine negative. (c) Phenylalanine positive. *(Javier Izquierdo/McGraw Hill)*

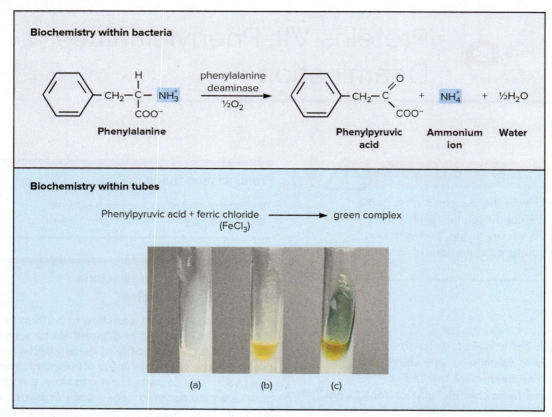

contains yeast extract to support growth of the bacteria, DL-phenylalanine, NaCl, Na_2HPO_4, and agar.

The **phenylalanine deaminase test** can be used to differentiate among enteric bacteria such as *E. coli* and *P. vulgaris*. *P. vulgaris* produces the enzyme phenylalanine deaminase, which deaminates phenylalanine, producing phenylpyruvic acid. When ferric chloride is added to the medium, it reacts with phenylpyruvic acid, forming a green compound. Since *E. coli* does not produce the enzyme, it cannot deaminate phenylalanine. When ferric chloride is added to an *E. coli* culture, there is no color change.

Procedure

First Period

1. Label two slants of phenylalanine deaminase agar with the name of the bacterium to be tested. Use another slant as a control. Add your name and date to each slant.
2. Using aseptic technique (*see Figure 15.3*), inoculate each of the slants with the respective bacteria.
3. Incubate aerobically at 37°C for 18 to 24 hours.
4. Alternatively, the cultures can be directly tested by the addition of KEY test tablets. Add a tablet to 1 ml distilled water, inoculate heavily with paste, and incubate for about 20 to 24 hours at 37°C. Add 3 or 5 drops of 10% $FeCl_3$ reagent. A yellow to green color that develops within 1 to 5 minutes is a positive test.

Second Period

1. With the Pasteur pipette, add 3 to 5 drops of the 10% $FeCl_3$ to the growth on the slant. Rotate each tube between your palms to wet and loosen the bacterial growth. The presence of phenylpyruvic acid is indicated by the development of a green color within 5 minutes and indicates a positive test for phenylalanine deamination (**Figure 29.1**). If there is no color change (slants remain yellow) after adding the reagent, the test is negative, and no deamination has occurred.

2. Based on your observations, determine and record in the report for this exercise which of the bacteria were able to deaminate phenylalanine.

Disposal

When you are finished with all of the tubes, discard them in the designated place for sterilization and disposal.

> **HELPFUL HINTS**
>
> - A positive phenylalanine test must be interpreted immediately after the addition of the $FeCl_3$ reagent (within 5 minutes) because the green color fades quickly.
> - Rolling the $FeCl_3$ over the slant aids in obtaining a faster reaction with a more pronounced color.
> - When not in use, the ferric chloride reagent should be kept in a dark bottle (no light exposure) and in the refrigerator.

NOTES

Laboratory Report 29

Name: _____

Date: _____

Lab Section: _____

Proteins VII: Phenylalanine Deamination

1. Complete the following table on phenylalanine deamination.

Bacterium	Color of the Slant	Deamination (+ or −)
E. coli	_____	_____
P. vulgaris	_____	_____

2. Describe the phenylalanine deamination reaction.

3. In summary, what is the major purpose of this exercise?

ASSESSMENT
Critical Thinking and Learning Outcomes Review

1. What are two ways that phenylalanine can be used by *P. vulgaris*?

2. What is the purpose of the ferric chloride in the phenylalanine deamination test?

3. When would you use the phenylalanine deamination test?

4. Name some bacteria that can deaminate phenylalanine.

5. Describe the process of deamination.

6. Why must the phenylalanine test be determined within 5 minutes?

EXERCISE 30
Proteins VIII: Dissimilatory Nitrate Reduction

SAFETY CONSIDERATIONS

- Be careful with the Bunsen burner flame.
- Since N, N-dimethyl-1-naphthylamine might be carcinogenic (nitrite test reagent B), wear disposable gloves and protective eyeglasses, and avoid skin contact or aerosols.
- The acids in nitrite test reagent A are caustic. Avoid skin contact and do not breathe the vapors.
- Be careful when working with zinc. Do not inhale or allow contact with skin.
- Keep all culture tubes upright in a test-tube rack or in a can.

MATERIALS

- suggested bacterial strains: *Escherichia coli*, *Pseudomonas fluorescens*, and *Staphylococcus epidermidis*
- garden soil
- Bunsen burner
- inoculating loop
- 1-ml pipette with pipettor
- nitrate broth tubes with Durham tubes
- nitrite test reagent A or Difco's SpotTest Nitrate Reagent A
- nitrite test reagent B or Difco's SpotTest Nitrate Reagent B
- zinc powder or dust or Difco's SpotTest Nitrate Reagent C
- test-tube rack
- incubator set at 37°C
- sterile test tubes
- wax pencil or lab marker
- container for biohazard waste
- safety glasses
- disposable gloves
- lab coat

LEARNING OUTCOMES

Upon completion of this exercise, students should demonstrate the ability to

1. Outline the biochemical process of dissimilatory nitrate reduction by bacteria
2. Describe how nitrate reduction can be determined from bacterial cultures
3. Perform a nitrate reduction test

SUGGESTED READING IN TEXTBOOK

1. Anaerobic Respiration Uses the Same Steps as Aerobic Respiration, section 11.5; see also figure 11.16 and table 11.2.

Pronunciation Guide

- *Escherichia coli* (esh-er-I-ke-a KOH-lie)
- *Pseudomonas fluorescens* (soo-do-MO-nas floor-ES-shens)
- *Staphylococcus epidermidis* (staf-il-oh-KOK-kus e-pee-DER-meh-diss)

Why Are the Following Bacteria Used in This Exercise?

In this exercise, you will learn how to perform the nitrate reduction test in order to differentiate between bacteria. Three different bacteria that give three different nitrate reduction results will be used. *Staphylococcus epidermidis* is unable to use nitrate as a terminal electron acceptor; therefore, it cannot reduce nitrate. *Escherichia coli* can reduce nitrate only to nitrite. *Pseudomonas fluorescens* often reduces nitrate completely to molecular nitrogen or even ammonia.

Medical Application

Most enteric bacteria are nitrate reducers. Pathogenic examples include *Escherichia coli* (opportunistic urinary tract infections), *Klebsiella pneumoniae* (bacterial pneumonia), *Morganella morganii,* and *Proteus mirabilis* (nosocomial infections). Nonenteric nitrogen-reducing pathogens include *Staphylococcus aureus* (staphylococcal food poisoning, bacteremia, various abscesses) and *Bacillus anthracis* (anthrax).

PRINCIPLES

Chemolithotrophic bacteria obtain energy through chemical oxidation and use inorganic compounds as electron donors with CO_2 as their main carbon source. **Chemoorganoheterotrophs** require organic compounds as their source of carbon and energy. Some chemolithotrophic and chemoorganoheterotrophic bacteria can use nitrate (NO_3^-) as a terminal electron acceptor during anaerobic respiration. In this dissimilatory process, nitrate is reduced to nitrite (NO_2^-) via a **nitrate reductase** as illustrated in **Figure 30.1**. Some of these bacteria possess the enzymes to further reduce the nitrite to either the ammonium ion (NH_3) or molecular nitrogen (N_2) as also illustrated in Figure 30.1.

The ability of some bacteria to reduce nitrate can be used in their identification and isolation. For example, *E. coli* can reduce nitrate only to nitrite, *P. fluorescens* reduces it completely to molecular nitrogen, and *S. epidermidis* is unable to use nitrate as a terminal electron acceptor.

The **nitrate reduction test** is performed by growing bacteria in a culture tube with a nitrate broth medium containing 0.5% potassium nitrate (KNO_3) and Durham tubes. After incubation, the culture is examined for the presence of gas (in the Durham tubes) and nitrite ions in the medium. The gas (a mixture of CO_2 and N_2) is released from the reduction of nitrate (NO_3^-) and from the citric acid cycle (CO_2) (**Figure 30.2**). The nitrite ions are detected by the addition of sulfanilic acid and N,N-dimethyl-1-naphthylamine to the culture. Any nitrite in the medium will react with these reagents to produce a pink or red color.

Bubbles indicate the presence of nitrogen gas for nonfermenters only; fermenters may also produce gas from carbohydrate sources. Even a small amount of gas in the Durham tube indicates a positive test for nonfermenters.

If a culture does not produce a color change, several possibilities exist: (1) the bacteria possess nitrate reductase and also reduce nitrite further to ammonia or

Figure 30.1 Dissimilatory Nitrate Reduction. After 24 to 48 hours of incubation, nitrate reagents are added to the culture tubes. Nitrite produced as a result of nitrate reduction will react with sulfanilic acid and N,N-dimethyl-1-naphthylamine to produce a red color as an indicator of a positive reaction.

Biochemistry within bacteria

$$NO_3^- + 2H^+ + 2e^- \xrightarrow{\text{nitrate reductase}} NO_2^- + H_2O$$
Nitrate Hydrogen Electrons Nitrite Water

$$2NO_3^- - 12H^+ + 10e^- \xrightarrow{\text{other enzymes}} N_2 + 6H_2O$$
Nitrate Hydrogen ions Electrons Nitrogen gas (bubbles in Durham tube) Water

Biochemistry within tubes

Sulfanilic acid + N,N-dimethyl-1-naphthylamine + nitrite ions
(colorless)   (colorless)

$\longrightarrow$ water + sulfobenzene azo-N,N-dimethyl-1-naphthylamine
(red color)

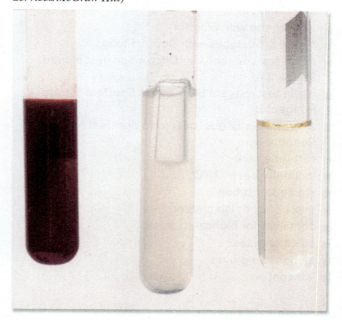

Figure 30.2 Nitrate Reduction Test. All three tubes contain nitrate broth. The tube on the left has turned red due to nitrate reduction and is a positive result for the reduction of nitrate to nitrite. The middle tube shows reduction of nitrate to nitrogen gas that is trapped in the Durham tube. The tube on the right is an uninoculated control. *(Auburn University Photographic Services/McGraw Hill)*

Microbial Biochemistry

molecular nitrogen; (2) they possess other enzymes that reduce nitrite to ammonia; or (3) nitrates were not reduced by the bacteria. To determine if nitrates were reduced past nitrite, a small amount of zinc powder or 5 to 10 drops of SpotTest Nitrate Reagent C is added to the culture containing the reagents. Since zinc reduces nitrates to nitrites, a pink or red color will appear and verifies the fact that nitrates were not reduced to nitrites by the bacteria (a negative test). If a red color does not appear, the nitrates in the medium were reduced past the nitrite stage to non-nitrite end products such as ammonia, hydroxylamine, or nitrogen gas (**Figure 30.3** and **Table 30.1**).

Procedure

First Period

1. Label three tubes of nitrate broth with the three respective bacteria (*E. coli, P. fluorescens,* and *S. epidermidis*); label the fourth tube "garden soil"

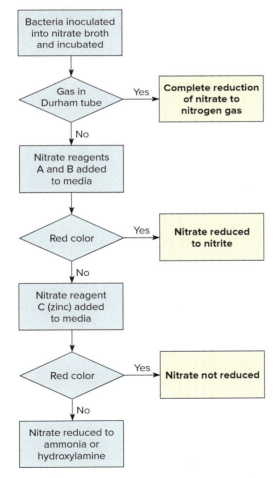

Figure 30.3 Flowchart for the Analysis of Nitrate Broth Tubes with Durham Tubes.

and the fifth tube "control." Add your name and date to each tube. The control tube serves two purposes: (1) to determine if the medium is sterile and (2) to determine if any O_2 comes out of the medium instead of out of the gas produced by the bacteria.

2. Using aseptic technique (*see Figure 15.3*), inoculate three tubes with the respective bacteria, and the fourth with about a gram of garden soil.

3. Incubate all five tubes for 24 to 48 hours at 37°C.

Second Period

1. Observe the tubes for the presence of growth, and the absence of growth in the control tube.

2. Examine the nitrate broth tubes. If there are gas bubbles in the Durham tubes, it means that the specific bacterium has reduced nitrate to a gaseous end product (nitrogen gas). If no bubbles are present, reduction may have resulted in the formation of nitrite or other nongaseous end products.

3. With a pipette and pipettor, while wearing disposable gloves and protective eyeglasses, add 0.5 ml of nitrate test reagent A and 0.5 ml of test reagent B to each of the culture tubes and mix. (Alternatively, about 5 to 10 drops of each reagent works well.) A distinct pink or red color indicates a positive test, provided the uninoculated control medium is negative.

4. Negative tests should be confirmed by adding several grains of zinc powder or 5 to 10 drops of Difco's nitrate reagent C and gently shaking the tube. If nitrate is present in the medium, it will turn red within 5 to 10 minutes; if it is absent, there will be no color change.

5. Record your results in the report for this exercise.

Disposal

When you are finished with all of the tubes and pipettes, discard them in the designated place for sterilization and disposal.

Table 30.1 Nitrate Broth Results and Interpretations

Interpretation (Symbol)	Test Results and Appearance
Nitrogen gas is produced by dissimilatory nitrate reduction; nitrogen gas +	If a bubble appears in the Durham tube, the bacterium is a known nonfermenter
If the source of gas is unknown, nitrate reagents A and B (and perhaps C) must be added	If a bubble appears in the Durhan tube, the bacterium may be a fermenter
Nitrate is reduced to nitrite; nitrite +	After reagents A and B are added, a red color appears
Nitrate reagent C (zinc powder) is added to continue the test	After adding nitrate reagents A and B, there is no color change (nitrate is not reduced)
Nitrate reduction to nongaseous, nonnitrite products; nitrite +	No color change after adding nitrate reagent C
No nitrate reduction; nitrate −	Red color appears after adding nitrate reagent C

> **HELPFUL HINTS**
> - Disposable gloves must be worn when using nitrite reagents A and B. If these solutions get on your hands, wash them immediately with soap and water for at least 15 minutes.
> - Bubbles indicate a positive test for nonfermenters only; fermenters may also produce gas from carbohydrates.
> - Even a small amount of gas production is a positive test for nonfermenters.
> - Keep the zinc powder away from open flames since it can explode.

Laboratory Report 30

Name: _____

Date: _____

Lab Section: _____

Proteins VIII: Dissimilatory Nitrate Reduction

1. On the basis of your observations, complete the following table.

Bacterium or Soil	Color with Reagents	Color with Zinc	Nitrate Reduction (+ or −)	End Products	Gas
E. coli	_____	_____	_____	_____	_____
P. fluorescens	_____	_____	_____	_____	_____
S. epidermidis	_____	_____	_____	_____	_____
Soil	_____	_____	_____	_____	_____
Control tube	_____	_____	_____	_____	_____

2. Illustrate or outline a complete test for the presence of nitrate reductase.

3. In summary, what is the major purpose of this exercise?

ASSESSMENT
Critical Thinking and Learning Outcomes Review

1. From your results, which bacteria are negative for nitrate reduction? Which are positive?

2. How do you explain the results from the soil sample?

3. Why is the development of a red color a negative test when zinc is added?

4. What are the end products that may result from the action of bacteria with nitrate-reducing enzymes?

5. What is the purpose of a control tube in this exercise?

NOTES

PART 5
Environmental Factors Affecting Growth of Microorganisms

The growth of microorganisms is greatly affected by the chemical and physical nature of their environment. An understanding of the environmental factors that promote microbial growth aids in understanding the ecological distribution of microorganisms. Therefore, the nature of some of these influences will be surveyed in this part of the lab manual, including temperature, pH, and osmotic pressure.

These same environmental factors that maximize microbial growth can also be manipulated to inhibit growth of unwanted microorganisms. Microbial control using antibiotics and disinfectants has become an important aspect of industrial and medical microbiology. Several exercises that cover control of microbial growth are also included in this part of the lab manual.

After completing the exercises in Part 5, you will be able to demonstrate proficiency in the following American Society for Microbiology Core Curriculum skills:

- collecting and organizing data in a systematic fashion
- presenting data in an appropriate form (graphs, tables, figures, or descriptive paragraphs)
- assessing the validity of data (including integrity and significance)
- drawing appropriate conclusions based on the results.

Forcellini Danilo/Shutterstock

Microbes are everywhere. Much of what we know about the microbial world is based on studying microbes that can be cultivated in the lab. However, modern sequence analysis technologies have revealed that microbes can colonize and thrive in vastly different environments. From the frigid waters of the arctic, to the crushing pressures of deep-sea hydrothermal vents, traces of microbial life cover our planet. Microbiologists are still working to understand how microbes can survive in such dissimilar settings, many of which seem quite extreme to human beings. Shown here is a hot spring from Yellowstone Park. Once thought to be too harsh to support life, we now know that hot springs are teeming with microbes. In fact, the beautiful colors of this hot spring are due to the microbes growing in it.

EXERCISE 31 Temperature

SAFETY CONSIDERATIONS

- Be careful with the Bunsen burner flame.
- Use caution with the hot water baths.
- Keep culture tubes upright in a test-tube rack.

MATERIALS

- suggested bacterial strains:
  Escherichia coli, Geobacillus stearothermophilus, Bacillus subtilis, Pseudomonas aeruginosa, Serratia marcescens, Staphylococcus aureus, and spore suspension of *Bacillus subtilis*.
- tryptic soy agar slants
- Bunsen burner
- inoculating loop
- tryptic soy broth dilution tubes
- test-tube rack
- sterile pipettes with pipettor
- sterile test tubes
- refrigerator set at 4°C
- incubators or water baths set at 4°, 23°, 30°, 37°, 55°, and 70°C
- wax pencil or lab marker
- sterile water
- container for biohazard waste
- safety glasses
- disposable gloves
- lab coat

LEARNING OUTCOMES

Upon completion of this exercise, students will demonstrate the ability to

1. Describe how microorganisms are affected by the temperature of their environment
2. Carry out an experiment that differentiates between several bacteria based on temperature sensitivity
3. Classify these same bacteria based on their temperature preference for growth
4. Determine the effects of heat on bacteria

SUGGESTED READING IN TEXTBOOK

1. Environmental Factors Affect Microbial Growth, Temperature, section 7.5; see also figures 7.16 and 7.19.
2. Microbes Can Be Controlled by Physical Means, section 8.2; see also table 8.2.

Pronunciation Guide

- *Geobacillus stearothermophilus* (geo-bal-SIL-lus ste-row-ther-MAH-fil-us)
- *Bacillus subtilis* (sub-TIL-us)
- *Escherichia coli* (esh-er-I-ke-a KOH-lie)
- *Pseudomonas aeruginosa* (soo-do-MO-nas a-ruh-jin-OH-sah)
- *Serratia marcescens* (se-RA-she-ah mar-SES-sens)
- *Staphylococcus aureus* (staf-il-oh-KOK-kus ORE-ee-us)

Why Are the Following Bacteria Used in This Exercise?

In this exercise, you will gain expertise in differentiating between bacteria based on temperature sensitivity and classifying bacteria based on their temperature preference for growth. *Bacillus stearothermophilus* is an endospore-forming rod that has an optimum growth temperature of 60°C to 65°C. Its spores are more heat resistant than those of any mesophilic species in the genus. *Bacillus subtilis* is an endospore-forming rod that has an optimum growth temperature of about 30°C to 40°C. All *Bacillus* species are Gram-positive. *Escherichia coli* is a Gram-negative, facultatively anaerobic rod that does not form spores and has an optimum growth temperature of 37°C. *Pseudomonas aeruginosa* is a Gram-negative rod with an optimum temperature of 37°C. *Staphylococcus aureus* is a Gram-positive coccus with an optimum growth temperature of 30°C to 37°C.

PRINCIPLES

Every microbe can be characterized based on their own unique temperature growth range. This temperature range is determined by the heat sensitivity of its particular enzymes, membranes, ribosomes, and other components. As a consequence, microbial growth has a fairly characteristic temperature dependence with distinct **cardinal temperatures**: minimum, maximum, and optimum (**Figure 31.1**). The **minimum growth temperature** is the lowest temperature at which growth will occur; the **maximum growth temperature** is the highest temperature at which growth will occur; and the **optimum growth temperature** is the temperature at which the rate of growth is most rapid. As you might suspect, the optimum temperature for the growth of a given microorganism is highly correlated with the temperature of the normal habitat of the microorganism. For example, the optimum temperature for the growth of bacteria pathogenic to humans is near that of the temperature of the body's tissues and organs (35°C to 37°C). Now you know why so many of our incubations take place at this temperature!

Microorganisms can be placed in one of five classes based on their temperature ranges for growth (**Figure 31.2**). **Psychrophiles** grow well at 0°C and have an optimum growth temperature of 10°C or lower. Psychrophiles are prominent microbes beneath the polar icecaps. Many species can grow at 0°C to 20°C and

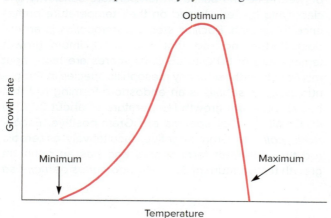

Figure 31.1 The Effect of Temperature on Growth Rate.
The three cardinal temperatures vary by microorganism.

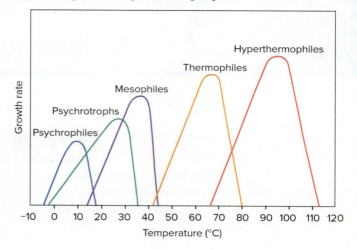

Figure 31.2 Temperature Requirements for Growth.
Microorganisms are commonly divided into groups based on their optimum growth temperature. This graph depicts the typical temperature ranges of these groups.

are called **psychrotrophs**. **Mesophiles** are microorganisms with growth optima around 20°C to 45°C. The majority of typical laboratory bacteria fall into this category. Some microorganisms are **thermophiles**; they can grow optimally at temperatures of 55°C or higher. These microbes flourish in hot environments such as compost near deep-sea hydrothermal vents, and in hot springs. A few thermophiles can grow between 85°C and 113°C or above and are called **hyperthermophiles.**

Recall that to sterilize microbiological media or certain foodstuffs, high temperatures are commonly used. Boiling is probably one of the easiest methods of ridding materials of harmful bacteria. However, not all bacteria are equally sensitive to high temperature. Some bacteria may be able to survive boiling even though they are unable to grow. These bacteria are termed **thermoduric.** Many of the spore formers (such as *B. subtilis*) can withstand boiling for 15 minutes because of their ability to form heat-resistant endospores.

Procedure

First Period

1. To economize on both time and media, work in groups. Each group will be assigned one temperature to study: 4°, 23°, 30°, 37°, 55°, or 70°C (**Table 31.1**).
2. Label each of the tryptic soy agar slants with the name of the test bacterium to be inoculated (*E. coli, G. stearothermophilus,* and *B. subtilis*), your name, and date.

Table 31.1	Temperature Assignments
Temperature	Group
4°C	A
23°C	B
30°C	C
37°C	D
55°C	E
70°C	F

3. Using aseptic technique, streak the surface of each slant with the appropriate bacterium. Incubate the slants for 24 hours at the temperature assigned to your group.
4. Obtain four sterile test tubes and label the first *S. aureus*, the second *B. subtilis spores*, the third *P. aeruginosa*, and the fourth *S. marcescens*. Make sure to also include your group name and date.
5. With a sterile pipette, aseptically add 1 ml of bacterial culture or spore suspension to the respective tubes.
6. Subject your tubes to the temperature you are assigned for 15 minutes.
7. After 15 minutes, let the samples adjust to room temperature. For each bacterial sample, make a dilution series as follows (*see the Appendix* for more information on dilutions):
   a. Pipette 0.1 ml of the incubated sample into the 0.9 ml tryptic soy broth in a sterile microfuge tube. Use a vortex to mix the tube thoroughly. With a fresh pipette tip, transfer 0.1 ml of this 10-fold dilution into another 0.9 ml of broth (yielding a 10^{-2} dilution) and mix. In the same way, prepare a series of tubes up to a 10^{-10} dilution.
8. Incubate all dilutions at 37°C for at least 24 hours.

Second Period
1. Observe the slants for the presence of growth. Record your observations and those of the other groups in your class; use a + for the presence of growth and a − for the absence of growth in the lab report for this exercise.
2. Observe your dilution series to see which tubes have growth as indicated by turbidity. You can also measure the optical density (at 600 nm) of each tube to obtain a quantitative measure of growth. The greater the number of bacteria present in a particular sample, the more this sample can be diluted and still contain bacteria in the aliquot transferred. For example, if bacterium A is more resistant to heat than bacterium B, bacterium A will require more dilutions in order to obtain a dilution tube free of growth.
3. Determine the last dilution in which growth occurred. Share your observations with your classmates and complete the lab report for this exercise.

Disposal

When you are finished with all of the tubes, pipette tips, and plates, discard them in the designated place for sterilization and disposal.

> **HELPFUL HINTS**
> - Make sure the spore suspension is diluted with sterile water to ensure that spores will not germinate prematurely.
> - Make sure to use a new pipette tip during each step of the dilution series.

NOTES

Laboratory Report 31

Name: _____

Date: _____

Lab Section: _____

Temperature

1. Based on your observations of bacterial growth and those of your classmates, complete the following table.

Bacterium	Growth (+) or (−)						Temperature Requirement Classification
	4°C	23°C	30°C	37°C	55°C	70°C	
E. coli							
S. marcescens							
B. subtilis							
G. stearothermophilus							

2. Based on your observations and those of your classmates, complete the following table, showing range of surviving bacteria.

Temperature for 15 Minutes	Last Dilution in Which Growth Occurred			
	S. marcescens	S. aureus	P. aeruginosa	B. subtilis spores
4°C				
23°C				
30°C				
37°C				
55°C				
70°C				

ASSESSMENT
Critical Thinking and Learning Outcomes Review

1. How can you be sure that the turbidity produced in the broth tubes was caused by the bacteria used for the inoculation?

2. How can you determine experimentally whether a bacterium is a psychrophile or a mesophile?

3. What limitations are there for using boiling water as a means of sterilizing materials?

4. Is *S. aureus* a mesophile? What about *P. aeruginosa*? Explain your answer.

5. Describe the three cardinal temperatures.

6. Of the bacteria tested in this exercise, which had the widest range of temperature tolerance? The narrowest range of temperature tolerance?

7. What are thermoduric bacteria?

8. Why is it unlikely for a psychrophile to be a human pathogen? Explain your answer.

9. Fever is a common symptom of an infection. Based on what you observed in this exercise, why is developing a fever useful for controlling a bacterial infection?

NOTES

EXERCISE 32   pH

SAFETY CONSIDERATIONS

- Be careful with the Bunsen burner flame.
- Hydrochloric acid and sodium hydroxide are extremely caustic; do not get these acids and bases on your skin or breathe the vapors.
- Keep culture tubes upright in a test-tube rack.

MATERIALS

- suggested bacterial and yeast strains: *Alcaligenes faecalis*, *Escherichia coli*, *Pseudomonas aeruginosa*, and *Saccharomyces cerevisiae*
- pH meter or pH paper
- tryptic soy broth tubes, pH 3.0
- tryptic soy broth tubes, pH 5.0
- tryptic soy broth tubes, pH 7.0
- tryptic soy broth tubes, pH 9.0
- Bunsen burner
- sterile pipettes with pipettor
- spectrophotometer
- cuvettes
- wax pencil or lab marker
- test-tube rack
- container for biohazard waste
- safety glasses
- disposable gloves
- lab coat

LEARNING OUTCOMES

Upon completion of this exercise, students will demonstrate the ability to

1. Predict how pH affects the growth of microbes
2. Perform an experiment that relates microbial growth to pH

SUGGESTED READING IN TEXTBOOK

1. Environmental Factors Affect Microbial Growth, pH, section 7.5; see also figure 7.18.

Pronunciation Guide

- *Alcaligenes faecalis* (al-kah-LIJ-e-neez fee-KAL-iss)
- *Escherichia coli* (esh-er-I-ke-a KOH-lie)
- *Saccharomyces cerevisiae* (sak-ah-ro-MI-seez ser-ah-VEES-ee-eye)
- *Pseudomonas aeruginosa* (soo-do-MO-nas a-ruh-jin-OH-sa)

Why Are the Following Microorganisms Used in This Exercise?

This exercise demonstrates the effect of pH on microbial growth. You will use three bacteria and one yeast that have different pH ranges for optimal growth. After performing an experiment that relates microbial growth to pH, you should appreciate the fact that these microbes have different pH optima.

PRINCIPLES

It is not surprising that **pH** (acidity; log $1/[H^+]$) dramatically affects microbial growth. The pH affects the activity of enzymes—especially those that are involved in biosynthesis and growth. Each microbial species possesses a definite pH **growth range** and a distinct pH **growth optimum. Acidophiles** have a growth optimum between pH 0.0 and 5.5; **neutrophiles** between 5.5 and 8.0; and **alkaliphiles** 8.0 to 11.5 (**Figure 32.1**). In general, different microbial groups have characteristic pH optima. The majority of bacteria and protozoa are neutrophiles. Most molds and yeasts occupy slightly

Figure 32.1 Classification of Microorganisms Grouped by Optimal Growth pH. Although some microorganisms can live at a very high or low pH, the internal pH of these organisms remains near neutrality.

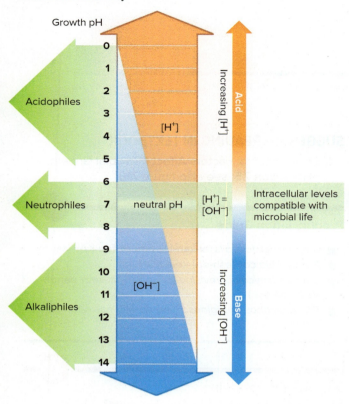

Table 32.1	Microbe Assignments
Microorganism	**Group**
A. faecalis	A
E. coli	B
P. aeruginosa	C
S. cerevisiae	D

acidic environments in the pH range of 4 to 6; algae also seem to favor acidity. Many archaea are acidophiles.

Many bacteria produce metabolic acids that may lower the pH and inhibit their growth. To prevent this, **buffers** that produce a pH equilibrium are added to culture media to neutralize these acids. For example, the peptones in complex media act as buffers. Phosphate salts are often added as buffers in chemically defined media.

In this exercise, you will work in groups to see how the pH affects the growth of several microorganisms.

Procedure

First Period

1. To economize on both time and media, work in groups. Each group of students will be assigned one microorganism to study. Label each of the tryptic soy broth tubes with the pH of the medium, your group name, date, and the microorganism to be inoculated (**Table 32.1**).

2. Using a sterile pipette, add 0.1 ml of your assigned microbial culture to the tube that has a pH of 3.0. Do the same for the tubes that have pH values of 5.0, 7.0, and 9.0.
3. Your classmates will repeat steps 1 and 2 for the remaining microbes.
4. Incubate the bacterial cultures for at least 24 hours at 37°C, and the *S. cerevisiae* culture for 48 hours at 30°C.

Second Period

1. Turn on the spectrophotometer and allow it to warm up (*see Figure 18.1*). Set the wavelength at 600 nm. Calibrate the spectrophotometer using a tryptic soy broth blank of each pH for each respective set of cultures.
2. Fill each cuvette with each of the cultures at the respective pH and read the absorbance.
3. Alternatively, if no spectrophotometer is available, examine the tubes by eye and score the growth in each tube on a scale of 1 (little to no growth) to 5 (abundant growth).
4. Record your results and those of your classmates in the lab report for this exercise.

Disposal

When you are finished with all of the tubes and pipettes, discard them in the designated place for sterilization and disposal. Do not throw away the cuvette tubes.

> **HELPFUL HINTS**
> - Be sure to thoroughly mix the microbial suspensions just before reading the absorbance.
> - Wipe any fingerprints off the cuvette before making a reading.

Laboratory Report 32

Name: _____

Date: _____

Lab Section: _____

pH

1. Summarize your results with respect to the pH and growth of each microorganism in the following table. Indicate the optimal pH with an asterisk.

Microorganism	pH	Visual Results	Absorbance	% Transmission	Optimal pH
A. faecalis	3				
	5				
	7				
	9				
E. coli	3				
	5				
	7				
	9				
P. aeruginosa	3				
	5				
	7				
	9				
S. cerevisiae	3				
	5				
	7				
	9				

2. Which microorganism grew best in the acid pH range?

3. Which microorganism grew best in the neutral pH range?

4. Which microorganism grew best in the alkaline pH range?

5. Which microorganism has the widest pH growth range?

6. Which microorganism has the narrowest pH growth range?

7. Using a different color line for each microorganism, graph the growth responses to pH variations.

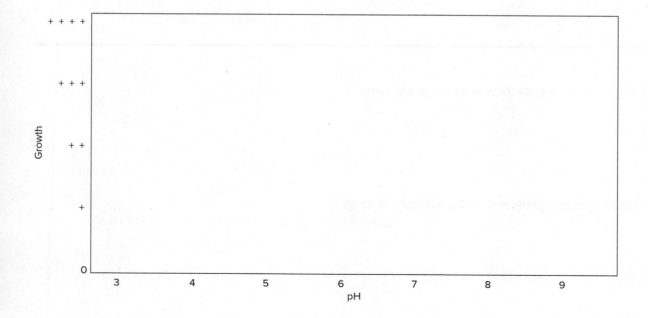

ASSESSMENT
Critical Thinking and Learning Outcomes Review

1. Why are buffers added to culture media?

2. Why do microorganisms differ in their pH requirements for growth?

3. What inhibits microbial growth at nonoptimal pHs?

4. What is the pH tolerance of bacteria compared to yeasts?

5. List and describe the chemistry of several common buffers used in microbiological media.

6. How is it possible that the byproducts of normal microbial metabolism can be growth inhibitory?

7. Once controversial, it is now well-established that bacteria can withstand the low pH environment found in the human stomach. One such organism, *Helicobacter pylori*, has been connected to peptic ulcer disease. Search the literature and describe how *H. pylori* is able to withstand the low pH in the gut.

EXERCISE 33 Osmotic Pressure

SAFETY CONSIDERATIONS
- Be careful with the Bunsen burner flame.

MATERIALS
- suggested bacterial strains:
  24- to 48-hour tryptic soy broth cultures of *Escherichia coli* and *Staphylococcus aureus*. A 48-hour salt broth culture of *Halobacterium salinarium*
- nutrient agar Petri plates with 0%, 0.5%, 5%, 10%, and 20% NaCl
- wax pencil or lab marker
- inoculation loop
- Bunsen burner
- *Difco Manual* or *BBL Manual* for the laboratory report
- container for biohazard waste
- safety glasses
- disposable gloves
- lab coat

LEARNING OUTCOMES
Upon completion of this exercise, students will demonstrate the ability to
1. Explain how osmotic pressure affects a bacterial cell
2. Predict how bacterial growth is related to osmotic pressure in the environment
3. Perform an experiment that differentiates among three different bacteria based on their tolerance for different salt concentrations (osmotic pressure)

SUGGESTED READING IN TEXTBOOK
1. Cell Walls Have Many Functions, section 3.4.
2. Environmental Factors Affect Microbial Growth, Solutes Affect Osmosis and Water Activity, section 7.5; see also figure 7.17.

Pronunciation Guide
- *Escherichia coli* (esh-er-I-ke-a KOH-lie)
- *Halobacterium salinarium* (hal-o-bak-TE-re-um sal-i-NAR-e-um)
- *Staphylococcus aureus* (staf-il-oh-KOK-kus ORE-ee-us)

Why Are the Following Bacteria Used in This Exercise?

In this exercise, you will differentiate three different bacteria based on their tolerance for different salt concentrations (osmotic pressure). The authors have thus chosen three bacteria that vary widely in osmotic tolerance. *Halobacterium salinarium* is an aerobic archaeon that requires a high salt concentration for growth. It is normally found in highly saline environments such as salt lakes and marine salterns. Mature cells retain their rod shape in 3.5 to about 5.2 M NaCl; at lower concentrations, pleomorphic forms appear, and at 1.5 M the cells are spherical because of loss of the cell wall. Suspensions are viscous at 1.5 M because of partial cell lysis; at 0.5 M, few, if any, cells can be detected. Sodium, chloride, and magnesium are required to maintain cell structure and rigidity. *Staphylococcus aureus* is a Gram-positive coccus. *S. aureus* can grow in the presence of 15% (about 2.5 M) sodium chloride. *Escherichia coli* is a Gram-negative rod that occurs as normal flora in the lower part of the intestine of warm-blooded animals. It is very sensitive to high salt concentrations—it lacks osmotic tolerance and cannot grow above 1.0 M sodium chloride.

PRINCIPLES
Since microorganisms are separated from their environment by a selectively permeable plasma membrane, they can be affected by changes in the osmotic pressure or water availability of their surroundings. **Osmotic pressure** is the force developed when two

solutions of different solute concentrations are separated by a membrane that is permeable only to the solvent. The **solvent** is the liquid, usually water, that dissolves a substance (the **solute**).

If a bacterium is placed in a **hypotonic solution** (low solute, high water content), water will enter the cell and cause it to burst (**Figure 33.1a**) unless something is done to prevent the influx. Most bacteria have rigid cell walls that maintain the shape and integrity of the cell; thus hypotonic solutions are not harmful to these bacteria. When bacteria are placed in a **hypertonic solution** (high solute, lower water content), water leaves, and the plasma membrane shrinks away from the wall (**Figure 33.1b**), a process known as **plasmolysis**. This process dehydrates the cell and prevents any further growth. In an **isotonic solution**, the concentration of solutes is the same (*iso* means "equal") outside and inside the bacterium. The bacterium is in osmotic equilibrium with its environment and does not change volume (**Figure 33.1c**).

A few bacteria, called **halophiles** (Gr. *hals*, salt, + *philic*, loving), are able to tolerate high (hypertonic) salt concentrations (**Figure 33.2**). Bacteria that can live in very salty environments are called **extreme**

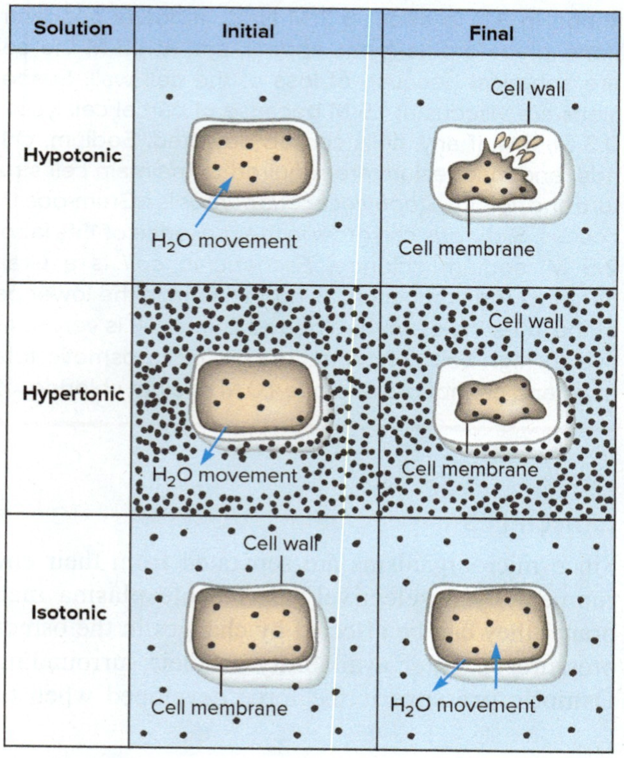

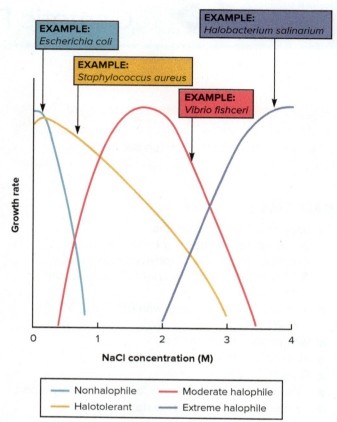

Figure 33.2 **The Effects of Sodium Chloride on Microbial Growth at Different Salt Tolerances or Requirements.** Four different patterns of microbial dependence on NaCl concentrations are shown. The curves are only illustrative and are not meant to provide precise shapes of salt concentrations required for growth.

Figure 33.1 **Effect of Osmotic Pressure on a Bacterial Cell.** Cells in this diagram will react differently when shifted to media with different solute concentrations. The dots represent solute (NaCl) molecules. The shaded light blue area represents water (solvent).

halophiles to distinguish them from the **moderate halophiles** that live in the sea. These extreme halophiles are mostly members of the Archaea and have significant modifications in their cell walls and membranes and will lyse in hypotonic environments.

All cells, even halophiles, must keep a relatively low intracellular Na^+ concentration because some solutes are moved into the cell by symport with Na^+. To achieve a low internal Na^+ concentration, halophilic microorganisms use special ion pumps to excrete Na^+ and replace it with cations, such as potassium (K^+), which is a comparable solute. In fact, the proteins and cell components (such as ribosomes) of halophiles require remarkably high intracellular potassium levels to maintain their structure.

In this exercise, you will examine the effect of various salt concentrations (osmotic pressure) on the growth of three species of bacteria. This tolerance (or lack of it) can be observed by the amount (or lack) of growth of the three bacterial species.

Procedure

First Period

1. With a wax pencil or lab marker, divide the bottom of each of the five Petri plates with increasing concentrations of NaCl into thirds. Place the name of the bacterium to be inoculated in each section. Add your name, salt concentration, and date.
2. Streak the respective bacteria onto the five different Petri plates.
3. Incubate the plates, inverted, for 48 hours at 37°C.

Second Period

1. Observe the relative amount of growth in each section at each salt concentration. Record this growth as − (none), +, ++, +++, and ++++ (the most).
2. Record your results in the report for this exercise.

Disposal

When you are finished with all of the plates, discard them in the designated place for sterilization and disposal.

NOTES

Laboratory Report 33

Name: _____

Date: _____

Lab Section: _____

Osmotic Pressure

1. Record the amount of growth of the three bacteria at the different salt concentrations in the following table. Use −, +, ++, +++, ++++ to indicate the relative amount of growth.

Medium	E. coli	H. salinarium	S. aureus
0% NaCl			
0.5% NaCl			
5.0% NaCl			
10% NaCl			
20% NaCl			

2. Which of these bacteria tolerates the most salt? _____
3. Which of these bacteria tolerates the least salt? _____
4. Which of these bacteria tolerates a broad range of salt? _____
5. How would you classify *H. salinarium* based on its salt needs? _____
6. Using the *Difco Manual* or *BBL Manual*, list the ingredients of mannitol salt agar.

Ingredient	Amount

7. Why is mannitol salt agar used to isolate staphylococci?

8. In summary, what is the major purpose of this exercise?

235

ASSESSMENT
Critical Thinking and Learning Outcomes Review

1. Compare isotonic, hypotonic, and hypertonic solutions and their effects on bacterial cells.

2. How is it possible for a bacterium to grow in a hypertonic environment?

3. What concentrations of NaCl are optimal for most bacteria?

4. What is unique about *H. salinarium*?

5. What foods can you think of that are protected from microbial destruction by salting?

EXERCISE 34

The Effects of Chemical Agents on Bacteria I: Disinfectants and Other Antimicrobial Products

SAFETY CONSIDERATIONS

- Be careful with the Bunsen burner flame.
- Always handle cultures with care since they may be potential pathogens.
- Keep all culture tubes upright in a test-tube rack or in a can.
- **Phenol is poisonous and caustic.** Do not handle with bare hands.

MATERIALS

- suggested bacterial strains: *Staphylococcus aureus* and *Pseudomonas aeruginosa*
- sterile screw-cap test tubes
- sterile 5-ml pipette with pipettor
- sterile 1-ml pipettes
- tryptic soy broth tubes (10 ml per tube)
- sterile water in Erlenmeyer flask
- sterile tubes for making dilutions
- Lysol®
- commercial disinfectants such as 3% hydrogen peroxide, 70% isopropyl alcohol, bleach, or other products (see **Table 34.1**). If commercial disinfectants are used, note the use-dilution and active ingredients. Dilute with tap water.
- phenol (carbolic acid)
- wax pencil or lab marker
- 37°C incubator
- test-tube rack
- Bunsen burner
- inoculating loop
- container for biohazard waste
- safety glasses
- disposable gloves
- lab coat

LEARNING OUTCOMES

Upon completion of this exercise, students will demonstrate the ability to

1. Investigate the effectiveness of some chemical disinfectants used in hospitals or homes as antimicrobial agents
2. Calculate a phenol coefficient

SUGGESTED READING IN TEXTBOOK

1. Microorganisms Are Controlled with Chemical Agents, section 8.3; see also table 8.3 and figures 8.1, 8.8 and 8.9.

Pronunciation Guide

- *Pseudomonas aeruginosa* (soo-do-MO-nas a-ruh-jin-OH-sah)
- *Staphylococcus aureus* (staf-il-oh-KOK-kus ORE-ee-us)

Why Are the Following Bacteria Used in This Exercise?

In this exercise, you will determine the antimicrobial effectiveness of some common disinfectants and calculate a phenol coefficient. The authors have chosen two common bacteria (*Staphylococcus aureus* and *Pseudomonas aeruginosa*) for use. Both of these bacteria give excellent positive or negative results depending on the disinfectant used. *S. aureus* and *Salmonella entericia* serovar Typhi are often used by laboratories as test microorganisms to determine phenol coefficients, although *Salmonella* will not be used in this particular exercise. We will instead use *Pseudomonas aeruginosa*, which is able to survive in many environments toxic to other bacteria.

PRINCIPLES

Many factors influence the effectiveness of chemical disinfectants and antiseptics. The **microbicidal** (to kill) or **microbiostatic** (to inhibit) **efficiency** of a chemical is often determined with respect to its ability to deter microbial growth. The types of chemical control agents and their uses are outlined in **Figure 34.1**. In general, to control microorganisms a **biocide** must be evaluated to determine the specific parameters under which it is to be effective. The first part of this exercise will investigate and evaluate the effect of several chemicals.

The microbicidal efficiency of a chemical is often determined with respect to phenol and is known as the **phenol coefficient (PC)**. The phenol coefficient is calculated by dividing the highest dilution of the antimicrobial of interest, which kills all organisms after incubation for 10 minutes but not after 5 minutes, by the highest dilution of phenol that has the same characteristics. Chemicals that have a phenol coefficient greater than 1 are more effective than phenol, and those that have a phenol coefficient less than 1 are less effective. However, this comparison should be used only for phenol-like compounds that do not exert bacteriostatic effects and are not neutralized by the subculture media used. The second part of this experiment will enable you to calculate a phenol coefficient for a select chemical. A list of commonly used antiseptics and disinfectants is shown in **Table 34.1**.

Procedure

First Period

Growth Inhibition

1. Each group of students should select one of the disinfectants or other antimicrobial products and, if necessary, dilute it according to the specifications on the label (the **use-dilution**).
2. Place 5 ml of disinfectant or antimicrobial into two sterile tubes. Add 0.05 ml of *P. aeruginosa* to one tube and 0.05 ml of *S. aureus* to the other.
3. Using the wax pencil or lab marker, label the tubes with your name and those of the respective bacteria. Mix each of the tubes in order to obtain a homogeneous suspension.
4. At intervals of 1, 2, 5, 10, and 15 minutes, transfer 0.1 ml of the mixture containing the bacteria and disinfectant or antimicrobial to separate tubes of tryptic soy broth. Do this for both bacteria. Also inoculate two tubes of broth with 0.1 ml of both bacteria and mark these "controls."
5. Incubate all tubes for 48 hours at 37°C.

Phenol Coefficient (See Safety Considerations)

1. Dilute phenol in sterile distilled water 1/80, 1/90, and 1/100; dilute the Lysol® 1/400, 1/450, and 1/500 so that the final volume in each tube is 5 ml.
2. Label 18 tryptic soy broth tubes with the name and dilution of disinfectant, the time interval of the subculture (e.g., 5 minutes, phenol 1/80), and your name. Each dilution should be tested after 5-, 10-, and 15-minute incubations.
3. Place in order in a test-tube rack, one test tube of each of the different Lysol® and phenol dilutions for each time interval.
4. Add 0.5 ml of *S. aureus* to each tube of disinfectant and note the time. Mix each of the tubes in order to obtain a homogeneous suspension and allow the disinfectant to come into contact with the bacteria.
5. Using aseptic technique, at intervals of 5, 10, and 15 minutes, transfer one loopful from each disinfectant tube into the appropriately labeled tryptic soy broth tube.
6. Incubate all tubes for 48 hours at 37°C.
7. If materials and time permit it, the experiment can be repeated with *P. aeruginosa* and other disinfectants for comparison.

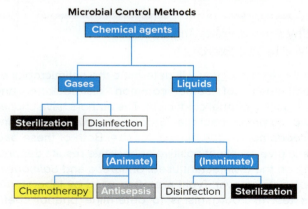

Figure 34.1 Chemical Microbial Control Measures.

Disinfection: The destruction or removal of vegetative pathogens but not bacterial endospores. Usually used only on inanimate objects.

Sterilization: The complete removal or destruction of all viable microorganisms. Used on inanimate objects.

Antisepsis: Chemicals applied to body surfaces to destroy or inhibit vegetative pathogens.

Chemotherapy: Chemicals used internally to kill or inhibit growth of microorganisms within host tissues.

Table 34.1	Active Ingredients of Various Commercial Antimicrobial Products	
Product	**Specific Chemical Agent**	**Antimicrobial Category**
Lysol® Sanitizing Wipes	Dimethyl benzyl ammonium chloride	Detergent (quat)
Clorox Disinfecting Wipes	Dimethyl benzyl ammonium chloride	Detergent (quat)
Tilex Mildew Remover	Sodium hypochlorites	Halogen
Lysol® Mildew Remover	Sodium hypochlorites	Halogen
Ajax Antibacterial Hand Soap	Triclosan	Phenolic
Dawn Antibacterial Hand Soap	Triclosan	Phenolic
Dial Antibacterial Hand Soap	Triclosan	Phenolic
Lysol® Disinfecting Spray	Alkyl dimethyl benzyl ammonium saccharinate/ethanol	Detergent (quats)/alcohol
ReNu Contact Lens Solution	Polyaminopropyl biguanide	Chlorhexidine
Wet Ones Antibacterial Moist Towelettes	Benzethonium chloride	Detergents (quat)
Noxzema Triple Clean	Triclosan	Phenolic
Scope Mouthwash	Ethanol	Alcohol
Purell Instant Hand Sanitizer	Ethanol	Alcohol
Pine-Sol	Phenolics and surfactant	Mixed
Allergan Eye Drops	Sodium chlorite	Halogen

Source: *Microbiology: A Systems Approach.* Cowan and Talaro, 2006. McGraw-Hill Higher Education.

Second Period

Growth Inhibition

1. Shake and observe each of the tubes for growth. Record the presence of growth as + and the absence of growth as −. Tabulate your results as well as the results of the class in part 1 of the report for this exercise.

Phenol Coefficient

1. Shake and observe all tryptic soy broth cultures for the presence (+) or absence (−) of growth.
2. Record your observations in part 2 of the report for this exercise.
3. From your data, calculate the phenol coefficient for Lysol®. For example, assume a 1/20 dilution of phenol (1 part phenol in a total of 20 parts liquid) kills *S. aureus* within 10 minutes. A 1/300 dilution of Lysol® also kills *S. aureus* within 10 minutes.

$$PC = \frac{300}{20} \text{ or } \frac{1/20}{1/300}$$

$$PC = 15$$

Thus Lysol® is 15 times more effective than phenol in killing *S. aureus*.

Disposal

When you are finished with all of the tubes and pipettes, discard them in the designated place for sterilization and disposal.

NOTES

Laboratory Report 34

Name: _____

Date: _____

Lab Section: _____

The Effects of Chemical Agents on Bacteria I: Disinfectants and Other Antimicrobial Products

1. Based on your observations of bacterial growth and those of your classmates, complete the following table.

Bacterium	Name of Disinfectant or Antimicrobial plus Active Ingredients*	Use Dilution	Control	Degree of Growth — Time of Exposure in Minutes				
				1	2	5	10	15

*Is triclosan present? Why is this important? _____

2. Based on your observations of bacterial growth and the observations of your classmates, complete the following table.

Disinfectant	Dilution	Growth in Subculture (Minutes)		
		5	10	15
Phenol	1/80			
	1/90			
	1/100			
Lysol®	1/400			
	1/450			
	1/500			

3. From the above data, calculate the phenol coefficient of Lysol®. For example, if the Lysol® dilution of 1/450 showed no growth at 10 minutes but growth at 5 minutes, and the phenol dilution of 1/90 showed no growth at 10 minutes but growth at 5 minutes, then:

$$\text{Phenol coefficient of Lysol}^® = \frac{1/450}{1/90} = 5$$

ASSESSMENT
Critical Thinking and Learning Outcomes Review

1. What are some limitations of a test like the one you performed on the evaluation of a disinfectant?

2. What are three criteria of a good disinfectant?

3. What is the phenol coefficient technique?

4. A disinfectant diluted 1/500 with water kills a bacterium after 10 minutes but not after 5 minutes. A 1/100 dilution of phenol kills the same bacterium after 10 minutes but not after 5 minutes. What is the phenol coefficient of the disinfectant?

5. What is the difference between microbicidal and microbiostatic?

6. What physical factors can influence the activity of a disinfectant?

7. Why do microorganisms differ in their response to disinfectants?

EXERCISE 35: The Effects of Chemical Agents on Bacteria II: Antibiotics

SAFETY CONSIDERATIONS

- Be careful with the Bunsen burner flame.
- Always handle cultures with care since they may be potential pathogens.
- The ethyl alcohol that is used to sterilize the forceps is flammable.

MATERIALS

- suggested bacterial strains: tryptic soy broth cultures of *Staphylococcus aureus*, *Escherichia coli*, *Pseudomonas aeruginosa*, and *Klebsiella pneumoniae*
- 150 × 15 mm Mueller-Hinton II agar plates poured to a depth of 4 mm
- antibiotic disk dispensers (BBL or Difco) or assorted individual vials containing antibiotic disks
- sterile swabs
- 37°C incubator
- forceps
- metric rulers
- wax pencil or lab marker
- 70% ethyl alcohol and beakers
- Bunsen burner
- container for biohazard waste
- safety glasses
- disposable gloves
- lab coat

LEARNING OUTCOMES

Upon completion of this exercise, students will demonstrate the ability to

1. Summarize the scope of antimicrobial activity of selected antibiotics
2. Perform the Kirby-Bauer method for determination of antibiotic sensitivity
3. Correctly interpret a Kirby-Bauer plate

SUGGESTED READING IN TEXTBOOK

1. Antimicrobial Activity Can Be Measured by Specific Tests, section 9.3; also see table 9.2 and figures 9.2–9.5.

Pronunciation Guide

- *Escherichia coli* (esh-er-I-ke-a KOH-lie)
- *Klebsiella pneumoniae* (kleb-se-EL-lah nu-MO-ne-ah)
- *Pseudomonas aeruginosa* (soo-do-MO-nas a-ruh-jin-OH-sah)
- *Staphylococcus aureus* (staf-il-oh-KOK-kus ORE-ee-us)

Why Are the Following Bacteria Used in This Exercise?

In this exercise, you will evaluate the antimicrobial activity of selected antibiotics using the Kirby-Bauer method. To accomplish this objective, the authors have chosen four bacteria that you have worked with in previous exercises and that often infect humans. The four bacteria are *Staphylococcus aureus*, *Escherichia coli*, *Pseudomonas aeruginosa*, and *Klebsiella pneumoniae*.

Medical Application

The number of antibiotics (and other antimicrobials) available today is larger than ever before. New antibiotics are continuously being developed and discovered; thus there is an increasing demand on the clinical laboratory to determine the antibiotic susceptibility or resistance of various pathogenic bacteria. In most clinical laboratories, the antibiogram has been replaced with molecular techniques.

PRINCIPLES

The **Kirby-Bauer** test for antibiotic susceptibility, also called the **disc diffusion** test, is a standard that has been used for over 70 years. First developed in the 1950s, it was refined by William Kirby and Alfred Bauer in 1966 and named after them. It was also standardized in 1966 by the World Health Organization. Although it has now been superseded in clinical labs by different automated tests, the Kirby-Bauer test is still used in some labs, particularly with certain bacteria for which automation does not work well for identification of antibiotic sensitivity.

In this method, antibiotics are impregnated onto paper disks and then placed on a seeded Mueller-Hinton II agar plate using a mechanical dispenser or sterile forceps. The plate is then incubated for 18 to 24 hours, and the diameter of the **zone of inhibition** around the disk is measured to the nearest millimeter. The size of this zone depends on the sensitivity of the specific bacterium to the specific antibiotic and the point at which the antibiotic's **minimum inhibitory concentration (MIC)** is reached. The inhibition zone diameter that is produced will indicate the susceptibility or resistance of a bacterium to the antibiotic (**Figure 35.1**). Antibiotic susceptibility patterns are called **antibiograms.** Antibiograms can be determined by comparing the zone diameter obtained with the known zone diameter size for susceptibility (**Table 35.1**). For example, a zone of a certain diameter indicates susceptibility; zones of a smaller diameter or no zone at all show that the bacterium is resistant to the antibiotic. Frequently, one will see colonies within the zone of inhibition when the strain is antibiotic resistant.

Many factors are involved in sensitivity disk testing and must be carefully controlled. These include size of the inoculum, distribution of the inoculum, incubation period, depth of the agar, diffusion rate of the antibiotic, concentration of antibiotic in the disk, and growth rate of the bacterium. If all of these factors are carefully controlled, this type of testing is highly satisfactory for determining the degree of susceptibility of a bacterium to a certain antibiotic.

The Kirby-Bauer method is not restricted to antibiotics. It may also be used to measure the sensitivity of any microorganism to a variety of antimicrobial agents such as sulfonamides and synthetic chemotherapeutics. Figure 35.1 and **Figure 35.2** illustrate the Kirby-Bauer method.

Procedure

First Period

1. With a wax pencil or lab marker, mark the lid of each Mueller-Hinton agar plate with your name, date, and the name of the bacterium to be inoculated. Each group of students will inoculate the surface of four Mueller-Hinton plates with *S. aureus, E. coli, P. aeruginosa,* and *K. pneumoniae,* respectively. Use a separate, sterile cotton swab for each bacterium. The swab is immersed in the culture tube (**Figure 35.2a**), and the excess culture is squeezed on the inner side of the test tube.

2. The swab is then taken and streaked on the surface of the Mueller-Hinton plate three times, rotating the plate 60° after each streaking (**Figure 35.2b**). Finally, run the swab around the edge of the agar. This procedure ensures that the whole surface has been seeded, resulting in a "lawn" of bacteria on the agar surface. Dispose of the swab in the biohazard container. Allow the culture to dry on the plate for 5 to 10 minutes at room temperature with the top in place.

3. Dispense the antibiotics onto the plate either with the multiple dispenser or individually with the single unit dispenser (**Figure 35.2c**). Make sure that contact is made between the antibiotic disk and the culture by gently pressing the disk with alcohol-flamed forceps (**Figure 35.2d**). Do not press the disk into the agar and do not move the disk once it is placed on the agar.

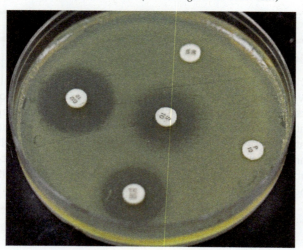

Figure 35.1 A Kirby-Bauer Plate. A Mueller-Hinton agar plate inoculated with *E. coli* and various antibiotics. Notice the diameter of the various zones of inhibition. The size of the zone of inhibition surrounding the disk reflects, in part, the sensitivity of the bacterium to the antibiotic. *(Lisa Burgess/McGraw Hill)*

Table 35.1   Interpretation of Inhibition Zones of Test Cultures

Disk Symbol	Antibiotic	Disk Content	Diameter of Zones of Inhibition (mm)		
			Resistant	Intermediate	Susceptible
AM	Ampicillin[a] when testing Gram-negative microorganisms and enterococci	10 μg	16 or less	–	17 or more
AM	Ampicillin[b] when testing staphylococci and penicillin G-susceptible microorganisms	10 μg	28 or less	–	29 or more
B	Bacitracin	10 units	8 or less	9–12	13 or more
CB	Carbenicillin when testing *Proteus* species and *E. coli*	50 μg	19 or less	18–22	23 or more
CB	Carbenicillin when testing *P. aeruginosa*	50 μg	13 or less	14–16	17 or more
C	Chloramphenicol (Chloromycetic®)	30 μg	12 or less	13–17	18 or more
CC	Clindamycin[c] when reporting susceptibility to clindamycin	2 μg	14 or less	15–20	21 or more
CC	Clindamycin[c] when reporting susceptibility to lincomycin	2 μg	16 or less	17–20	21 or more
CL	Colistin[d] (Coly-mycin®)	10 μg	8 or less	9–10	11 or more
E	Erythromycin	15 μg	13 or less	14–22	23 or more
GM	Gentamicin	10 μg	12 or less	13–14	15 or more
K	Kanamycin	30 μg	13 or less	14–17	18 or more
ME	Methicillin[e]	5 μg	9 or less	10–13	14 or more
N	Neomycin	30 μg	12 or less	13–16	17 or more
NB	Novobiocin[f]	30 μg	17 or less	18–21	22 or more
OL	Oleandomycin[g]	15 μg	11 or less	12–16	17 or more
P	Penicillin G. when testing staphylococci[h]	10 units	28 or less	–	29 or more
P	Penicillin G. when testing other microorganisms[h,i]	10 units	14 or less	15–21	22 or more
PB	Polymyxin B[d]	300 units	8 or less	9–11	15 or more
R	Rifampin when testing *N. meningitidis* susceptibility only	5 μg	16 or less	17–19	20 or more
S	Streptomycin	10 μg	6 or less	7–9	10 or more
S	Sulfonamides	300 μg	12 or less	13–16	17 or more
T(TE)	Tetracycline[j]	30 μg	14 or less	15–18	19 or more
VA	Vancomycin	30 μg	14 or less	15–16	17 or more

Source: Based on data from the National Committee for Clinical Laboratory Standards (NCCLS). NCCLS frequently updates the interpretive tables (such as this table) through new editions of the standards and supplements. Users should refer to the most recent edition. The current standards can be obtained from NCCLS, 940 West Valley Road, Suite 1400, Wayne, PA 19087–1898, USA. (www.clsi.org).
[a]The ampicillin disk is used for testing susceptibility of both ampicillin and betacillin.
[b]Staphylococci exhibiting resistance to the penicillinase-resistant penicillin class disks should be reported as resistant to cephalosporin class antibiotics. The 30 mcg cephalothin disk cannot be relied upon to detect resistance of methicillin-resistant staphylococci to cephalosporin class antibiotics.
[c]The clindamycin disk is used for testing susceptibility to both clindamycin and lincomycin.
[d]Colistin and polymyxin B diffuse poorly in agar, and the accuracy of the diffusion method is thus less than with other antibiotics. Resistance is always significant, but when treatment of systemic infections due to susceptible strains is considered, it is wise to confirm the results of diffusion test with a dilution method.
[e]The methicillin disk is used for testing susceptibility of all penicillinase-resistant penicillins; that is, methicillin, cloxacillin, dicloxacillin, oxacillin, and nafcillin.
[f]Not applicable to medium that contains blood.
[g]The oleandomycin disk is used for testing susceptibility to oleandomycin and trioleandomycin.
[h]The penicillin G. disk is used for testing susceptibility to all penicillinase-susceptible penicillins except ampicillin and carbenicillin: that is, penicillin G., phenoxymethyl penicillin, and phenethicillin.
[i]This category includes some organisms such as enterococci and Gram-negative bacilli that may cause systemic infections treatable with high doses of penicillin G. Such organisms should be reported susceptible only to penicillin G. and not to phenoxymethyl penicillin or phenethicillin.
[j]The tetracycline disk is used for testing susceptibility to all tetracyclines; that is chlorotetracycline, demeclocycline, doxycycline, methacycline, oxytetracycline, minocycline, and tetracycline.

4. Incubate the plates for 24 hours at 37°C. DO NOT INVERT THE PLATES. The bacteria will grow into what is called a "lawn." During the incubation period, as the antibiotics diffuse outward from the disk, the growth or survival of the bacteria in the lawn may be affected and this is one of the major objectives of this experiment.

Second Period

1. Measure the zones of inhibition to the nearest mm for each of the antibiotics tested. (**Figure 35.2d** and **Figure 35.3**). Do not open the lid to measure the zones of inhibition; measure on the underside of the plate. This will eliminate the possibility of contaminating the surroundings.

Figure 35.2 Antimicrobic Sensitivity Testing. (a) Immerse swab in the culture to test. (b) Cover the entire surface of the plate. (c) Dispense disks on the surface of the agar. (d) Gently press disk onto the surface of the agar. (e) Make observations after 24 hours of incubation. Note the safety gloves on the hands.

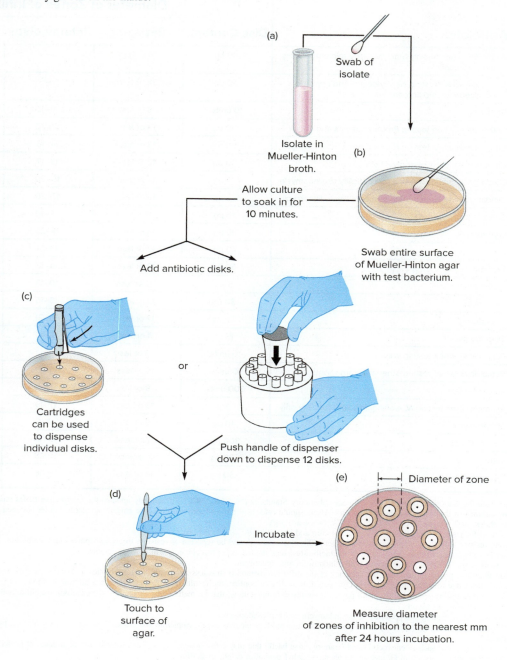

2. Using Table 35.1, determine whether the bacteria are resistant (R), of intermediate susceptibility (I), or susceptible (S). (If there is no inhibition, growth extends up to the rim of the disks on all sides, and the bacterium is reported as resistant to the antibiotic in the disk.) Record the results in the report for this exercise.

Disposal

When you are finished with all of the plates, discard them in the designated place for sterilization and disposal.

Environmental Factors Affecting Growth of Microorganisms

Figure 35.3 Zone of Inhibition (ZOI). (a) Zone of inhibition for gentamicin on *Pseudomonas aeruginosa*. (b) Measuring the zone of inhibition. The diameter of the zone of inhibition is measured in mm, from the edge of the growth, across the center of the antimicrobic disk. In this photo, the ZOI measures 18 mm across. *(a–b: Lisa Burgess/McGraw Hill)*

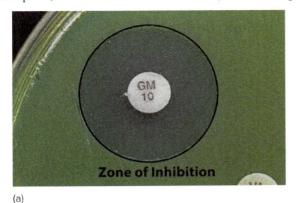
(a)

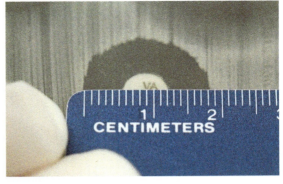
(b)

HELPFUL HINTS

- If the plate is correctly inoculated, symmetrically aligned inhibition ellipses will be seen against an even lawn of confluent growth. If growth is weak or uneven, the test should be repeated.
- Colonies growing within the inhibition ellipses may represent resistant subpopulations and the manufacturer has provided reading guidelines to interpret different types of results.
- Do not place the cotton swab on the benchtop. Place it in the proper container for disposal. Remember it contains bacteria that could get on the benchtop.

NOTES

Laboratory Report 35

Name: _____

Date: _____

Lab Section: _____

The Effects of Chemical Agents on Bacteria II: Antibiotics

1. Based on your measurements, complete the following table on the susceptibility of each test bacterium to the antibiotic by using an R (resistant), I (intermediate), or S (sensitive). Exchange data with other students in the class in order to complete the entire table.

Antibiotic or Antimicrobial	Disk Code	S. aureus		E. coli		P. aeruginosa		K. pneumoniae	
		Zone Size (mm)	S, I, R	Zone Size (mm)	S, I, R	Zone Size (mm)	S, I, R	Zone Size (mm)	S, I, R
1.									
2.									
3.									
4.									
5.									
6.									
7.									
8.									
9.									
10.									
11.									
12.									

2. From the above table, which antibiotic would you use against each of the following?

   S. aureus _____

   E. coli _____

   P. aeruginosa _____

   K. pneumoniae _____

3. Pick two antibiotics that you used and list the class of drug, spectrum of use, and their mechanism of action.

ASSESSMENT
Critical Thinking and Learning Outcomes Review

1. Propose how you would determine whether the zone of inhibition is due to death or to inhibition of a bacterium.

2. What factors must be carefully controlled in the Kirby-Bauer method?

3. What are the media components of the Mueller-Hinton media that are relevant to this procedure?

4. If the clinical laboratory reports bacterial susceptibility to an antibiotic but the patient is not responding to it, what could have gone wrong?

5. What are the similarities and differences in response to plates with Gram-positive and Gram-negative bacteria? Between enterics and nonenterics?

6. What is the difference between an antibiotic and an antimicrobial?

7. Discuss one of the antibiotics you used and how bacteria are becoming more resistant to it.

EXERCISE 36 Bacterial Growth Curve

SAFETY CONSIDERATIONS

- Be careful with the Bunsen burner flame.

MATERIALS

- 10- to 12-hour (log phase) tryptic soy broth cultures of *Escherichia coli*
- 100 ml of brain-heart infusion (for rapid growth) in a 250-ml Erlenmeyer flask
- 99-ml saline blanks
- 100-ml bottles of tryptic soy agar or individual 15-ml tubes of TSA
- 37°C water bath with shaker or temperature-controlled shaker incubator
- spectrophotometer
- cuvettes
- Nephelo culture flasks
- colony counter
- 28 Petri plates
- 1-ml and 10-ml sterile pipettes with pipettor
- Bunsen burner
- wax pencil or lab marker
- 1,000-ml beaker
- ruler
- container for biohazard waste
- safety glasses
- disposable gloves
- lab coat

LEARNING OUTCOMES

Upon completion of this exercise, students will demonstrate the ability to

1. Explain the growth dynamics of a bacterial culture
2. Identify the typical phases of a bacterial growth curve
3. Use a spectrophotometer
4. Measure bacterial growth and turbidity
5. Plot a growth curve and determine the generation time of a culture of *E. coli*

SUGGESTED READING IN TEXTBOOK

1. Growth Curves Consist of Five Phases, section 7.4; see also figures 7.11–7.15.

Pronunciation Guide

- *Escherichia coli* (esh-er-I-ke-a KOH-lie)

Why Are the Following Bacteria Used in This Exercise?

In this experiment, you will learn how to measure bacterial growth, plot a bacterial growth curve, and determine the generation time. To accomplish this, the authors have chosen the common bacterium *Escherichia coli*, which has a generation time of approximately 21 minutes at 37°C.

PRINCIPLES

The four phases (**lag, logarithmic or exponential, stationary,** and **death or decline**) of growth (**Figure 36.1**) of a bacterial population can be determined by measuring the **turbidity** of the population in a broth culture. Turbidity is not a direct measure of bacterial numbers but an indirect measure of biomass, which can be correlated with cell density during the log growth phase. Since about 10^7 bacterial cells per milliliter must be present to detect turbidity with the unaided eye, a spectrophotometer can be used to achieve increased sensitivity and obtain quantitative data (**Figure 36.2**).

The construction of a complete **bacterial growth curve** (increase and decrease in cell numbers versus time) requires that aliquots of a shake-flask culture be measured for population size at intervals over an extended period. Because this may take many hours, such a procedure does not lend itself to a regular laboratory session.

Figure 36.1 Bacterial Growth Curve in a Closed System. The four phases of the growth curve are illustrated on the curve. On this graph, the number of viable cells expressed as a logarithm (log) is plotted against time in hours. Note that with a generation time of 30 minutes, the bacterial population has risen from 10 (10^1) cells to 1,000,000,000 (10^9) cells in 16 hours. During the lag phase, bacteria are preparing their cell machinery for growth. During the exponential (log or logarithmic) growth phase, bacterial growth approximates an exponential curve (straight line on a logarithmic scale). In the stationary phase, bacteria stop growing and shut down their growth machinery while turning on stress responses to help retain viability. Finally, in the death or decline phase, cells are dying in a negative exponential curve.

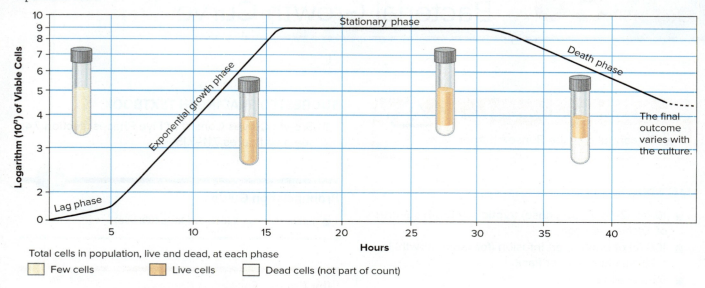

Figure 36.2 Turbidity Measurements as Indicators of Bacterial Growth. (a) Holding broth culture tubes up to a light source is one method of checking for differences in cloudiness (turbidity). Notice that the broth culture tube on the left is transparent, indicating little or no bacterial growth; the broth culture tube on the right is cloudy and opaque, indicating heavy bacterial growth. (b) Highly sensitive and accurate measurements can be made with a spectrophotometer. (1) A broth culture tube with no bacterial growth will allow light to easily pass through. Therefore more light will reach the photo-detector and give a higher transmittance value. (2) In a broth culture tube with growth, the bacterial cells scatter the light, resulting in less light reaching the photo-detector and, therefore, giving a lower transmittance value. Both a newer digital and an older analog spectrophotometer are shown. *(a: Javier Izquierdo/McGraw Hill)*

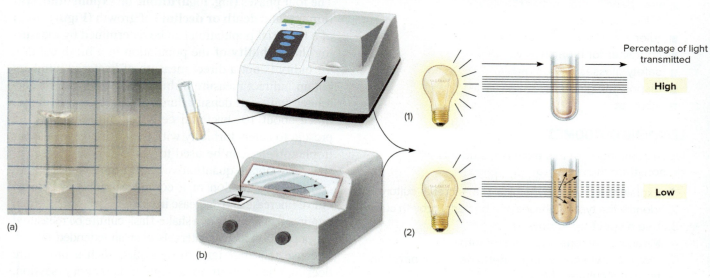

Environmental Factors Affecting Growth of Microorganisms

Therefore, this exercise has been designed to demonstrate only the lag and log phases of a bacterial growth curve. The bacterial population will be plotted on graph paper by using both an indirect and a direct method for the measurement of growth. The resulting growth curve can be used to delineate stages of the growth cycle. It also makes possible the determination of the growth rate of a particular bacterium under standardized conditions in terms of its **generation time**—the time required for a bacterial population to double (**Figure 36.3**).

The indirect method uses spectrophotometric measurements of the developing turbidity in a bacterial culture taken at regular intervals. These samples serve as an index of increasing cellular mass. The graphical determination of generation time is made by extrapolation from the log phase, as illustrated in Figure 36.3. For example, select two points (0.2 and 0.4) on the absorbance (A) scale that represent a doubling of turbidity. Using a ruler, extrapolate by drawing a line between each absorbance on the ordinate and the exponential phase of the growth curve. From these two points, draw perpendicular lines to the time intervals on the abscissa. From these data, the generation time can be calculated as follows:

Generation time = t (A of 0.4) − t (A of 0.2)

Generation time = 90 minutes − 60 minutes

= 30 minutes

The same graphical generation time determination can be done with a plot of population counts.

The growth rate constant can also be determined from the data. When the $\log_{10}$ of the cell numbers or absorbance is plotted versus time, a straight line is obtained, the slope of which can be used to determine the value of g and k (**Figure 36.4**). The dimensions of k are reciprocal hours or per hour. The growth rate constant will be the same during exponential growth regardless of the component measured (e.g., cell biomass, numbers). The growth rate constant provides the microbiologist with a valuable tool for comparison between different microbial species when standard growth and environmental conditions are maintained.

Once the growth rate constant is known, the **mean generation time (doubling time)** can be calculated from the following equation:

$$g = \frac{1}{k}$$

This equation also allows one to calculate the growth rate constant from the generation time.

As mentioned previously, the generation time can be read directly from the bacterial growth curve plot, and the growth rate constant can then be determined.

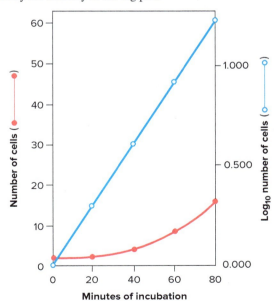

FIGURE 36.4 Exponential Microbial Growth. Four generations of growth are plotted directly (—) and in the logarithmic form (—). The growth curve is exponential, as shown by the linearity of the log plot.

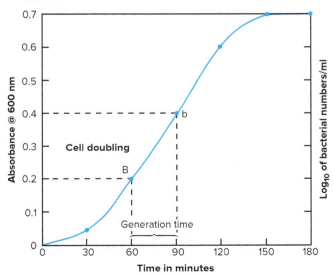

Figure 36.3 Indirect Method for Determining Bacterial Generation Time.

To calculate the generation time (g) from these data with an equation, use the following formula:

$$\text{Generation time} = \frac{0.301t}{\log_{10}N_t - \log_{10}N_0}$$

where N_0 = bacterial population at point B or any other point at the beginning of the log phase

N_t = bacterial population at point b or any other point at or near the end of the log phase

t = time in minutes between b and B (Figure 36.3).

From the previous equation, one can also determine the specific **mean growth rate constant** (k) for any culture during unrestricted growth. During this time, the rate of increase of cells is proportional to the number of cells present at any particular time. In mathematical terms, the growth rate is expressed as

$$K = \frac{n}{t}$$

where n is the number of generations per unit time. The symbol k represents the mean growth rate constant, converting the equation to logarithms:

$$k = \frac{\log N_t - \log N_0}{0.301t}$$

Procedure

First Period

1. Separate 21 sterile saline blanks (99 ml each) into seven sets of three each. Using the wax pencil or lab marker, label each set as to the time of inoculation ($t = 0$, $t = 30$, $t = 60$, $t = 90$, $t = 120$, $t = 150$, and $t = 180$) and the dilution in each blank (10^{-2}, 10^{-4}, and 10^{-6}) (**Figure 36.5**).
2. Place the flask containing the brain-heart infusion broth in the 37°C water bath or incubator for 15 minutes. (Warming the brain-heart infusion broth before adding the bacteria will decrease both the duration of the lag phase and drop in absorbance that is generally visualized as the bacteria adjust in the medium and to the temperature, allowing for a shift into the log phase.)
3. Using the wax pencil or lab marker, label seven sets of four plates each with your name, the time of inoculation (use the same times as in step 1), and the dilution (10^{-4}, 10^{-5}, 10^{-6}, 10^{-7}) to be plated.
4. Melt three 100-ml bottles or 20 15-ml tubes of tryptic soy agar in a water bath, cool to 45°C, and leave in the 45°C water bath until needed (see step 8).
5. Using a sterile pipette, transfer 5 ml of the log phase *E. coli* culture to the flask containing 100 ml of brain-heart infusion broth. Label with your name, time, and date. The approximate absorbance (A) of this broth should be about 0.1 at 600 nm (*see Exercise 18 for proper use of the spectrophotometer*).
6. After the initial A has been determined, shake the culture and aseptically transfer 1 ml to the 99-ml water blank labeled $t = 0$ and 10^{-2} and continue to serially dilute to 10^{-6}.
7. Place the culture flask in the shaker water bath or incubator, set at 37°C and 120 rpm. If a shaker bath is not available, the flask should be shaken periodically.
8. Plate the 0 time dilutions into the appropriately labeled Petri plates, using the amounts indicated in Figure 36.5. Pour 15 ml of the melted agar into each plate and mix by gently swirling the plate on a flat surface.
9. Thereafter, at 30-minute intervals, transfer 5 ml of the broth culture to a cuvette and determine the A of the culture at 600 nm. Be sure to suspend the bacteria thoroughly each time before taking a sample.
10. At the same time interval, transfer 1 ml of the culture into the 10^{-2} water blank of the set labeled with the appropriate time (see step 1). Complete the serial dilution once again as indicated in Figure 36.5 and plate into the labeled (see step 2) Petri plates. Add melted agar as per step 8.
11. When the media in the Petri plates harden, incubate them in an inverted position for 24 hours at 37°C.

Second Period

1. Perform colony counts on all plates.
2. Record all measurements and corresponding bacterial counts in the table in the report for Exercise 36.
3. On the graph paper provided, plot the following:

    a. Absorbances on the ordinate, and incubation times on the abscissa. Use Figure 36.3 as an example.

Figure 36.5 Dilution Plating and Spectrophotometric Procedure for Constructing Bacterial Growth Curves. These steps will be needed for each time point.

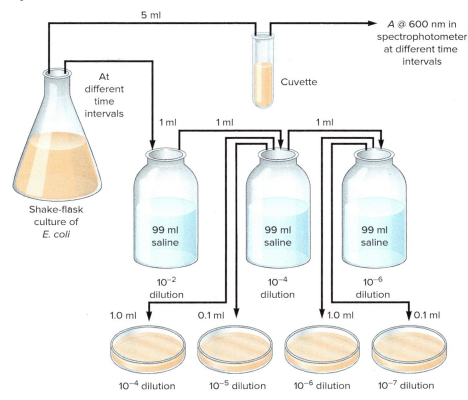

b. Log_{10} values of the bacterial counts on the ordinate, and incubation times on the abscissa. Connect the points with a ruler.
c. In addition to using the graph paper, you should also construct a graph of the data using semilog graph paper at the end of the exercise. Calculate generation time and mean growth rate constant. Employ both the graphical method and growth equations.

Disposal

When you are finished with all of the tubes, pipettes, cuvettes, and plates, discard them in the designated place for sterilization and disposal.

HELPFUL HINTS

- Be sure to maintain good aseptic technique when making transfers and report any spills to your instructor. Carefully clean and decontaminate your work area at the end of the experiment.
- The bacterial growth curve study can be very easily run using Nephelo culture flasks, which have side arms that fit into the spectrophotometer's cuvette compartment so that growth can be followed without removing samples.

NOTES

Laboratory Report 36

Name: _____

Date: _____

Lab Section: _____

Bacterial Growth Curve

Classical Method

1. Based on your data on absorbance and plate counts (bacterial cells per milliliter), complete the following table.

Incubation Time in Minutes	Absorbance @ 600 nm	Plate Counts, Bacteria/ml	Dilution Factor	Log of Bacteria/ml
0				
30				
60				
90				
120				
150				
180				

2. Calculate the generation time for this *E. coli* culture by the indirect method, using the formula given in the Principles section and by the indirect method using your growth curve and extrapolations from the absorbances for doubling. Show all calculations in the space provided.

    a. From formula

    b. From growth curve

3. What is the *k* value for your *E. coli* culture?

4. In summary, what is the major purpose of this experiment?

257

ASSESSMENT
Critical Thinking and Learning Outcomes Review

1. Define generation time.

2. When following bacterial growth, why is absorbance plotted instead of percent transmission?

3. Can generation time be calculated from any phase of the growth curve? Explain your answer.

4. What is occurring in a bacterial culture during the lag phase? During the growth phase?

5. What is the significance of a k value?

6. How can the mean generation time be determined for a bacterial culture?

Graph paper

Absorbance

Log₁₀ cell number

Time

Graph paper

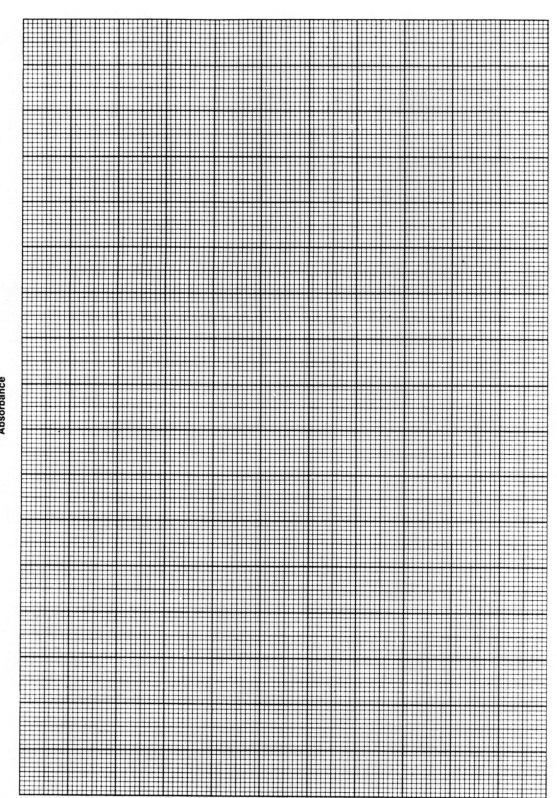

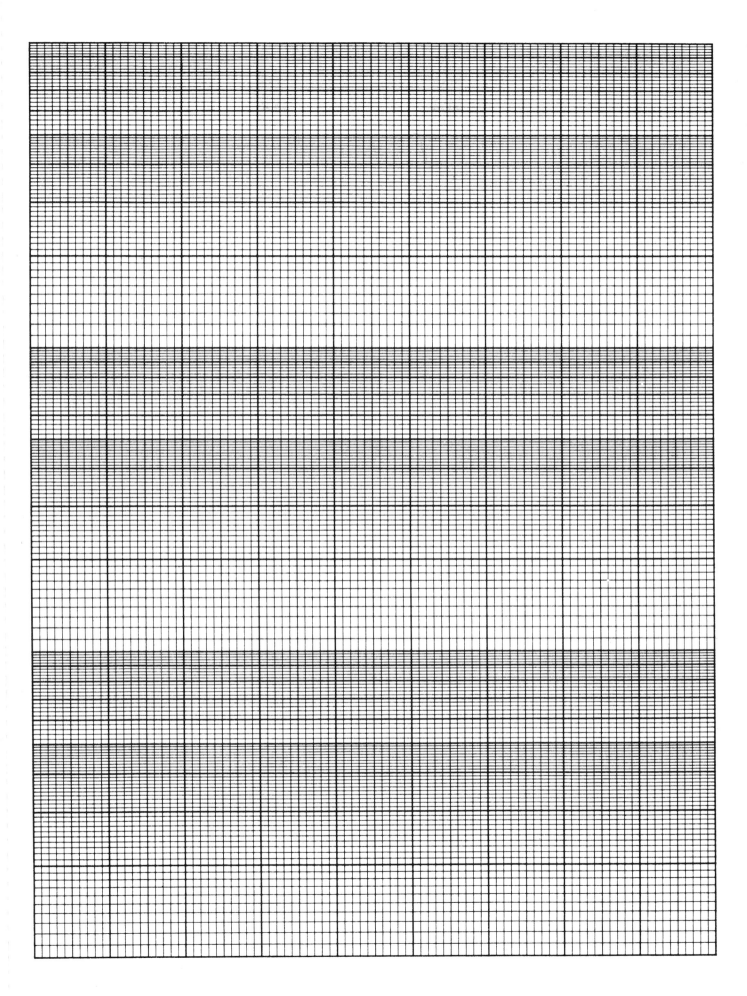

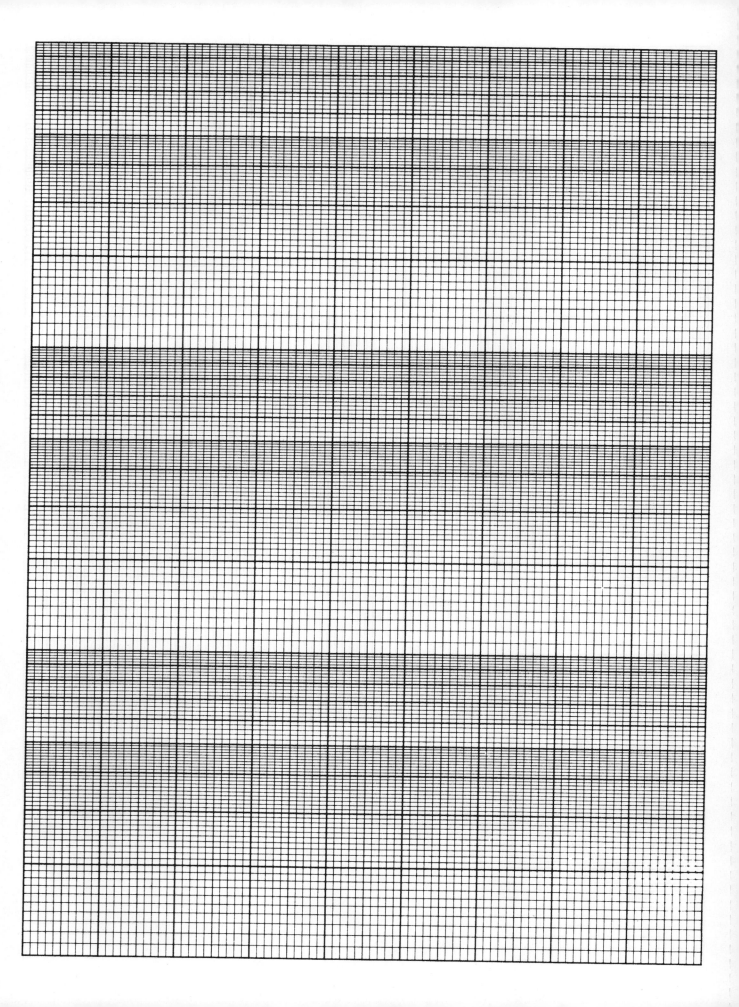

NOTES

PART 6
Environmental and Food Microbiology

This part of the manual contains exercises dealing with environmental and food microbiology.

The quality of **potable** (drinking) **water** available for public use is of major concern to everyone. Municipal and rural water supplies can transmit human diseases such as cholera, typhoid fever, and gastroenteritis. The amount of fecal contamination is monitored by counting the number of nonpathogenic, or **indicator,** bacteria, present in human sewage in large numbers, and that survive in water long enough to be satisfactorily counted. The bacterium that satisfies these criteria best is *Escherichia coli*. It is a member of the **coliform group** of bacteria. Even though all coliforms do not come from human feces, the **total coliform count** is used as an index of sewage pollution. The classical approach to count coliforms is to calculate the **most probable number (MPN)** of coliforms by using lactose or lauryl tryptose broth fermentation tubes. Commercially available alternatives make use of synthetic substrates to determine if water samples are contaminated.

Soils are complex environments in which microorganisms play a major role, contributing to nutrient cycling and decomposition. Complex microbial interactions with other microorganisms, macroorganisms, and nutrients influence degradation processes. In this section, you will isolate and quantify microbes found in a soil sample. You will also cultivate your own soil microbiome by constructing Winogradsky columns.

Most of the foods we consume contain microorganisms. In many instances, these microorganisms have been introduced into the food as part of the production process (e.g., cheese, pickles, yogurt, and sausage). Once in the food product and when a suitable temperature exists, the microorganisms will use the food product as an energy source, metabolizing it, and excreting waste products. Some of these waste products may make the food product unpalatable, and, as a result, the food source is said to be spoiled. At other times, the waste products from the microorganisms may cause **food poisonings** or **food intoxications**. In other instances, the waste products of the microorganisms may be desirable since their products aid in the production of a specific food. In this section, you will isolate and enumerate bacteria associated with food.

Budgetstockphoto/Getty Images

It is no secret that human activity has a huge impact on the environment. This is also true at the microscopic level. The water and soil around us are filled with microbes who play a vital role in biogeochemical cycles that are needed to maintain a healthy planet. Disruption of these microbial communities can not only disrupt such cycles, but in some cases can also lead to human disease. A prime example of this is when sewage spills contaminate drinking water and seafood supplies. In contrast, microbes can also play a positive role in our food supply. Production of foods like cheese and bread depends on microbial metabolism. In this part of the lab manual, you will perform a variety of exercises that will provide an introduction to environmental and food microbiology.

After completing the exercises in Part 6, you will meet the American Society for Microbiology Core Curriculum Laboratory Skills:

- using biochemical test media and accurately recording macroscopic observations;
- performing aseptic transfers;
- isolating colonies;
- correctly spreading appropriate dilutions;
- estimating the number of microorganisms in a sample using serial dilution techniques; and
- extrapolating plate counts to obtain correct microbial count in the starting sample.

EXERCISE 37 Monitoring Water for Coliforms

SAFETY CONSIDERATIONS

- Be careful with the Bunsen burner flame.
- Dispose of all water samples in the biohazard waste container.
- Keep all culture tubes upright in a test-tube rack.
- UV light can cause damage to the eyes if looked at directly; use of a face shield is required.

MATERIALS

- 1-ml single-strength lauryl tryptose broth fermentation tubes
- 0.1-ml single-strength lauryl tryptose broth fermentation tubes
- 10-ml double-strength lauryl tryptose broth fermentation tubes
- 100 ml water sample
- sterile saline
- longwave ultraviolet lamp
- Gram staining reagents
- Levine's eosin methylene blue (EMB) agar plates
- tryptic agar slants
- brilliant green lactose bile broth tubes
- sterile 10-ml pipettes with pipettor
- sterile 1-ml pipettes
- wax pencil or lab marker
- test-tube rack
- 37°C incubator
- inoculating loop and needle
- 37°C water bath
- 150-ml sterile nonfluorescent sample bottles
- ONPG/MUG substrate
- Bunsen burner
- container for biohazard waste
- safety glasses
- disposable gloves
- lab coat

LEARNING OUTCOMES

Upon completion of this exercise, students will demonstrate the ability to

1. Determine the presence of coliform bacteria in a water sample
2. Calculate the possible number of coliform bacteria present in a water sample
3. List and explain each step (presumptive, confirmed, completed) in the multiple-tube technique for determining coliforms in the water sample
4. Perform a coliform test using ONPG/MUG

SUGGESTED READING IN TEXTBOOK

1. Purification and Sanitary Analysis Ensure Safe Drinking Water, section 42.1; see also figure 42.1 and table 42.1.

Pronunciation Guide

- *Citrobacter* (SIT-ro-bac-ter)
- *Escherichia* (esh-er-I-ke-a)
- *Enterobacter* (en-ter-oh-BAK-ter)
- *Klebsiella* (kleb-se-EL-lah)

Medical Application

Many important human pathogens colonize living organisms other than humans, including many wild animals and birds. Many of these pathogens can be shed in water and infect humans. Among the most prominent water-borne pathogens of recent times are the protozoans *Giardia* and *Cryptosporidium*; the bacteria *Campylobacter, Salmonella, Shigella, Vibrio, Escherichia,* and *Mycobacterium*; and hepatitis A and Norwalk viruses.

265

Figure 37.1 The Most Probable Number (MPN) Procedure. This method enables detection of coliforms in a water sample. Note the difference between presumptive, confirmed, and completed tests.

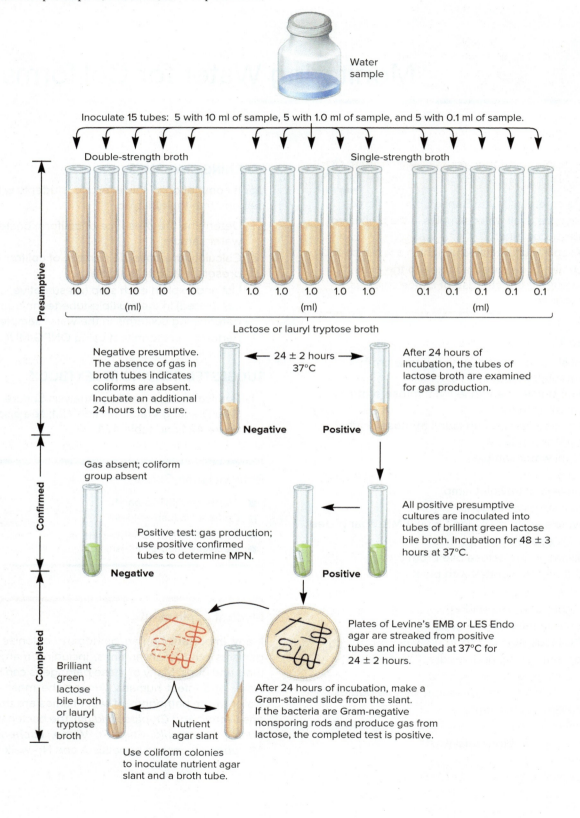

266 Environmental and Food Microbiology

When waters are used for drinking or recreation, and if they are contaminated with human or animal waste, the possibility for disease transmission exists. This includes direct consumption of contaminated water, or indirectly through consumption of contaminated seafood. Thus the microbial content of waters used by humans must be continuously monitored to ensure that the water is free of infectious agents.

Attempting to survey water for specific pathogens can be very difficult, expensive, and time-consuming, so most assays of water purity are focused on detecting contamination by fecal coliforms (a collection of Gram-negative non-spore-forming rods that ferment lactose to acid and gas). High levels of fecal contamination typically indicates that the water contains pathogens and is consequently unsafe. One coliform, *Escherichia coli,* is found almost exclusively in human and animal waste. Its presence in water provides strong evidence that other intestinal microorganisms also may be present and that the water may be unfit for human consumption.

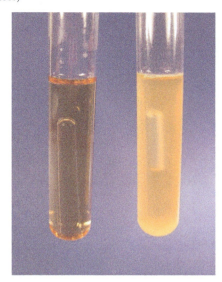

Figure 37.2 **Lauryl Tryptose Broth.** Within 24 hours of incubation at 37°C, notice the gas bubble in the Durham tube on the right. This indicates a positive presumptive test, "presumably" due to the presence of coliforms. The tube on the left is negative. *(Barry Chess)*

PRINCIPLES

The number of total coliforms (*Enterobacter, Klebsiella, Citrobacter, Escherichia*) in a water sample can be determined by a statistical estimation called the ***most probable number* (MPN) test** (**Figure 37.1**). This test involves a series of multiple fermentation tubes and is divided into three parts: the **presumptive, confirmed, and completed tests.** Note that each test is based on one or more of the characteristics of coliforms.

In the presumptive test, dilutions from the water sample are added to lauryl tryptose broth fermentation tubes. After 24 to 48 hours of incubation at 37°C, the tubes are examined for the presence of bacteria capable of fermenting lactose with gas production, which "presumably" are coliforms (**Figure 37.2**). Gas is detected by appearance of a bubble in the Durham tube (recall the carbohydrate fermentation experiment with phenol red broth). Note that the lauryl tryptose broth is selective for Gram-negative bacteria due to the presence of lauryl sulfate.

In the confirmed test, culture material is transferred from the highest dilution of those lactose broth tubes that showed growth and gas production into brilliant green lactose bile broth, which is selective and differential for coliforms. The tube is incubated for 48 hours at 37°C. Gas formation in the Durham tube is a confirmed test for total coliforms (**Figure 37.3**).

In the completed test, a sample from the positive green lactose bile broth is streaked onto Levine's EMB agar and incubated for 18 to 24 hours at 37°C. This medium contains lactose for the coliform bacteria. EMB has methylene blue, which is inhibitory for Gram-positive bacteria.

On EMB agar, coliforms produce small colonies with dark centers (**Figure 37.4**). Samples are then inoculated into brilliant green lactose bile broth or lauryl tryptose broth, and onto a nutrient agar slant. These tubes are incubated for 24 hours at 37°C. If gas is produced in the lactose broth, and the isolated bacterium is a Gram-negative (based on a Gram stain) nonsporing rod, the completed test is positive.

Figure 37.3 **Brilliant Green Lactose Broth.** Within 48 hours of incubation at 37°C, notice the gas bubble in the Durham tube on the right. This indicates a positive confirmed test, "presumably" due to coliforms. The tube on the left is negative. *(Barry Chess)*

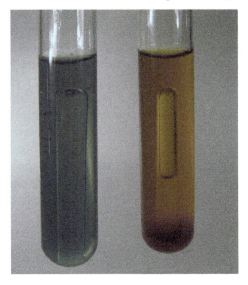

Figure 37.4 EMB Agar with Isolated Colonies of *Escherichia coli*. Notice the distinct, metallic green color. *(Lisa Burgess/McGraw Hill)*

Figure 37.5 ONPG/MUG Defined Substrate Test. The uninoculated control bottle is on the left **(a)**. Coliforms hydrolyze ONPG yielding a yellow-colored compound **(b)**. *E. coli* hydroyzes MUG generating a blue fluorescent compound **(c)**. *(a-b: Donald A. Klein; c: Ethan Shelkey and Joanne M. Willey, Ph.D.)*

An estimate of the number of coliforms (most probable number) can also be done in the presumptive test. In this procedure, 15 lactose broth tubes are inoculated with the water sample. Five tubes receive 10 ml of water, 5 tubes receive 1 ml of water, and 5 tubes receive 0.1 ml of water. A count of the number of tubes showing gas production is then made, and the figure is compared to a table (**Table 37.1**) developed by the American Public Health Association. The number is the most probable number (MPN) of coliforms per 100 ml of the water sample. It should be noted that the MPN index usually comes from the presumptive test if raw sewage is being tested and comes from confirmed or completed tests for other types of samples.

More recently, a simple alternative to the classical MPN procedure has been developed. This test is based on two key reagents, *o*-Nitrophenyl-β-D-Galactopyranoside (ONPG) and 4-Methylumbelliferyl-β-D-Glucuronide (MUG). When the enzyme galactosidase hydrolyzes the colorless ONPG, it produces galactopyranose and *o*-nitrophenol, which is a bright yellow color (**Figure 37.5a**). If the test bottle is as yellow as or more yellow than the control, this portion of the test is positive and reported as "total coliforms present." The enzyme beta-glucuronidase cleaves MUG into its components; the product 4-methylumbelliferone is fluorescent when viewed under UV light (**Figure 37.5b**). If the test bottle fluorescence is equal to or greater than that of the control, the presence of *E. coli* has been confirmed. This part of the test is reported as "*E. coli* present." If the water sample is not as yellow and does not fluoresce, it is reported as negative and the test is reported as "total coliforms absent" and "*E. coli* absent."

Procedure for the MPN Test

First Period

Presumptive Test

1. Mix the bottle of water to be tested 25 times. Inoculate five of the double-strength lauryl tryptose broth tubes with 10 ml of the water sample; five single-strength tubes with 1 ml of the water sample; and five single-strength tubes with 0.1 ml of the water sample. Carefully mix the contents of each tube without spilling any of the broth by rolling the tubes between the palms of your hands. Using the wax pencil or lab marker, label all tubes with your name, date, and the amount of water added.
2. Incubate the three sets of tubes for 24 to 48 hours at 37°C.
3. Observe after 24 and 48 hours. The presence of gas in any tube after 24 hours is a positive presumptive test. The formation of gas during the next 24 hours is a doubtful test. The absence of gas after 48 hours is a negative test.
4. Determine the number of coliforms per 100 ml of water sample (see Principles section and use Table 37.1). For example, if gas were present in all five of the 10 ml tubes, in only one of the 1 ml series, and in none of the 0.1 ml series, your test results would read 5-1-0. Table 37.1 indicates that

Table 37.1	Most Probable Number (MPN) Index for Five Tubes Each of 10 ml, 1 ml, and 0.1 ml						
10 ml	1 ml	0.1 ml	MPN Index per 100 ml	10 ml	1 ml	0.1 ml	MPN Index per 100 ml
0	0	0	<2	4	2	1	26
0	0	1	2	4	3	0	27
0	1	0	2	4	3	1	33
0	2	0	4	4	4	0	34
1	0	0	2	5	0	0	23
1	0	1	4	5	0	1	31
1	1	0	4	5	0	2	43
1	1	1	6	5	1	0	33
1	2	0	6	5	1	1	46
2	0	0	5	5	1	2	63
2	0	1	7	5	2	0	49
2	1	0	7	5	2	1	70
2	1	1	9	5	2	2	94
2	2	0	9	5	3	0	79
2	3	0	12	5	3	1	110
3	0	0	8	5	3	2	140
3	0	1	11	5	3	3	180
3	1	0	11	5	4	0	130
3	1	1	14	5	4	1	170
3	2	0	14	5	4	2	220
3	2	1	17	5	4	3	280
3	3	0	17	5	4	4	350
4	0	0	13	5	5	0	240
4	0	1	17	5	5	1	350
4	1	0	17	5	5	2	540
4	1	1	21	5	5	3	920
4	1	2	26	5	5	4	1,600
4	2	0	22	5	5	5	≥2,400

the MPN for this reading would be 33 coliforms per 100 ml of water sample.

Second Period

Confirmed Test

1. Record your results of the presumptive test in the report for this exercise.
2. Using an inoculating loop, from the tube that has the highest dilution of water sample and shows gas production, transfer one loopful of culture to the brilliant green lactose bile broth tube. Incubate for 48 hours at 37°C. The formation of gas at any time within 48 hours constitutes a positive confirmed test.

Third Period

Completed Test

1. Record your results of the confirmed test in the report for this exercise.
2. From the positive brilliant green lactose bile broth tube, streak an EMB plate.
3. Incubate the plate inverted for 24 hours at 37°C.
4. If coliforms are present, select a well-isolated colony and inoculate a single-strength, brilliant green lactose bile broth tube and streak a nutrient agar slant.
5. Gram stain any bacteria found on the slant.

Monitoring Water for Coliforms

6. The formation of gas in the lactose broth and the demonstration of Gram-negative, nonsporing rods in the agar culture is a satisfactorily completed test revealing the presence of coliforms and indicating that the water sample was polluted. This is a positive completed test.

Procedure for MUG Assay

First Period

1. Add 50 ml of the water sample to two sample bottles and ONPG/MUG substrate reagent to each. In a third control bottle, combine sterile saline with the ONPG/MUG reagent. Place the screw lids tightly on the bottles.
2. Mix thoroughly by inverting the bottle ten times to achieve even distribution of the reagent throughout the sample.
3. Incubate the bottles for 18 hours at 37°C.

Second Period

1. Examine the bottles. A clear, colorless solution is negative for coliforms. A yellow color indicates a positive reaction for total coliforms.
2. Place each of the bottles in front of an ultraviolet lamp in a dark room. If the bottle is fluorescent, it is positive for *E. coli*.

Disposal

When you are finished with all of the tubes, pipettes, and plates, discard them in the designated place for sterilization and disposal.

> **HELPFUL HINTS**
>
> - Do not confuse the appearance of an air bubble in a clear Durham tube with actual gas production; the broth medium will become cloudy if gas is formed as the result of fermentation.
> - When collecting your water sample, the upper 40 cm of most waters usually contains the greatest numbers of live bacteria.
> - Use sterile containers to collect water samples.

Laboratory Report 37

Name: _____

Date: _____

Lab Section: _____

Monitoring Water for Coliforms

1. Results of presumptive test:

Number of Positive Tubes			
10 ml H$_2$O added	1.0 ml H$_2$O added	0.1 ml H$_2$O added	MPN
24 hours 48 hours	24 hours 48 hours	24 hours 48 hours	

2. Results of confirmed test:

Positive		Negative	
24 Hours	48 Hours	24 Hours	48 Hours

3. Results of completed test:

Lactose Fermentation Results	Morphology

4. Sketch the appearance of each bottle from the ONPG/MUG test, noting any color change.

ASSESSMENT
Critical Thinking and Learning Outcomes Review

1. Why are coliforms selected as the indicator of water potability?

2. Does a positive presumptive test indicate that water is potable?

3. Why is the MPN test qualitative rather than quantitative?

4. What is the function of the following in the MPN test?
    a. lactose broth
    b. Levine's EMB agar
    c. nutrient agar slant
    d. Gram stain

5. What does a metallic green sheen indicate on an EMB plate? Is EMB considered a selective and/or differential medium? Explain your answer.

6. What bacterial diseases can be transmitted by polluted water?

7. What does a positive ONPG test indicate? What does a positive MUG test indicate?

8. Describe the biochemistry behind the ONPG and MUG tests. What enzymes are being assayed? What products are being produced?

EXERCISE 38: Enumeration of Soil Microorganisms

SAFETY CONSIDERATIONS

- Be careful with the Bunsen burner flame.
- Be careful when growing fungi as they can form spores which will easily contaminate bacterial cultures.

Pronunciation Guide

- *Aspergillus* (as-per-JIL-us)
- *Mucor* (MU-kor)
- *Penicillium* (pen-a-SIL-ee-um)
- *Rhizopus* (rye-ZOH-pus)

MATERIALS

- 1 g of soil
- tryptic soy agar
- glycerol yeast extract agar supplemented with 50 mg/l cycloheximide
- Sabouraud dextrose agar
- 99 ml bottle of sterile water
- 0.9 ml sterile saline dilution tubes
- wax pencil or lab marker
- Bunsen burner
- micropipettor and sterile tips
- cell spreaders
- vortex mixer
- container for biohazard waste
- safety glasses
- disposable gloves
- lab coat

LEARNING OUTCOMES

Upon completion of this exercise, students will demonstrate the ability to

1. Describe the diversity of the microorganisms present in garden soil
2. Determine the number of bacteria, actinomycetes, and fungi in the garden soil using the plate count method

SUGGESTED READING IN TEXTBOOK

1. Soils Are an Important Microbial Habitat, section 30.1, and Diverse Microorganisms Inhabit Soil, section 30.2; see also table 30.1 and figure 30.3.

PRINCIPLES

The level of microbial diversity in soil exceeds that of almost any other habitat on Earth, with density estimates of 10^9 to 10^{10} cells per gram of soil. The soil environment supports active microbial growth that contributes to global cycling of elements like carbon and nitrogen and makes nutrients available to plants. Simply put, without soil microbes, there would be no plants, which would eliminate herbivores, and so on up the food chain.

Actinomycetes (including actinoplanetes, nocardioforms, and streptomycetes), other bacteria, and filamentous fungi (*Rhizopus, Mucor, Penicillium,* and *Aspergillus*) are all important members of the soil microbial community. Protozoa, algae, cyanobacteria, nematodes, insects and other invertebrates, and viruses are also important members but will not be studied in this exercise (**Figure 38.1**).

Since soils vary greatly with respect to their physical features (e.g., pH, mineral content, temperature, and other factors), the microorganisms present will also vary. For example, acid soils will have a higher number of fungi compared to alkaline soils, and rich garden soil will contain more actinomycetes than either the other bacteria or fungi. Not surprisingly, no single culture technique is available to count the microbial diversity found in an average soil sample. Thus in this exercise, you will try to determine only the relative number of fungi, actinomycetes, and other bacteria in a soil sample using the serial dilution agar plating method.

Figure 38.1 The Soil Habitat. A typical soil habitat contains a mixture of clay, silt, and sand along with soil organic matter. Roots and animals (e.g., nematodes and mites), as well as protozoa and bacteria, consume oxygen, which rapidly diffuses into soil pores where the microbes live. Note that two types of fungi are present: mycorrhizal fungi, which derive their organic carbon from plant roots; and saprophytic fungi, which help degrade organic matter.

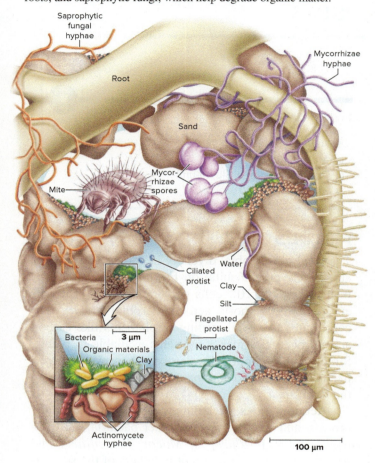

By counting each colony forming unit (CFU) and multiplying by the dilution, you will then obtain a census of the microbes in your soil sample. To support the three different groups of microorganisms, you will use three types of media: (1) Sabouraud dextrose agar for the isolation of fungi, (2) glycerol yeast extract agar with cycloheximide for the isolation of actinomycetes, and (3) tryptic soy agar for the isolation of other bacteria.

Procedure

1. The dilution scheme for enumeration of soil microorganisms is illustrated in **Figure 38.2**.

2. Place 1 g of soil in a 99 ml sterile water blank. Mix the soil and water thoroughly by shaking the water-soil mixture vigorously for 1 minutes. This is now a 10^{-2} dilution of soil.

3. Using a wax pencil or lab marker, label three sets of four Petri plates each as follows: glycerol yeast extract for **actinomycetes** (10^{-5}, 10^{-6}, 10^{-7}, and 10^{-8}), Sabouraud dextrose agar for **fungi** (10^{-4}, 10^{-5}, 10^{-6}, and 10^{-7}), and tryptic soy agar for **other bacteria** (10^{-6}, 10^{-7}, 10^{-8}, and 10^{-9}). Be sure to use the correct medium for each type of microorganism.

4. Using a micropipette and aseptic technique, perform the dilution scheme illustrated in **Figure 38.2a**, transferring 1 ml between tubes. Then, distribute the proper amount of each soil dilution to the respective Petri plate to achieve the desired dilution. For example, for the 10^{-5} plate, transfer 0.1 ml from the 10^{-4} dilution tube.

5. Using a cell spreader, evenly distribute the diluted soil sample over the surface of the agar plate. Rotate the plate on the bench as you move the cell spreader in circles.

6. Invert the Petri plates and incubate for 7 days at room temperature. Observe daily for the appearance of colonies. Count the plates with fewer than 250 colonies but more than 25. Designate plates with over 250 colonies as too numerous to count (TNTC) and those with less than 25 colonies as too few to count (TFTC). Record your data in the report for this exercise.

7. Determine the number of respective microorganisms per milliliter of original culture (or in this case, gram of soil) as follows:

$$\text{Microorganisms per gram of soil} = \frac{\text{CFU/plate}}{\text{dilution used}}$$

For example, if 200 colonies were present on the 10^{-7} plate, the calculation would be

$$\text{Microorganisms per gram of soil} = \frac{200}{10^{-7}}$$

$$= 2.0 \times 10^9 \text{ microorganisms/gram of soil}$$

Disposal

When you are finished with all of the pipettes and plates, discard them in the designated place for sterilization and disposal.

Figure 38.2 Dilution scheme for Enumerating Soil Microorganisms. (a) Serial dilutions can be made in any sterile test tube. Make sure to use a fresh micropipette tip between each transfer. Mix the dilution tubes vigorously, but be careful to avoid spills. (b) Use the pictured scheme to plate the serial dilutions. When plating the dilution, transfer 0.1 ml from the dilution tube to the appropriate plate.

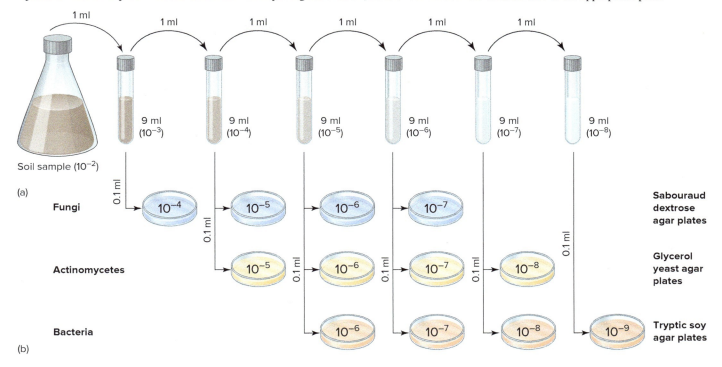

HELPFUL HINTS

- Optimal results in this exercise depend on thoroughly mixed soil samples containing evenly distributed microorganisms.
- Be sure to use the correct medium for each type of microorganism.
- Be careful when using the cell spreaders that you do not press too hard and gouge the surface of the agar plates.

Enumeration of Soil Microorganisms

NOTES

Laboratory Report 38

Name: _____

Date: _____

Lab Section: _____

Enumeration of Soil Microorganisms

1. Complete this table to determine the number of microorganisms per gram of soil.

Microorganism	Dilution	Number of Colonies	Microorganisms/Gram of Garden Soil
Actinomycetes	10^{-5}		
	10^{-6}		
	10^{-7}		
	10^{-8}		
Other bacteria	10^{-6}		
	10^{-7}		
	10^{-8}		
	10^{-9}		
Fungi	10^{-4}		
	10^{-5}		
	10^{-6}		
	10^{-7}		

2. For each type of medium, sketch the appearance of a plate that has well-isolated colonies.

3. In summary, what is the major purpose of this experiment?

ASSESSMENT
Critical Thinking and Learning Outcomes Review

1. Why were three different media used in this experiment? Look up the ingredients of each medium. What components make them selective and/or differential?

2. Would the same results be obtained in this experiment if the soil sample was collected during a different time of the year? Explain your answer.

3. What general group (s) of soil bacteria cannot be determined using the media and procedures in this exercise? (*Hint:* Think about how you incubated your plates.)

4. What generalizations can you make from this exercise with respect to your soil sample?

5. What are some physical features of soil that influence microbial populations?

6. How do actinomycetes differ from other groups of bacteria found in soil?

7. Why are different dilutions used for bacteria, fungi, and actinomycetes?

EXERCISE 39 Winogradsky Columns

SAFETY CONSIDERATIONS

- Slides and coverslips are made of glass. Dispose of any broken glass in the appropriate container.
- Dispose of all slides, coverslips, and syringes when finished, since they will contain enriched microorganisms from an environmental sample.

MATERIALS

- anaerobic mud
- soil
- pond water
- a thick glass 2-l graduated cylinder (or colorless plastic bottle cut to serve as a cylindrical container)
- calcium sulfate ($CaSO_4$)
- dipotassium phosphate (K_2HPO_4)
- cellulose
- aluminum foil
- spatulas
- syringes
- 50 ml Burk's nitrogen-free media broth in a 100-ml flask
- 50 ml *Beggiatoa* media broth in a 100-ml flask
- 50 ml *Chlorobium* media broth in a 100-ml flask
- 50 ml *Chromatium* media broth in a 100-ml flask
- microscopes
- glass slides
- coverslips
- container for biohazard waste
- safety glasses
- gloves
- lab coat

LEARNING OUTCOMES

Upon completion of this exercise, students will demonstrate the ability to

1. Distinguish the effect of multiple environmental factors on the metabolic needs and diversity of microbes
2. Apply classical microbial enrichment techniques using environmental samples

SUGGESTED READING IN TEXTBOOK

1. Photosynthetic Bacteria Are Diverse, section 20.4; see also table 20.1 and figures 20.3, and 20.4.
2. Class *Alphaproteobacteria* Includes Many Oligotrophs, section 21.1; see also table 21.1, and figures 21.2, 21.3, 21.20 and 21.21.
3. Microorganisms in Marine Ecosystems, section 29.2; see also figure 29.4.

Pronunciation Guide

- *Beggiatoa* (be-jee-ah-TO-ah)
- *Chlorobium* (klo-RO-be-um)
- *Chromatium* (kro-MA-te-um)
- *Desulfovibrio* (de-sul-fo-VEEB-reo)

PRINCIPLES

Specific microbes thrive under specific environmental conditions, and these environments can be dramatically changed by the microbes that inhabit them. In the 1880s, Russian microbiologist Sergei Winogradsky, a pioneer in the field of microbial ecology, came up with a very simple experimental set-up that would allow him to study and culture a broad diversity of microorganisms involved in the sulfur cycle. In this exercise you will get a chance to replicate this experiment by constructing a **Winogradsky column**. By combining sediment and water from a pond or marsh in a sealed glass column facing a light source, it is possible to create a vertical gradient of oxygen that will lead to very different processes depending on light and oxygen availability (**Figure 39.1a**).

At the top of the column, oxygenic photosynthesis will be the dominant process, mostly carried out by Cyanobacteria and diatoms (**Figure 39.1b**). Also close to the surface, a variety of sulfur-oxidizing bacteria will

Figure 39.1 **The Winogradsky Column.** (**a**) Schematic representing the main types of metabolism and groups of organisms responsible for these reactions, including organisms that we will enrich for in this exercise. (**b**) Cyanobacteria, like the genus *Anabaena* shown here, are among the main photoautotrophs found at the top of the column. (**c**) *Beggiatoa* is a genus of filamentous sulfur-oxidizing bacteria also found toward the top of the column. (**d**) *Chromatium* is a genus of purple sulfur bacteria that is able to store sulfur in intracellular globules. (**e**) *Chlorobium* is a group of anaerobic green sulfur bacteria able to harvest light through vesicles attached to the membrane called chlorosomes. *(b: Science History Images/Alamy Stock Photo; c: ©Roger Burks/University of California at Riverside and Mark Schneegurt/Wichita State University, and Cyanosite (www-cyanosite.bio.purdue.edu); d: James T. Staley)*

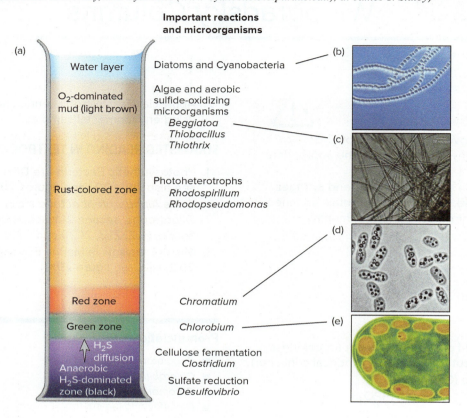

be able to use hydrogen sulfide (H_2S) produced by sulfate reduction at the bottom of the column. These include bacteria belonging to the genera *Thiobacillus*, *Thiothrix*, and the filamentous *Beggiatoa* (**Figure 39.1c**), which tends to produce a white biofilm. In the middle of the column a variety of photoheterotrophs and photolithotrophs will thrive on the side of the column that is facing the light producing a rust colored zone. These include sulfur-oxidizing bacteria of the genus *Chromatium* and *Chlorobium* that will create distinct red and green layers in the column (**Figure 39.1d,e**), respectively. Cellulose added to the column is used as a carbon source for anaerobic cellulose degradation and fermentation carried mostly by members of the genus *Clostridium* at the bottom of the column. Sulfate reduction is another important anaerobic process that takes place at the bottom of the column. This form of anaerobic respiration is carried out by heterotrophs like *Desulfovibrio*, which are able to use the added sulfate as a terminal electron acceptor and generate hydrogen sulfide that diffuses to the top of the column.

At the end of this exercise, you will have created a self-sustaining stratified microbial ecosystem able to recycle sulfur that can be maintained for years with the proper exposure to light and addition of water. Considering the metabolic diversity of microorganisms enriched by the Winogradsky column, we will also be able to use enrichment techniques targeting some of these specific types of metabolism and observe their morphological diversity under the microscope.

Procedure

First Period

1. Label a thick glass 2-l graduated cylinder or a clear, topless plastic bottle with your initials.
2. Weigh 5.0 g of cellulose, 2.5 g of dipotassium phosphate, and 5.0 g of calcium sulfate. Combine them with approximately 0.5 l of anaerobic mud and 0.5 l of soil and mix into a thick slurry with a large spatula.

3. Pack the mixture tightly to the bottom of the tube until you reach about half of the height of the tube.
4. Pour 2 to 3 inches of pond water over the top of the mud slurry.
5. Cover the top with a piece of aluminum foil to minimize evaporation.
6. Incubate in front of a window with direct sunlight or in front of a lamp. Over the next 5–10 weeks, take breaks during the lab to observe our columns and record any changes that occur.

Second Period

1. Carefully examine your Winogradsky column periodically on the weekly basis, until you and your instructor determine that clear layers have developed along the column (approximately 5-10 weeks).
2. Examine the different layers that have formed and record your observations in the report for this exercise. Make a note of any changes in color that you find in the microaerophilic and anaerobic sections of the column, particularly of the side facing the sun. In particular, look for white, red, purple, and green zones.
3. With a syringe collect a little water from the very top of the column in the area that was set facing the sun. Transfer 1 ml of the water from the top of the column to a flask with Burk's nitrogen-free medium and incubate in front of the light. Then, transfer just a drop of water from the top of the column you collect to a slide and place a cover slip to make observations under the microscope.
4. With sampling spatulas, collect samples from the zone you identified in step 2 that has a white film on the interface between the sediment and the glass. Transfer about 1 ml of this mud to a flask with *Beggiatoa* medium using aseptic technique and incubate at room temperature in the dark. Then prepare slides with *very small* portions of the white biofilm.
5. With different sampling spatulas, collect samples from the zone you identified as red (or purple) and green toward the bottom of the column.

Transfer about 1 ml of mud from the red/purple zone to a flask with *Chromatium* medium, and transfer 1 ml of mud from the green zone to a flask with the *Chlorobium* medium. Both flasks should be incubated in front of a light source. If possible, prepare slides with **very small** portions from these samples to observe under the microscope. You can resuspend them in deionized water on a separate slide before transferring them to the slide you will use for visualizing.
6. Record all your observations.

Third Period

1. Observe for changes in color in all of your media each lab period.
2. Once each medium shows increases in turbidity accompanied by appropriate changes of color, collect a sample with a loop and place in a microscope slide with a cover slip for each of the four enrichments.
3. Record all your microscopic observations in the lab report for this exercise.

Disposal

Flasks, pipettes, spatulas, and syringes should be discarded in the designated place for sterilization and disposal.

> **HELPFUL HINTS**
> - If you are making the column using a colorless plastic bottle, the cut should be made at the widest spot of the bottle.
> - Make sure to supplement the top water with deionized water during the enrichment process if any evaporation has occurred.
> - Given the metabolic diversity of organisms you will be enriching for in this experiment, growth rates will vary greatly and cell density will develop at a different pace between all the enrichment flasks.

NOTES

Laboratory Report 39

Name: _____

Date: _____

Lab Section: _____

Winogradsky Columns

1. Indicate in the Winogradsky column below the changes you observed in your column by indicating color and possible organisms found in each zone.

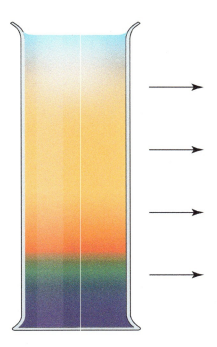

→ _____

→ _____

→ _____

→ _____

2. Draw a representative field of your microscopic observations for each of the enrichment media.

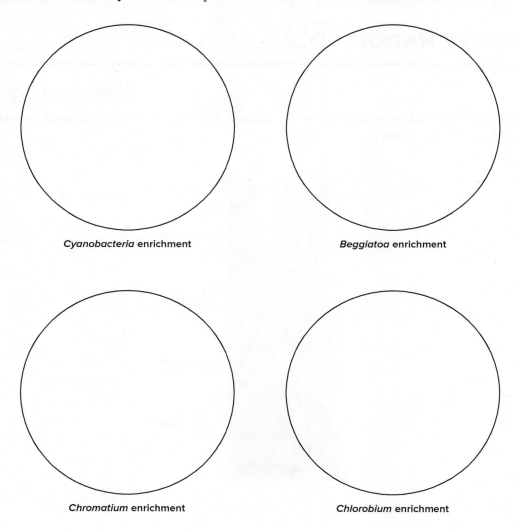

Cyanobacteria enrichment

Beggiatoa enrichment

Chromatium enrichment

Chlorobium enrichment

3. In summary, what is the major purpose of this exercise?

ASSESSMENT
Critical Thinking and Learning Outcomes Review

1. Are other columns from the lab different from yours? What could account for these differences?

2. Based on where specific microorganisms grow, can you estimate where oxygen is absent on your Winogradsky column? Explain.

3. What is the role of cellulose and calcium sulfate in the column?

4. If you could use ^{13}C-labeled cellulose, provide at least two examples of where you would expect to find ^{13}C-labeled carbon in the column.

5. Why are the light or dark conditions important for the enrichment flasks you prepared?

EXERCISE 40 Bioluminescence

SAFETY CONSIDERATIONS
- Be careful with the Bunsen burner flame.

MATERIALS
- 24 hour tryptic soy broth cultures of *Vibrio fischeri*
- sea water complete medium agar plates
- sea water complete medium broth
- screw-cap bacteriological culture tubes, 20 × 150 mm, 34 ml
- parafilm
- Bunsen burner
- 5 ml pipette with pipettor
- inoculating loops
- spreaders
- container for biohazard waste
- safety glasses
- disposable gloves
- lab coat

LEARNING OUTCOMES
Upon completion of this exercise, students will demonstrate the ability to
1. Observe bioluminescence by a marine bacterium
2. Explain the role of cell density in cell-cell communication and quorum sensing

SUGGESTED READING IN TEXTBOOK
1. Microbial Growth in Natural Environments, section 7.6: see also figures 7.26 and 7.27.
2. Bacteria Combine Several Regulatory Mechanisms to Control Complex Cellular Processes, section 14.6: see also figures 14.23 and 14.24.
3. Class *Gammaproteobacteria* is the Largest Bacterial Class, section 21.2: see also table 21.3.

Pronunciation Guide
- *Vibrio fischeri* (VIB-rio FISH-er-i)

Why Are the Following Microorganisms Used in This Exercise?

By culturing the marine bacterium *Vibrio fischeri* on complete medium sea water agar plates, the biochemical reaction known as bioluminescence can be observed through the presence of glowing green and bluish-green colors. Marine bioluminescent *Vibrio* species are model organisms for studying quorum sensing and have improved our understanding of quorum sensing in pathogenic *Vibrio* species, like *V. cholerae*.

PRINCIPLES

Several species in the marine genera *Vibrio* and *Photobacterium* can emit light of a green to blue-green color. This phenomenon is called **bioluminescence** and plays an important role in the symbiotic relationship between these bacteria and specific types of marine fauna. For example, the genus *Photobacterium* lives in association with the flashlight fish, allowing it to receive nutrients from the fish while it provides a unique device for frightening off would-be predators.

The enzyme luciferase catalyzes the reaction and uses reduced flavin mononucleotide, in the presence of molecular oxygen, and a long-chain aldehyde (glyceraldehyde) as substrates as indicated in the equation below. In the process, outer electrons surrounding

FMNH$_2$ become excited. Light is emitted when the electronically excited FMNH$_2$ returns to its ground state (FMN).

$$FMNH_2 + O_2 + RCHO \xrightarrow{\text{luciferase}} FMN + H_2O + RCOOH + \text{light}$$

Studies on the regulation of luciferase synthesis have shown that the genes encoding it are only expressed when the density of the bacterial population in the host organism reaches a critical point. This phenomenon, called **quorum sensing**, is now recognized as an important mechanism used by a variety of Gram-negative and Gram-positive bacteria to regulate the expression of certain genes.

In this lab, we will work with *Vibrio fischeri*, a bioluminescent bacterium that inhabits the light organ of the bobtail squid and some species of fish. In the case of *V. fischeri*, cell-cell communication through quorum sensing allows for the increased expression of the genes involved in the production of luciferase and bioluminescence at high cell densities (**Figure 40.1**). An autoinducer (AI) molecule produced by *V. fischeri* called N-acyl homoserine lactone (AHL) activates a transcriptional regulator (LuxR), increasing the expression of the necessary structural genes for the production of luciferase (*luxCDABEG*) and the expression of *luxI* for the production of additional AHL molecules. Quorum sensing has many roles in other species. For example, other *Vibrio* species, like *V. cholerae* and *V. parahemolyticus*, are important pathogens that regulate the production of virulence factors through very similar quorum sensing mechanisms.

Procedure

First Period

1. Obtain an overnight culture of *Vibrio fischeri*. Using an inoculating loop, aseptically transfer some of this culture onto 6 sea water complete medium agar plates. On 5 of the plates, streak for isolated colonies. On the remaining plate, draw a design. Be creative! Use the agar as your canvas and the bacteria as your paint.
2. Incubate the plates for 12–24 hours at 30°C.

Second Period

1. Remove the plates from the incubator and take them into a dark room. Observe the plates for light emission. It may take a while for your eyes to adjust to the darkened room and be able to see the bioluminescence. Try to look at the plates frequently in the first 24 hours of incubation because sometimes the bacteria are luminous for only a few hours (**Figure 40.2**).
2. After making your observations, transfer 2 ml of sea water complete medium broth to the plates

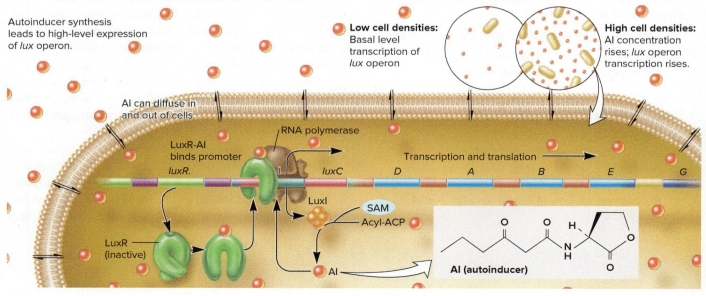

Figure 40.1 Quorum Sensing in *Vibrio fischeri*. Changes in expression of luciferase genes in response to the concentration of an autoinducer (AI). At high cell densities, transcription of the *lux* operon increases, increasing the expression of luciferase.

288   Environmental and Food Microbiology

Figure 40.2 Bioluminescence on an Agar Plate. A plate culture of a bioluminescent *Vibrio*. (*Javier Izquierdo/McGraw Hill*)

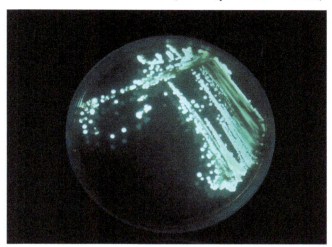

where you have streaked *V. fischeri* for single colonies. Use a sterile cell spreader to mix the streaked culture with the newly added liquid media.

3. Label two screw-cap culture tubes with the two levels of cell density you will use, one as 1× (low density) and the other as 4× (high density). Using a pipettor, transfer 2 ml of the broth-bacteria suspension from one plate to the 1× screw-cap culture tube. From the remaining 4 plates to the 4× screw-cap culture tube (a total of 8 ml).
4. Bring the volume of each tube to 22 ml, by adding 20 ml of sterile sea water complete medium broth to the 1× tube and 14 ml of sterile sea water complete medium broth to the 4× tube. Close tightly and seal the culture tubes with parafilm.
5. In a dark room, take note of the relative levels of brightness in each tube. After 1 minute, invert the tubes gently back and forth to observe how bioluminescence changes in the presence of oxygen and how different concentrations of cells affect bioluminescence. Any observable difference should be apparent in the first 5–10 minutes. Record your observations in the report for this exercise.

Disposal

When you are finished with all of the plates, discard them in the designated place for sterilization and disposal.

> **HELPFUL HINTS**
>
> - Make observations of your plates at multiple time points in the first 24 hours of incubation since bioluminescence will decrease over time.
> - Make sure to securely close the cap and seal with parafilm before inverting the tubes to avoid spills.

NOTES

Laboratory Report 40

Name: _____

Date: _____

Lab Section: _____

Bioluminescence

1. Make a drawing of the patterns of bioluminescence in one of your plates in the plate below. Are your cells equally bioluminescent throughout the whole plate?

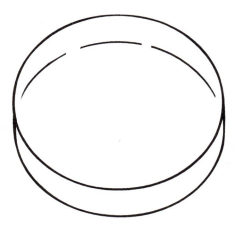

2. Compare the brightness observed in both screw-cap tubes in the following table.

	Brightness?	Does Brightness Change with Tube Inversion?
Low density (1×)		
High density (4×)		

3. What is the purpose of this exercise?

291

ASSESSMENT
Critical Thinking and Learning Outcomes Review

1. What is the relationship between bioluminescence and quorum sensing?

2. How do some species of *Vibrio* and *Photobacterium* emit a blue-green light?

3. Why is bioluminescence brighter closer to the bubble in the sealed screw-cap tube?

4. If your instructor provided you with a mutant *V. fischeri* strain without the ability to make the autoinducer AI-1 but an intact LuxR and *lux* operon, what would you expect to see in a plate with wild-type *V. fischeri* and this mutant plated side by side?

5. What if the LuxR protein was mutated and was not able to bind DNA?

EXERCISE 41

Plate Counts and Quality Assessment of Milk

SAFETY CONSIDERATIONS

- Be careful with the Bunsen burner flame and the boiling water bath.
- Keep all culture tubes upright in a test-tube rack.

MATERIALS

- pasteurized milk sample
- unpasteurized milk may be obtained from any dairy farm. If a sample is not available, pasteurized milk incubated at 37°C may be used or the instructor may wish to contaminate the sample with coliforms such as *Escherichia coli, Klebsiella pneumoniae,* or *Enterobacter aerogenes*
- 30°C incubator
- 15-ml and 5-ml violet red bile agar (VRBA) plates
- 9-ml and 99-ml saline blanks
- wax pencil or lab marker
- methylene blue solution (1/25,000)
- screw-cap test tubes
- sterile 10-ml and 1-ml pipettes with pipettor
- 37°C water bath
- Bunsen burner
- test-tube rack
- colony counter
- container for biohazard waste
- safety glasses
- disposable gloves
- lab coat

LEARNING OUTCOMES

Upon completion of this exercise, students will demonstrate the ability to

1. Explain the purpose of pasteurization
2. Determine the sanitary quality of milk by performing a coliform analysis and methylene blue reductase test

SUGGESTED READING IN TEXTBOOK

1. Microbes Can Be Controlled by Physical Means, section 8.2.
2. Microbiology of Fermented Foods: Beer, Cheese, and Much More, section 40.5; see also figure 40.6.
3. Food-Borne Disease Outbreaks, section 40.3.

Pronunciation Guide

- *Enterobacter aerogenes* (en-ter-oh-BAK-ter a-RAH-jen-eez)
- *Escherichia coli* (esh-er-I-ke-a KOH-lie)
- *Klebsiella pneumoniae* (kleb-se-EL-lah nu-MO-ne-ah)
- *Lactotoccus lactis* (lak-toh-KOK-kus LAK-tees)

Medical Application

Milk is normally sterile as secreted by the lactating glands of healthy animals. From that point on, it is subjected to contamination from two major sources: (1) the normal flora of the mammary ducts, and (2) flora of the external environment, including the hands of milkers, milking machinery, utensils, and the animal's coat.

There are a number of human diseases that might become milkborne via contaminated milk handlers, including streptococcal infections, shigellosis, and salmonellosis. In addition, unpasteurized milk has also been reported to cause tuberculosis, brucellosis, listeriosis and yersiniosis.

PRINCIPLES

Pasteurization is a means of processing raw milk, before it is distributed, to assure that it is relatively free of bacteria and safe for human consumption. It is a heating process gentle enough to preserve the physical and nutrient properties of milk, but sufficient to destroy pathogenic microorganisms. The two methods most commonly used for pasteurization of milk are (1) heating at 62.9°C (145°F) for 30 minutes or (2) heating to 71.6°C (161°F) for a minimum of 15 seconds.

The presence of coliforms in milk (and milk products) is a major indicator of the sanitary quality of milk. Their presence can be determined by a coliform plate count. A high count means that there is the possibility of the presence of disease-causing bacteria (*Salmonella*, *Campylobacter*, *E.coli* O157:H7, *Listeria*). A low count decreases this possibility but does not completely guarantee the absence of disease-causing bacteria. In the first part of this exercise, each group of students will perform a plate count on a pasteurized and on an unpasteurized milk sample.

Milk that contains a large number of growing bacteria will have a lower concentration of O_2 compared to milk with few bacteria. This is because growing aerobic and facultatively anaerobic bacteria (*see Exercise 17*) use oxygen as a final electron acceptor in cellular respiration. The dye, **methylene blue**, is a redox indicator. It loses its blue color in an anaerobic environment and is reduced to leuco-methylene blue (**Figure 41.1**). As a result, the **methylene blue reductase test** can be used to rapidly screen the quality of milk for the load of coliforms and *Lactococcus lactis*, which are strong reducers of methylene blue and indicators of contamination. The larger the bacterial load, the more quickly the milk will spoil. The speed at which the reduction occurs (**Methylene *B*lue *R*eduction *T*ime [MBRT]**) and the blue color disappears indicates the quality of milk as follows:

a. Reduction within 30 minutes—very poor milk quality (class 4 milk)
b. Reduction between 30 minutes and 2 hours—poor milk quality (class 3 milk)
c. Reduction between 2 and 6 hours—fair quality (class 2 milk)
d. Reduction between 6 and 8 hours—good quality (class 1 milk)
e. No reduction of blue color over 8 hours—excellent quality of milk (class 1 milk)

In the second part of this exercise, each group of students will perform a methylene blue reductase test on a pasteurized and unpasteurized milk sample.

Procedure

Coliform Analysis

1. Shake the milk sample 25 times. Make dilutions of the pasteurized and unpasteurized milk samples as indicated in **Figure 41.2a**.
2. Use the wax pencil or lab marker to label the violet red bile agar (VRBA) plates with your name, date, the respective dilution, and either pasteurized or unpasteurized milk.
3. Pipette 0.1-ml milk aliquots of each dilution into the appropriate plates and spread with a sterile L-shaped spreader.
4. Incubate all plates at 30°C for 24 hours.

Figure 41.1 Methylene Blue Reduction Test for Bacteria.

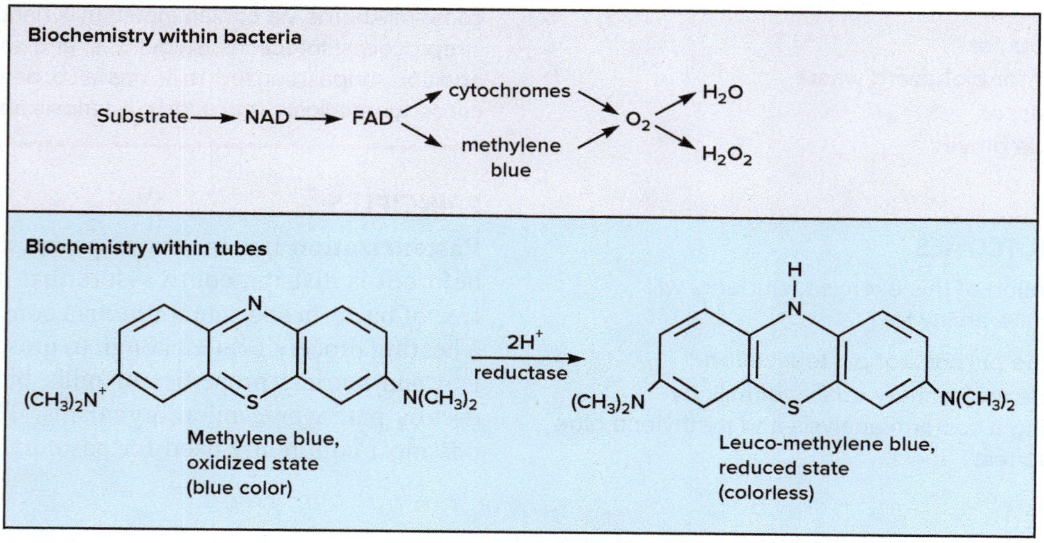

Figure 41.2 Examination of Milk for Bacteria. (a) Dilution series for coliform analysis of pasteurized and unpasteurized milk samples. (b) Methylene blue reductase test.

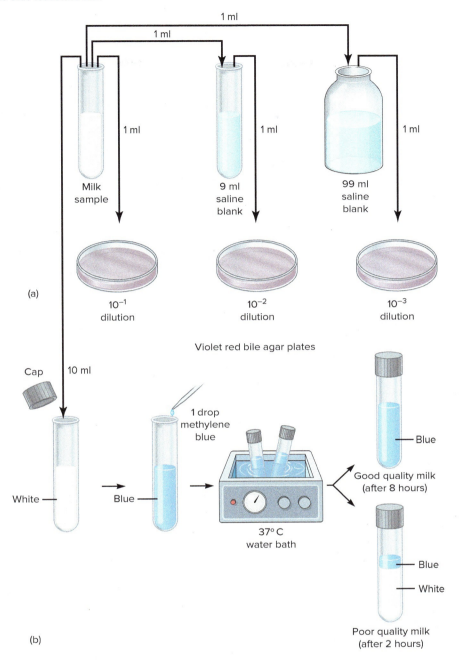

Second Period

1. For accuracy, select the plate that has between 25 and 250 colonies (Food and Drug Administration standards), which are deep red, and surrounded by a pink halo. Record these as the coliform count per milliliter of milk.

Methylene Blue Reductase Test

1. Label the two screw-cap tubes with your name and "pasteurized" and "unpasteurized," respectively.
2. Using the 10-ml pipette, transfer 10 ml of unpasteurized milk to one screw-cap tube and, with another pipette, 10 ml of pasteurized milk to the other tube (**Figure 41.2b**).
3. Add 1 drop of methylene blue to each tube.
4. Cap tightly and invert the tubes several times.
5. Place the tubes in a test-tube rack and place the rack in the 37°C water bath. After a 5-minute incubation, remove the tubes from the water bath and invert several times to mix again.

6. Observe the tubes at 30-minute intervals for 8 hours. Reduction is demonstrated by a change in color of the milk sample from blue to white. When at least four-fifths of the tube has turned white, the end point of reduction has been reached, and the time should be recorded (**Figure 41.3**).

7. Record your results and the class of milk in the report for this exercise.

Disposal

When you are finished with all of the pipettes, tubes, and saline blanks, discard them in the designated place for sterilization and disposal.

Figure 41.3 **Methylene Blue Reductase Test.** The tube on the left illustrates what can be expected when this test has just begun. The blue color indicates that dissolved oxygen is present in the milk. The lack of color in the middle tube indicates that the oxygen has been depleted due to bacterial metabolism. The tube on the right shows the endpoint of the reaction, when 80% of the tube has turned white. *(Barry Chess)*

Laboratory Report 41

Name: _____

Date: _____

Lab Section: _____

Plate Counts and Quality Assessment of Milk

1. Data from coliform analysis.

Milk Sample	Number of Colonies		
	Dilution		
	10^{-1}	10^{-2}	10^{-3}
Pasteurized	_____	_____	_____
Unpasteurized	_____	_____	_____

Number of colonies per milliliter of pasteurized milk _____

Number of colonies per milliliter of unpasteurized milk _____

Calculations:

2. Data from methylene blue reductase analysis.

	Unpasteurized Milk	Pasteurized Milk
Methylene blue reduction time	_____	_____
Milk class	_____	_____

3. In summary, what is the major purpose of this experiment?

ASSESSMENT
Critical Thinking and Learning Outcomes Review

1. What is the function of the methylene blue in the reductase test for milk quality?

2. Why does milk sour when it is not refrigerated?

3. How can a milk sample be contaminated by humans?

4. Why is milk pasteurized and not sterilized?

5. What are the differences in methylene blue reduction time between the different classes of milk? What do these differences signify?

6. As a bacteriological medium, how does milk differ from water?

7. What are some bacteria normally found in milk?

PART 7
Medical Microbiology

Medical microbiology deals with the diagnosis, treatment, and prevention of infectious diseases caused by all groups of microorganisms. As such, a complete coverage of these groups is not feasible in a few exercises. This part of the manual has, therefore, been constructed to enable you to perform some of the routine techniques that are used in a clinical microbiology laboratory. For example, you will isolate some of the microorganisms that are part of the normal microbiota by using aseptic technique and selective and differential media. In subsequent exercises, you will use various techniques to examine, isolate, and identify some of the more prevalent and potentially pathogenic microorganisms. As these exercises are performed, keep in mind that very similar if not identical procedures are routinely done for the many other microorganisms that are associated with the human body.

When carrying out the following exercises, view all microorganisms as potential pathogens. Pay attention to the safety considerations for each exercise in this section and listen carefully to the protocols outlined by your instructor. It is absolutely essential to follow all safety precautions to prevent infecting yourself or other members of your group with the microbes used in the following exercises.

After completing the exercises in Part 7, you will meet the following American Society for Microbiology Core Curriculum Laboratory Skills:

- using a bright-field microscope to view and interpret slides;
- properly preparing slides for microbiological examination;
- proper use of aseptic technique for transferring and handling of microorganisms and instruments;
- use of appropriate microbiological media and test systems; and
- the correct use of standard laboratory equipment.

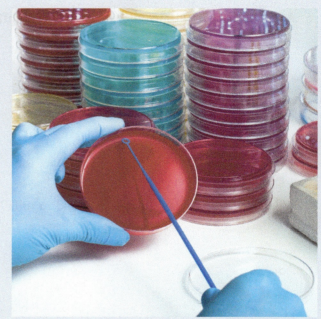

angellodeco/Shutterstock

To cultivate and identify infectious microbes from patient samples, clinical microbiologists rely on selective and differential media. These media formulations are designed to rapidly and clearly report on an organism's unique metabolism. The metabolic properties of infectious microbes have been catalogued through years of study. By comparing the metabolic signature of a clinical isolate to this catalogue, identification of an infectious microbe is possible, which allows a physician to design an appropriate treatment plan.

EXERCISE 42 Staphylococci

SAFETY CONSIDERATIONS

- **Staphylococci are potential pathogens (BSL2);** use aseptic technique throughout this experiment.
- Keep your hands scrupulously clean.
- If you have any minor cuts or scratches, they should be protected with bandages.
- While in the laboratory, keep your hands away from your face.
- Keep all tubes upright in a test-tube rack.

MATERIALS

- suggested bacterial strains: *Staphylococcus aureus, Staphylococcus epidermidis,* and *Staphylococcus saprophyticus*
- Vogel-Johnson agar plates
- Mueller-Hinton plate
- mannitol salt agar plate
- sterile cotton swabs
- tryptic soy broth tube
- blood agar plates
- citrated rabbit plasma tubes
- DNase test agar plate
- nutrient gelatin deep
- novobiocin antibiotic discs
- forceps
- 37°C incubator
- Bunsen burner
- wax pencil or lab marker
- inoculating loop
- container for biohazard waste
- safety glasses
- disposable gloves
- lab coat

LEARNING OUTCOMES

Upon completion of this exercise, students will demonstrate the ability to

1. Describe the medical significance of the staphylococci
2. Distinguish the pathogenic species of staphylococci from the nonpathogenic species using biochemical tests

SUGGESTED READING IN TEXTBOOK

1. Phylum Firmicutes, Class *Bacilli*: Aerobic Endospore Forming Bacteria, section 22.2; see also table 22.3 and figure 22.16.
2. Direct Contact Diseases Can Be Caused By Bacteria, section 38.3, Staphylococcal Diseases; see also figures 38.21–38.24 and table 38.2.

Pronunciation Guide

- *Staphylococcus aureus* (staf-il-oh-KOK-kus ORE-ee-us)
- *Staphylococcus epidermidis* (staf-il-oh-KOK-kus e-pee-DER-meh-diss)
- *Staphylococcus saprophyticus* (staf-il-oh-KOK-kus sa-pro-FIT-e-kus)

Medical Application

The pathogenic staphylococcus most often encountered is *Staphylococcus aureus,* which causes common food poisoning, osteomyelitis, bacterial pneumonia, and other diseases. The coagulase test is used to distinguish between pathogenic and nonpathogenic members of the genus *Staphylococcus*. Strains of *S. aureus* are coagulase positive, whereas the nonpathogenic *S. epidermidis* and *S. saprophyticus* are coagulase negative. In some cases, the coagulase-negative staphylococci can cause infections, particularly in persons with compromised immune systems.

PRINCIPLES

The genus *Staphylococcus* consists of Gram-positive cocci, about 1 μm in diameter, usually in irregular clusters (**Figure 42.1**). They are nonmotile and nonsporing. Members of this genus are facultatively anaerobic. Colonies are round, convex, mucoid, and display a variety of pigments (**Figure 42.2**). They are chemoorganotrophic, requiring nutritionally rich media. Staphylococci have respiratory and fermentative metabolism, producing acid but no gas from carbohydrates. They are able to grow on nutrient agar with 5% NaCl and are usually positive for catalase. Oxidase-negative members contain cytochromes and are Voges-Proskauer positive. Most species reduce nitrate to nitrite. Optimum growth is at 37°C. Staphylococci are commensals on the skin and in the human mouth and upper respiratory tract. They can be human pathogens.

S. aureus is the most important clinical member of this genus. It may be isolated from the skin or mucous membranes of the body. It can cause various infections (e.g., carbuncles, abscesses, pneumonia, endocarditis, food poisoning, toxic shock syndrome) throughout the body. Increasingly, strains of *S. aureus* are resistant to antibiotics. This includes resistance to beta-lactam antibiotics like methicillin. Methicillin-resistant *S. aureus* (MRSA) strains are particularly problematic in clinical settings.

Figure 42.1 **Morphology of staphylococci.** Gram stained smear prepared from a staphylococcal culture illustrating the characteristic grape-like clusters of Gram-positive cocci. *(Toeytoey2530/iStock/Getty Images)*

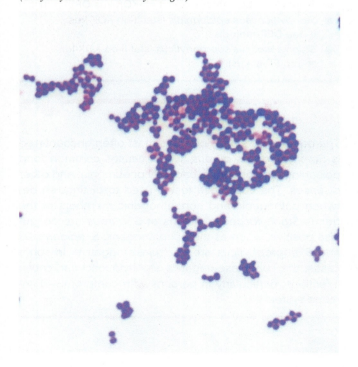

Figure 42.2 ***Staphylococcus aureus* on Chocolate Agar.** Chocolate agar is a rich medium that supports growth of even the most fastidious bacteria. Colonies of *S. aureus* (top half) are typically smooth and have a yellow pigmentation, while colonies of *S. epidermidis* (bottom half) are white. *(Nathan Rigel/McGraw Hill)*

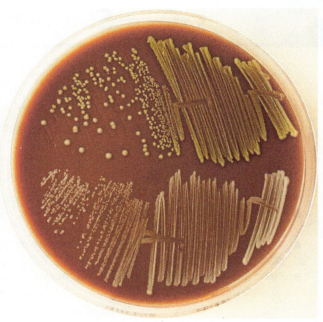

Other species of *Staphylococcus* can also cause disease. *S. epidermidis* is an opportunistic pathogen causing skin lesions and endocarditis. It is part of the normal microbiota of the skin. Additionally, *S. saprophyticus* can cause urinary tract infections, especially in females. Both of these species are coagulase negative, but they can be differentiated from each other by using different biochemical tests.

Table 42.1 lists characteristics used to differentiate between the three most commonly isolated *Staphylococcus* species. One of the simplest ways to isolate and identify *S. aureus* and other Gram-positive cocci of medical importance is through the use of blood agar plates. Observing hemolysis patterns (i.e., destruction of red blood cells) is particularly useful in identification (**Figure 42.3**). In **alpha-hemolysis,** a zone of greenish coloration with an indistinct margin forms around colonies growing on blood agar. The green color results from partial decomposition of hemoglobin. **Beta-hemolysis** is a sharply defined zone of clear hemolysis with no greenish tinge surrounding the colony. In this case, hemolysis is complete. Of course, some bacteria cannot cause lysis of red blood cells and thus produce no pattern of hemolysis on blood agar. This is called **gamma-hemolysis**.

Table 42.1	Some Characteristics of *Staphylococcus* Species		
	Species		
Characteristic	*S. aureus*	*S. epidermidis*	*S. saprophyticus*
Coagulase	+	–	–
Blood agar lysis	Beta	Gamma	Gamma
Nitrate reduction	+	+	–
Acid produced aerobically			
Mannitol	+	–	d
D-trehalose	+	–	+
Sucrose	+	+	+
D-xylose	–	–	–
Novobiocin resistance at an MIC of 5 µg/ml	S	S	R
DNase activity	+	–/weak	–
Gelatinase activity	+	–	–
Pigmentation	Yellow	White	White, yellow

Figure 42.3 **Patterns of Hemolysis Observed on Blood Agar.** Blood agar plates demonstrating (**a**) alpha, (**b**) beta, and (**c**) gamma hemolysis. Note the difference in the greenish appearance of incomplete hemolysis in (a) versus complete destruction of red blood cells in (b). *(a–c: Nathan Rigel/McGraw Hill)*

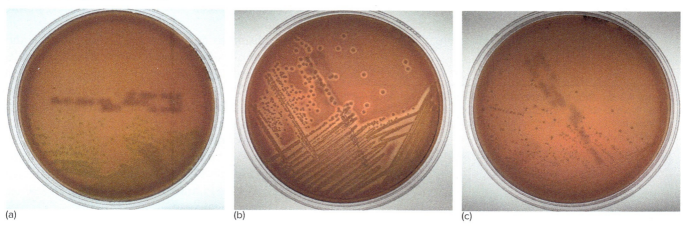

(a)   (b)   (c)

Another important tool used to identify staphylococci is mannitol salt agar (MSA). Mannitol (a 6-carbon sugar alcohol formed by the reduction of mannose) is fermented by *S. aureus*, which turns the pH indicator Phenol Red from a pink to a yellow color due to the decrease in pH. (**Figure 42.4**). Staphylococci that do not use mannitol (e.g., *S. epidermidis* and *S. saprophyticus*) produce no color change (**Figure 42.4**).

Vogel-Johnson agar is another selective and differential medium used to identify *S. aureus* (**Figure 42.5**). *S. aureus* grows on this medium and produces black colonies from the reduction of potassium tellurite to free tellurium. A yellowing of the medium indicates fermentation of mannitol and the production of an acid pH. The high concentration of lithium chloride, tellurite, and glycine inhibits the growth of most other organisms, including other species of staphylococci.

Coagulases are enzymes that cause blood plasma to clot. Although coagulase activity is not required for pathogenicity, this enzyme is a good indicator of the potential of staphylococci to cause severe disease. Inside a human host, coagulase-producing staphylococci form a fibrin clot around themselves to avoid attack by the host's defenses. In the coagulase test used in this

Figure 42.4 *Staphylococcus aureus* and *Staphylococcus epidermidis* on **Mannitol Salt Agar.** Notice that the medium has turned yellow around the growing bacteria since the bacteria are able to ferment the mannitol and produce an acid pH. *(Javier Izquierdo/McGraw Hill)*

Figure 42.5 *Staphylococcus aureus* on **Vogel-Johnson agar.** Note the black pigmentation of the colonies and the yellow color of the medium in the area nearest the inoculum. *(Nathan Rigel/McGraw Hill)*

Figure 42.6 **Coagulase Test.** Coagulase-producing staphylococci form a clot when grown in rabbit plasma, whereas coagulase-negative staphylococci (*S. saprophyticus* [inset]) do not form a clot. *(Lisa Burgess/McGraw-Hill)*

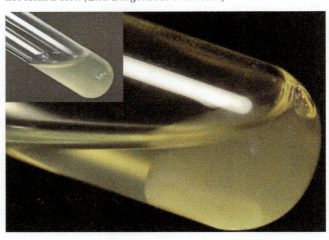

exercise (**Figure 42.6**), coagulase-positive staphylococci will cause rabbit plasma to clot by using coagulase to initiate the clotting cascade. Citrate and EDTA are usually added to act as anticoagulants to help prevent false-positive results. Cultures should be considered coagulase negative if they fail to induce clotting after 4 hours.

In addition to coagulase production, most pathogenic strains of staphylococci produce a nuclease called **DNase**. DNase degrades host DNA into its constituent nucleotides. To demonstrate the presence of DNase, DNase agar plates containing toluidine blue are streaked with staphylococci. After 24 hours of incubation, DNase-positive cultures capable of DNA hydrolysis will show a faintly pink or clear halo around the inoculum (**Figure 42.7**). The absence of a pink/clear halo indicates the culture does not produce DNase and is thus DNase negative.

This exercise is designed to illustrate (1) methods of culturing the staphylococci and (2) the clinical laboratory techniques that are used to distinguish pathogenic from nonpathogenic species.

Procedure

First Period

Mannitol Salt, Vogel-Johnson, Blood, and DNase Agars

1. Label your agar plates using a wax pencil or lab marker. Add your name, date, and organism being inoculated.

Figure 42.7 DNase Agar. DNase degradation on the left side of the plate and no degradation on the right side of the plate that has been inoculated with two different species of *Staphylococcus*. Notice the clear zone around the growth on the left. *(Javier Izquierdo/McGraw Hill)*

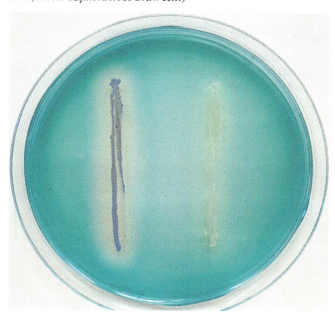

2. On one plate of each variety, streak *S. aureus* for single colonies. Do the same with *S. epidermidis* and *S. saprophyticus* on the other plates. Incubate the plates for 24 hours at 37°C.

Gelatinase Activity

1. Label three gelatin deeps with your name, date, and the bacterium to be inoculated. Label a fourth tube control.
2. Using aseptic technique and an inoculating needle, inoculate three of the deeps with the appropriate bacterium by stabbing the medium three-fourths of the way to the bottom of the tube.
3. Incubate the four tubes for at least 24 hours at 37°C.

Coagulase Test

1. Add 0.5 ml of citrated rabbit plasma to three small test tubes. With the wax pencil or lab marker, label the tubes with the respective bacteria, your name, and date.
2. Inoculate one tube with enough *S. aureus* colony material to make a cloudy suspension. Inoculate another tube with *S. epidermidis* or *S. saprophyticus*. (Alternatively, you can add about 5 drops of 18- to 24-hour broth culture to each tube.) Leave the last tube uninoculated and label it "control."
3. Incubate the tubes at 37°C for at least 3 hours. Afterward, examine the tubes for the presence or absence of clotting. A positive coagulase test is represented by any degree of clotting, from a loose clot suspended in plasma to a solid clot (Figure 42.7). For comparison, you should carefully observe the uninoculated control tube.

Novobiocin Sensitivity

1. Label a Mueller-Hinton agar plate with your name, date, and bacterium to be inoculated.
2. Inoculate the surface of a Mueller-Hinton plate with a sterile swab for susceptibility testing. Repeat the procedure for all three staphylococci.
3. Flame the tips of the forceps to sterilize them. Then, place a sterile disc impregnated with novobiocin onto the center of each plate.
4. Incubate each plate, disc side up, for 18 hours at 37°C.

Second Period

Mannitol Salt and Vogel-Johnson Agars

1. Observe your plates for growth. Note any color change to the surrounding medium.
2. Record your observations in the lab report section for this exercise.

Blood Agar

1. Observe your plates for colony pigmentation and hemolysis patterns. It might help to hold your plates up so that the light is in the background. If available, you can also use a lightbox.
2. Record your observations in the lab report section for this exercise.

DNase Agar

1. Examine your plates. Note that the DNase-producing colonies will be surrounded by a pink/clear zone and are thus DNase positive. The absence of a zone indicates the organism is DNase negative.
2. Based on your observations, determine the DNase reaction of each organism and record the results in the report for this exercise.

Gelatinase Activity

1. Remove the gelatin deep tubes from the incubator and place them in the refrigerator (at 4°C) for 30 minutes. Alternatively, you may place them in an ice water bath for 5 minutes.
2. When the bottom resolidifies, remove the tubes and gently slant them. Check to see if the surface of the medium is fluid or liquid. If the gelatin is liquid, this indicates that gelatin has been hydrolyzed by the bacterium. If no hydrolysis has occurred, the medium will remain semi-solid (i.e., will appear like the uninoculated control).
3. Record your observations in the lab report section for this exercise.

Novobiocin Sensitivity

1. Measure and record the diameter of the zone of growth inhibition around each antibiotic disc.
2. Use the following parameters to score sensitivity to novobiocin: 17 mm or less = resistant, 18–21 mm = intermediate, 22 mm or more = sensitive. Record your findings in the lab report for this exercise.

Disposal

When you are finished with all of the tubes, pipettes, swabs, and plates, discard them in the designated place for sterilization and disposal.

> **HELPFUL HINTS**
>
> - Be sure to carefully measure the diameter of inhibition zones to get accurate results.

Laboratory Report 42

Name: _____

Date: _____

Lab Section: _____

Staphylococci

1. Biochemical Tests

	S. aureus	S. epidermidis	S. saprophyticus
Mannitol fermentation			
Hemolysis pattern			
Coagulase activity			
DNase activity			
Gelatinase activity			
Colony pigmentation			
Novobiocin resistance			

2. Describe the appearance of each species of *Staphylococcus* on Vogel-Johnson agar plates. What do your observations tell you about the metabolic properties of each organism?

ASSESSMENT
Critical Thinking and Learning Outcomes Review

1. What are the three patterns of hemolysis observed on blood agar, and how are they different?

2. What test can differentiate the three major species of *Staphylococcus*?

3. How would you treat a person who has a carbuncle? Would this approach be similarly useful in treating staphylococcal food poisoning? Explain your answer.

4. Describe the cellular morphology and arrangement of staphylococci.

5. What is the principle behind the coagulase test?

6. How could you distinguish *S. epidermidis* from *S. saprophyticus*?

7. Why are staphylococcal infections frequent among hospital patients?

8. In summary, what is the major purpose of this experiment?

Medical Microbiology

EXERCISE 43 Pneumococcus

SAFETY CONSIDERATIONS

- Be careful with the Bunsen burner flame.
- **Streptococcus pneumoniae is a pathogen (BSL2);** use aseptic technique throughout this experiment.
- Keep all culture tubes upright in a test-tube rack.

MATERIALS

- suggested bacterial strains: *Streptococcus mitis* and *Streptococcus pneumoniae*
- blood agar plates
- chocolate agar plates
- glass slides
- 3% hydrogen peroxide
- inoculating loops
- optochin discs
- wax pencil or lab marker
- 10% sodium deoxycholate
- 0.85% saline (NaCl)
- Bunsen burner
- container for biohazard waste
- safety glasses
- disposable gloves
- lab coat

LEARNING OUTCOMES

Upon completion of this exercise, students will demonstrate the ability to

1. Culture pneumococci
2. Perform biochemical and serological diagnostic tests to identify pneumococci and differentiate them from other alpha-hemolytic streptococci

SUGGESTED READING IN TEXTBOOK

1. Phylum Firmicutes Class *Bacilli:* Aerobic Endospore Forming Bacteria, section 22.2; see also table 22.3 and figure 22.20.
2. Opportunistic Diseases Can Be Caused by Bacteria, section 38.1, Streptococcal Pneumonia.

Pronunciation Guide

- *Streptococcus mitis* (strep-to-KOK-us MY-tiss)
- *Streptococcus pneumoniae* (strep-to-KOK-us new-MOH-nee-eye)

Medical Application

Pneumococci are among the most important agents of bacterial pneumonia. This bacterium can cause an acute inflammation of the bronchial and/or alveolar membranes. When the alveoli are involved, fluid accumulation may rupture the alveolar membranes. Laboratory diagnosis is often made by isolating the causative agent from the patient's sputum sent for culture. *S. pneumoniae* can also be distinguished from other alpha-hemolytic streptococci (e.g., *Streptococcus mitis*) because it is lysed by bile salts and does not grow in the presence of optochin.

PRINCIPLES

S. pneumoniae (pneumococci) is the causative agent of many types of diseases (e.g., bacterial lobar pneumonia, conjunctivitis, otitis media, meningitis, peritonitis). Many normal individuals harbor this bacterium in their nasopharynx. *S. pneumoniae* is a Gram-positive

Figure 43.1 **Morphology of Pneumococci.** Gram stain of a smear illustrating several short chains and diplococci, characteristic of *Streptococcus pneumoniae*. *(Source: Dr. Mike Miller/CDC)*

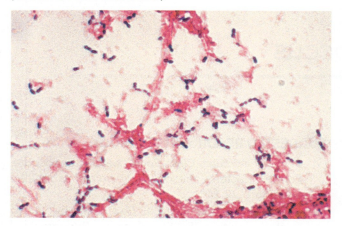

Table 43.1	Properties of *S. pneumoniae* and *S. mitis*	
Test	*S. pneumoniae*	*S. mitis*
Hemolysis	α	α
Bile solubility	+	−
Optochin sensitivity	+	−

coccus, usually arranged in pairs or short chains (**Figure 43.1**). The pneumococci are somewhat fastidious in their nutritional needs and do not survive well in competition with other microbes. The medium of choice for culturing pneumococci is blood agar or chocolate agar (**Figure 43.2**). Chocolate agar gets its name based on its appearance; this is misleading since it is blood agar where the red blood cells have been lysed by heating prior to casting the plate. It is a very rich source of nutrients and is used to cultivate fastidious organisms. Colonies of *S. pneumoniae* that form on this medium appear small and shiny with a dense center and a lighter, raised margin, giving them a dimpled appearance. Blood agar is also used to cultivate fastidious microbes, and it can also be used to detect the production of hemolytic enzymes. The hemolysis pattern on blood agar exhibited by pneumococci is **alpha-hemolysis** (a zone of greenish coloration occurs around the colonies). Some strains of *S. pneumoniae* produce a significant amount of capsule, which makes the colonies appear wet or slimy. This capsule is an important virulence factor, and is used for epidemiological studies to track disease outbreaks.

In contrast, *S. mitis* colonies are smaller, gray to whitish gray, and opaque with smooth edges. **Table 43.1** differentiates *S. pneumoniae* from *S. mitis*. *S. mitis* is part of the viridans streptococci. These commensals are part of the oral microbiota. They can cause a range of diseases, from dental caries to endocarditis. Since they display alpha-hemolysis on blood agar, it is important to be able to distinguish these streptococci from *S. pneumoniae*. These tests include **bile solubility** and **optochin sensitivity.** In the bile solubility test, pneumococci will dissolve (undergo lysis) in bile (10% sodium deoxycholate), whereas other alpha-hemolytic streptococci will not. In the optochin test, the pneumococci exhibit sensitivity to the reagent optochin (ethylhydrocupreine) and their growth will be inhibited, but other alpha-hemolytic streptococci are unaffected (Figure 43.2).

This exercise is designed to acquaint you with methods of culturing pneumococci and performing biochemical tests used to identify them.

Figure 43.2 **Optochin sensitivity on Blood Agar.** The organism on the left is sensitive to optochin, while the organism on the right is resistant. *(Source: Dr. Richard Facklam/CDC)*

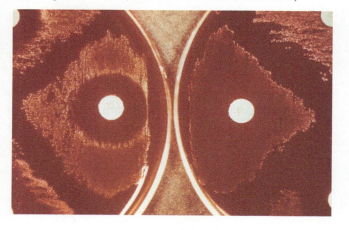

Procedure

First Period

Blood and Chocolate Agars

1. Label your blood and chocolate agar plates using a wax pencil or lab marker. Add your name, date, and organism being inoculated.
2. On one blood agar plate, streak *S. mitis* for single colonies. Do the same with *S. pneumoniae* on the other blood agar plate. Repeat this procedure for two chocolate agar plates. Incubate the plates for 24 hours at 37°C.

Optochin Sensitivity

1. Label a blood agar plate with your name, date, and bacterium to be inoculated.
2. Inoculate the surface of the plate with a sterile swab for susceptibility testing. Repeat the procedure for both streptococci.
3. Flame the tips of the forceps to sterilize them. Then, place a sterile disc impregnated with optochin onto the center of each plate.
4. Incubate each plate, disc side up, for 24 hours at 37°C.

Second Period

Blood and Chocolate Agar

1. Observe your plates for colony pigmentation (on chocolate agar) and hemolysis patterns (on blood agar). When scoring the hemolysis, it might help to hold your plates up so that the light is in the background. If available, you can also use a lightbox.
2. Record your observations in the lab report section for this exercise.

Optochin Sensitivity

1. Measure and record the diameter of the zone of growth inhibition around each optochin disc.
2. *S. pneumoniae* should have a zone of growth inhibition ≥14 mm in diameter. *S. mitis* should display resistance to optochin.

3. Record your findings in the lab report for this exercise.

Bile Solubility

1. Using the chocolate agar plates, place a drop of 10% deoxycholate on a well-isolated colony of *S. mitis* and *S. pneumoniae*.
2. Observe the colony carefully. If the colony dissolves or flattens within 5 minutes, it is bile-soluble. As a control, repeat this experiment using a drop of saline instead of a drop of deoxycholate.
3. Record your observations in the lab report for this exercise.

Disposal

When you are finished with all of the tubes, pipettes, swabs, and plates, discard them in the designated place for sterilization and disposal.

HELPFUL HINTS

- A candle jar (*Figure 45.4*) can be used to create a microaerophilic environment preferred by many pneumococci.

NOTES

Laboratory Report 43

Name: _____

Date: _____

Lab Section: _____

Pneumococcus

1. Describe the morphology of an *S. pneumoniae* colony.

2. Describe the morphology of an *S. mitis* colony.

3. Optochin susceptibility:
    a. *S. pneumoniae*

    b. *S. mitis*

4. Bile solubility:
    a. *S. pneumoniae*

    b. *S. mitis*

5. In summary, what is the major purpose of this experiment?

ASSESSMENT
Critical Thinking and Learning Outcomes Review

1. What differentiates alpha- and beta-hemolytic streptococci?

2. What does the optochin test demonstrate?

3. Do some research and describe the difference between blood and chocolate agar.

4. How can you differentiate between *S. mitis* and *S. pneumoniae*?

5. What does the bile solubility test indicate?

6. What type of hemolysis is produced by *S. pneumoniae*?

7. What role does a bacterial capsule play in an infection?

EXERCISE 44 Streptococci

SAFETY CONSIDERATIONS

- Be careful with the Bunsen burner flame.
- ***Streptococcus* spp. are potential pathogens (BSL-2).** Use aseptic technique throughout this exercise.
- Keep all culture tubes upright in a test-tube rack.

Legal Considerations

Either the course director or the laboratory instructor should check with the appropriate college or university officials (legal department) to see if it is considered an invasion of students' privacy to ask students to provide throat, skin, or rectal swabs as well as other personal information concerning the microbiota of their bodies.

MATERIALS

- suggested bacterial strains:
  1 tryptic soy broth culture of *Streptococcus equi* (Group C beta-hemolytic)
  1 tryptic soy broth culture of *Streptococcus bovis* (Group D, alpha-hemolytic)
  1 tryptic soy broth culture of *Streptococcus pyogenes* (Group A beta-hemolytic streptococci)
  1 tryptic soy broth culture of a non–Group A beta-hemolytic streptococci (e.g., *Streptococcus agalactiae*. *S. agalactiae* is a member of Group B).
  1 tryptic soy broth culture of *Staphylococcus aureus*
- serological test tubes
- 1-ml pipettes with pipettor
- 5-ml pipettes
- 10-ml pipette
- sheep blood agar plates
- bacitracin disks
- SXT disks
- bile esculin agar slants
- Gram stain materials
- Bunsen burner
- inoculating loop
- forceps
- REMEL BactiCard™ Strep rapid test for the presumptive identification of streptococci and related genera and Enterococcus/Group A Strep screen rapid tests; RIM® A.R.C. STREP A TEST
- container for biohazard waste
- safety glasses
- disposable gloves
- lab coat

LEARNING OUTCOMES

Upon completion of this exercise, students will demonstrate the ability to

1. Compare and contrast morphological and colonial characteristics of the streptococci as well as their medical significance
2. Perform several biochemical and serological tests to identify different *Streptococcus* species
3. Compare and contrast the alpha streptococci *S. bovis* with *S. pyogenes*
4. Demonstrate how bacitracin sensitivity, the CAMP reaction, and SXT sensitivity are used to differentiate the Lancefield groups of beta-hemolytic streptococci

SUGGESTED READING IN TEXTBOOK

1. Phylum Firmicutes, Class *Bacilli:* Aerobic Endospore-Forming Bacteria, section 22.2; see also table 22.3 and figures 22.19 and 22.20.
2. Bacteria Can Be Transmitted by Airborne Routes, section 38.1, Streptococcal Diseases (Group A); see also figures 38.8 and 38.9.

Pronunciation Guide

- *Staphylococcus aureus* (staf-il-oh-kok-kus ORE-ee-us)
- *Streptococcus* (strep-to-KOK-us)
- *S. agalactiae* (a-gal-ACT-e-a)

- *S. bovis* (bo-VIS)
- *S. equi* (e-QY)
- *S. mutans* (MYOO-tans)
- *S. pyogenes* (pi-OH-gen-eez)
- *S. salvarius* (sal-vah-REE-uss)

Medical Application

In the clinical laboratory, observation of hemolysis is an important first step in the identification among streptococcal species. Most alpha-hemolytic streptococci are members of the normal throat microbiota and do not need to be identified further when they are isolated from respiratory cultures. However, when beta-hemolytic streptococci are found in throat cultures, the laboratory must proceed with further testing to determine the antigenic group. The most important streptococcal group is Group A, which is responsible for streptococcal pharyngitis ("strep throat") and a variety of other serious skin and deep-tissue infections.

PRINCIPLES

The streptococci are Gram-positive cocci arranged in chains (**Figure 44.1**). Streptococci are classified according to their hemolytic activity, immunologic properties (the serological classification of Lancefield), and resistance to chemical and physical factors.

Figure 44.1 Morphology of Streptococci. Gram stain smear illustrating long chains (since cell division occurs in one plane) of Gram-positive cocci characteristic of streptococci. *(Javier Izquierdo/McGraw Hill)*

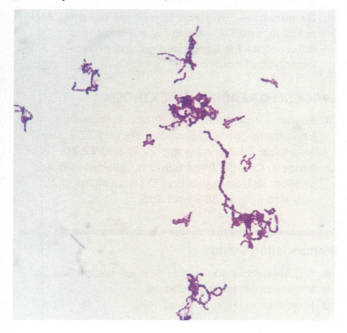

Figure 44.2 Streak Plate of Streptococci on Blood Agar. This blood agar plate illustrates the small, semitranslucent gray-white colonies of streptococci surrounded by zones of beta-hemolysis. *(Javier Izquierdo/McGraw Hill)*

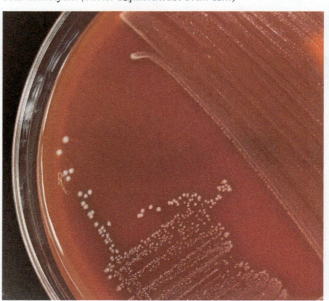

One of the streptococci you will study in this exercise is *S. pyogenes*. Streptolysins produced by *S. pyogenes* cause the lysis of red blood cells in vitro, producing **beta-hemolysis** (a clear zone of hemolysis with no color change) on blood agar (**Figure 44.2**). Two types of beta lysins are produced: streptolysin O and streptolysin S. The former is oxygen-labile, while the latter is oxygen-stable. Streptolysin O is demonstrated only in deep colonies on the blood agar medium. Since most strains of *S. pyogenes* produce both types of lysins, surface hemolysis is generally observed.

S. pyogenes is classified as Group A, based on a chemical substance known as C carbohydrate (an antigenic, group-specific hapten) found in the cell wall. There are 18 to 20 different immunologic groups of streptococci (A to O, excluding I and J) based on the presence of C factor. Members of Group A, such as *S. pyogenes*, are the streptococci most often responsible for human infections. Some diseases in humans caused by Group A streptococci include tonsillitis, septic sore throat, scarlet fever, otitis media, rheumatic fever, meningitis, erysipelas, acute endocarditis, and glomerulonephritis. *S. pyogenes* produces many enzymes and toxins such as streptokinase, leukocidins, streptodornase, hyaluronidase, hemolysins (streptolysin O and S), nucleases, and erythrogenic toxin.

The medium of choice for *S. pyogenes* is blood agar. On blood agar, the colonies are opaque, domed, about 0.5 mm in diameter, and surrounded by a zone of beta-hemolysis. The optimal temperature for growth is 37°C.

Figure 44.3 The Group A and Group C Streptococcal Responses to Bacitracin and SXT.

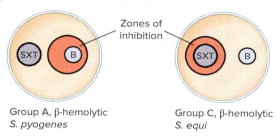

Figure 44.4 Bacitracin Sensitivity Test. The left half of this blood agar plate was inoculated with a pure culture of a Group A beta-hemolytic streptococcus. The right half of the plate was inoculated with *Streptococcus agalactiae*. The two disks contain 10 μg bacitracin. After 24 hours of incubation, the inhibition of growth around the antibiotic (disk on the left) is a positive bacitracin test. The bacteria on the right are bacitracin resistant. (*Javier Izquierdo/McGraw Hill*)

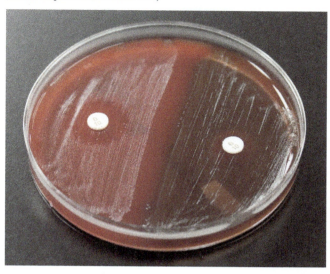

S. pyogenes may be found in the throat and nasopharyngeal areas of humans. Also found in these areas are the viridans streptococci (e.g., *S. salvarius, S. mutans*) that produce either alpha-hemolysis (as evidenced by the greenish to gray pigmentation produced around the colony growth on blood agar) or no hemolysis. The viridans streptococci do not produce C carbohydrate and are usually nonpathogenic opportunists. *S. bovis* (L. *bos,* cow) is an example of a member of the viridans streptococci group.

In addition to hemolysis, *S. pyogenes* can be distinguished from viridans streptococci by means of the **bacitracin sensitivity test.** In this test, a filter paper disk impregnated with bacitracin is applied to the surface of a blood agar plate that has been previously streaked with the bacteria to be identified. The appearance of a zone of inhibition surrounding the disk is a positive test for Group A streptococci (**Figures 44.3** and **44.4**). An absence of a zone of inhibition suggests non–Group A bacteria (e.g., Group C). *S. pyogenes* is also resistant to SXT disks, which contain a mixture of 1.25 μg trimethoprim and 27.75 μg of sulfamethoxazole. Viridans streptococci are inhibited by these agents.

Group B streptococci can be distinguished from other beta-hemolytic streptococci by their production of a substance called the **CAMP factor.** CAMP is an acronym for the names of the investigators (**C**hristie, **A**tkins, and **M**unch-**P**etersen) who first described the factor. This factor is a peptide that acts together with the β-hemolysin produced by some strains of *S. aureus,* enhancing the effect of the latter on a sheep blood agar plate (**Figure 44.5**).

Group D streptococci (*S. bovis*) and enterococci can be differentiated from other streptococci by using bile esculin agar slants. Group D streptococci grow readily on the bile esculin agar and hydrolyze the esculin, imparting a dark brown color to the medium (**Figure 44.6**). This reaction denotes their bile tolerance and ability to

Figure 44.5 The CAMP Test for Distinguishing Group B Streptococci from Other beta-Hemolytic Streptococci. Group B streptococci display arrowhead zone with increased hemolysis (CAMP positive). Group A streptococci display no increase in hemolysis (CAMP negative). (*Javier Izquierdo/McGraw Hill*)

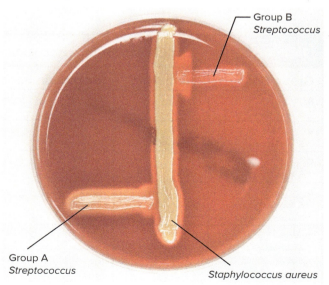

Figure 44.6 Esculin Reaction. The *S. pyogenes* on the left does not hydrolyze esculin, the tube remains brown, and this constitutes a negative reaction. The *E. faecalis* on the right has hydrolyzed esculin with resulting blackening of the medium and constitutes a positive reaction. *(Javier Izquierdo/McGraw Hill)*

hydrolyze esculin, and constitutes a positive reaction. Gram-positive bacteria (other than group D streptococci and enterococci) are inhibited by the bile salts.

This exercise is designed to familiarize the student with the morphological and colonial characteristics, diagnostic tests (**Table 44.1**), and serological identification of *S. pyogenes*. Also included is the comparison of the usually nonpathogenic alpha streptococci (*S. bovis*) with the pathogenic *S. pyogenes*.

Procedure

First Period

1. Each group of students is to inoculate one blood agar plate with Group A beta-hemolytic *S. pyogenes* and another plate with Group C beta-hemolytic *S. equi*. Use a loop to inoculate the plates. A viridans streptococcus such as *S. salvarius* can be used in place of *S. equi*.
2. Place a bacitracin disk and an SXT disk on the plates. Be sure to place the disks on opposite sides of the plates, in well-inoculated areas. Incubate the plate for 24 hours at 37°C. Do not invert.
3. With an inoculating loop, streak a strain of *S. aureus* down the center of a blood agar plate (as shown in Figure 44.5). On one side of the plate, inoculate a strain of Group B streptococci (*S. agalactiae*) by making a streak at a 90° angle, starting 5 mm away from the *S. aureus* and extending outward to the edge of the agar. On the other side of the plate, inoculate the strain of Group A streptococci (*S. pyogenes*). This streak should not be directly opposite the Group B streak.
4. Incubate the plates at 37°C for 24 hours.
5. Inoculate one bile esculin agar slant with *S. equi*, one with *S. bovis*, one with *S. agalactiae*, and one with *S. pyogenes*. Incubate the slants at 37°C for 24 hours.
6. REMEL has a BactiCard™ Strep rapid test for the presumptive identification of streptococci and related genera using enzyme technology and an Enterococcus/Group A Strep screen using esculin hydrolysis and PYR tests to identify *Enterococcus* and Group A *Streptococcus*. The RIM® A.R.C. STREP A TEST (REMEL) is a one-step rapid test

Table 44.1	Diagnostic Tests for Streptococcal Differentiation				
Group Species	**Hemolysis**	**Bacitracin Sensitivity**	**CAMP**	**Bile Esculin**	**SXT Sensitivity**
A *S. pyogenes*	Beta	Sensitive	–	–	Resistant
B *S. agalactiae*	Beta	Resistant	+	–	Resistant
C *S. equi*	Beta	Resistant	–	–	Sensitive
D *E. faecalis* *S. bovis*	Alpha, beta, or none	Resistant	–	+	Resistant (some sensitive)
Viridans *S. salvarius* *S. mitis* *S. mutans*	Alpha or none	Resistant (some sensitive)	–	–	Sensitive

for the detection of Group A streptococcal antigen from throat swabs. It is one of the most common rapid tests used in clinical facilities. These rapid tests can be used per group of students or for demonstration purposes.

Second Period

1. Examine the plates for bacitracin and SXT susceptibility. A zone of bacitracin inhibition greater than or equal to 14 mm in diameter indicates the presence of Group A streptococci. There will not be a zone around the non–Group A streptococci. Positive SXT results for the presence of Group C streptococci are indicated by a zone of inhibition greater than or equal to 16 mm in diameter. There will not be a zone around the non–Group C.
2. Observe the area of hemolysis around the *S. aureus* streak. At the point adjacent to the streak of Group B streptococci, you should be able to see an arrowhead-shaped area of increased hemolysis indicating the production of the CAMP factor. There should be no change in the hemolytic zone adjacent to the streak of the Group A streptococci since most strains do not produce the CAMP factor.
3. Gram stain each group of streptococci from the blood agar plates.
4. Examine the bile esculin agar slants for any color change. A dark brown color change is a positive reaction for group D streptococci. No color change is a negative reaction.
5. Complete the report for this exercise.

Disposal

When you are finished with all of the tubes, pipettes, swabs, and plates, discard them in the designated place for sterilization and disposal.

> **HELPFUL HINTS**
>
> - It should be noted that the zone of hemolysis around the bacitracin disk is so distinct because a pure culture is being used. If this is not so, then your stock culture is probably contaminated.
> - Sometimes it is easiest to observe hemolysis by examining the zone with the 4× objective of a compound microscope. Any intact erythrocytes are readily seen.

NOTES

Laboratory Report 44

Name: _____

Date: _____

Lab Section: _____

Streptococci

1. Observations from streptococci.

Bacterium	Gram Stain	Group	CAMP Factor	Colony on Blood Agar	Hemolysis	Bacitracin Sensitivity	SXT Sensitivity	Bile Esculin Hydrolysis
S. bovis	_____	_____	_____	_____	_____	_____	_____	_____
S. pyogenes	_____	_____	_____	_____	_____	_____	_____	_____
S. equi	_____	_____	_____	_____	_____	_____	_____	_____
S. agalactiae	_____	_____	_____	_____	_____	_____	_____	_____

2. In summary, what is the major purpose of this experiment?

ASSESSMENT
Critical Thinking and Learning Outcomes Review

1. How could you determine if a sore throat was caused by *S. pyogenes* (Group A, beta-hemolytic)?

2. What types of diseases are caused by *S. pyogenes?*

3. Do all Group A streptococci produce erythrogenic toxin? Explain your answer.

4. What is SXT sensitivity? CAMP factor? What organisms are identified by those tests?

5. How would you differentiate between alpha- and beta-hemolysis?

EXERCISE 45 Neisseriae

SAFETY CONSIDERATIONS

- Be careful with the Bunsen burner flame.
- The catalase reagent (3% hydrogen peroxide) and 1 N HCl are caustic.
- Work with them in the fume hood.
- **Neisseria spp. are potential pathogens (BSL-2).** Use aseptic technique throughout this experiment.
- Keep all culture tubes upright in a test-tube rack.

MATERIALS

- suggested bacterial strains:
  Neisseria subflava biovar *subflava*
  Neisseria flavescens
  Moraxella nonliquefaciens
- tryptic soy agar plate
- modified Thayer-Martin plate
- chocolate agar plate
- tubes enriched nitrate broth containing Durham tubes
- candle jar
- CTA (cysteine tryptic agar) glucose medium tubes
- CTA fructose medium tubes
- CTA maltose medium tubes
- CTA sucrose medium tubes
- indophenol oxidase reagent or Difco SpotTest Oxidase Reagent
- catalase reagent (3% hydrogen peroxide)
- nitrate test reagents
- DNase test agar plate
- 1-ml pipettes with pipettor
- Gram-stain reagents
- glass slides
- wax pencil or lab marker
- 37°C incubator
- prepared slides of *Neisseria gonorrhoeae* exudate
- 1 N HCl
- Bunsen burner
- REMEL BactiCard™ Neisseria rapid test for the presumptive identification of *Neisseria* spp.
- container for biohazard waste
- safety glasses
- disposable gloves
- lab coat

LEARNING OUTCOMES

Upon completion of this exercise, students will demonstrate the ability to

1. Explain the clinical isolation and identification of *Neisseria* species
2. Perform isolation and biochemical tests to differentiate *N. subflava, N. flavescens,* and *M. nonliquefaciens*

SUGGESTED READING IN TEXTBOOK

1. Class *Gammaproteobacterias* the Largest Bacterial Class, Order *Burkholderiales* Includes Chemoheterotrophs and Chemolithotrophs, section 21.2, see also figure 21.16 and table 21.3.
2. Direct Contact Diseases Can Be Caused by Bacteria, section 38.3, Sexually Transmitted Infections, Gonorrhea; see also table 38.1 and figure 38.18.

Pronunciation Guide

- *Moraxella nonliquefaciens* (mo-rak-SEL-ah non-li-kwe-FA-sens)
- *Neisseria flavescens* (nis-SE-re-ah flav-ES-sens)
- *N. gonorrhoeae* (go-nor-REE-ah)
- *N. lactamica* (lak-TAM-ica)
- *N. mucosa* (MU-cosa)
- *N. sicca* (SIK-ah)
- *N. subflava* (sub-FLA-vah)

Medical Application

Recently, nucleic acid probe tests and gene amplification assays for detecting *N. gonorrhoeae* directly from patient urogenital specimens have become available. These tests can be completed in 2 to 4 hours and make the diagnostic results available the same day the specimen is taken. However, these methods cannot be used for medical-legal cases. This is because there is a very small risk that the sample could be contaminated and result in a false-positive report and the conviction of an innocent individual. Thus cultural isolation (such as is done in this experiment) and confirmatory identification of *N. gonorrhoeae* are the only test results admissible in court. In addition, strict protocols in the collection, transport, and laboratory testing of specimens must be strictly followed. These protocols are referred to as the chain of custody.

PRINCIPLES

The family *Neisseriaceae* includes bacteria that are Gram-negative cocci occurring in pairs or groups. The two species that are pathogenic to humans are *N. gonorrhoeae*, which causes gonorrhea, and *N. meningitidis*, which causes cerebrospinal meningitis. Since these bacteria are very fastidious and sensitive to changes in the atmosphere, the following precautions must be taken in isolating and culturing them: (1) highly enriched media must be used, and (2) the bacteria must be incubated in an increased CO_2 and water atmosphere. It is of historical interest that the genus name *Neisseria* was named after Albert Ludwig Sigesmund Neisser, who discovered the cause of gonorrhea in the 1880s.

Other nonpathogenic *Neisseria* may be isolated from the respiratory tract of humans. These include *N. sicca, N. subflava, N. flavescens, N. mucosa,* and *N. lactamica. Moraxella nonliquefaciens* is also commonly found in the respiratory tract. However, it is a Gram-negative rod that produces cocci only in a stationary culture. All of these bacteria can be distinguished from the pathogenic species by their fermentation patterns, nitrate and nitrite reduction capabilities, and growth patterns on enriched and nutrient agars (**Table 45.1**).

The purpose of this exercise is to acquaint you with methods of clinical laboratory identification of the *Neisseria* species. *M. nonliquefaciens* will also be studied because of its practical laboratory implications, since it is the only species that produces DNase.

Table 45.1	Some Distinguishing Characteristics of *Neisseria* and *Moraxella*		
	Bacteria		
Characteristic	*N. subflava*	*N. flavescens*	*M. nonliquefaciens*
Growth on soy agar at 35°–37°C	−	+	+
Growth on modified Thayer-Martin agar at 37°C	+	−	−
Acid produced from			
Glucose	+	−	−
Fructose	−	−	−
Maltose	+	−	−
Sucrose	−	−	−
Lactose	−	−	−
Reduction of			
NO_3	−	−	+
NO_2	+	+	−
Colony pigmentation	None	Yellow	None
Colony morphology	Gray to white, smooth	Opaque, smooth	Opaque, smooth
DNase	−	−	+
Catalase	+	+	+
Oxidase	+	+	+

Figure 45.1 Morphology of Neisseria. Gram stain of a culture of *N. gonorrhoeae* showing the Gram-negative cocci occurring most in pairs; some have formed clumps during the staining procedure (×1,000). *(Javier Izquierdo/McGraw Hill)*

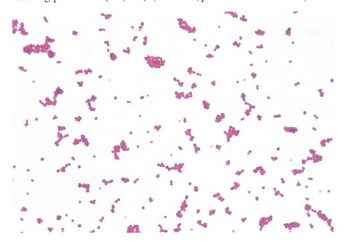

Figure 45.2 Gram Stain of a Urethral Discharge Showing Acute Gonococcal Urethritis. Note the presence of the intracellular Gram-negative diplococci (seen in pairs with adjacent sides flattened) within the cells. *(Source: Joe Miller/Centers for Disease Control and Prevention)*

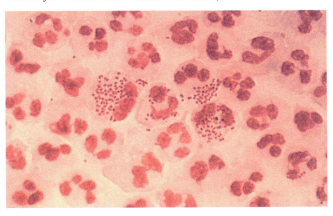

Figure 45.3 Chocolate Agar: An Enriched Medium. Culture of the fastidious *Neisseria* spp. on chocolate agar. The brown color is the result of heating red blood cells and lysing them before adding them to the medium. It is called chocolate agar because of its brown color. Notice the off-white colonies with no discoloration of the agar. *(Kathy Park Talaro)*

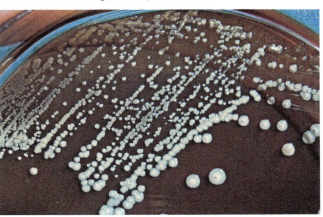

Procedure

First Period

1. Gram stain the known stock cultures of *N. subflava* (**Figure 45.1**), *N. flavescens,* and *M. nonliquefaciens.* Note the resistance of *M. nonliquefaciens* to decolorization.
2. Carefully examine the prepared slide of a *N. gonorrhoeae* exudate. Notice the many white blood cells that are present (**Figure 45.2**).
3. Warm the media to be used to room temperature. With a wax pencil or lab marker, divide the bottom of a tryptic soy agar plate, modified Thayer-Martin plate, and DNase plate into three sections. Modified Thayer-Martin agar is one of the recommended media when *Neisseria* spp. are suspected. However, only a few species are able to grow on this medium, including non-pathogenic *N. subflava* and pathogenic *N. gonorrhoeae* and *N. meningitidis* (Table 45.1). Chocolate agar can also be used (**Figure 45.3**). With the wax pencil or lab marker, place your name, date, and the name of the respective bacteria to be inoculated on the plates. Streak each section with the respective bacteria.
4. Incubate the plates at 37°C in a candle jar that contains a dampened paper towel for providing a humid atmosphere (**Figure 45.4**) for 24 to 48 hours. (The candle jar provides an anaerobic environment as the burning candle uses up the O_2 in the sealed jar.)
5. Inoculate the enriched nitrate tubes with the respective bacteria. Label with your name, date, and the name of the bacteria. Incubate for 24 to 48 hours at 37°C (*see Figure 30.2*).
6. Heavily inoculate the following cysteine tryptic agar (CTA) carbohydrate-containing media tubes just below the surface (**do not go deep into the medium**): glucose, maltose, fructose, and sucrose. Tighten the caps and incubate at 37°C. Examine daily for 5 days or until acid production is seen (*see Figure 19.2*).
7. Add 1 to 2 drops of oxidase reagent to each stock culture. A positive oxidase test is indicated by a darkening of the color of the colony within

Figure 45.4 A Candle Jar. Agar plates are placed in a jar, a candle is lit, and the jar is sealed. When the O_2 has been consumed, the candle is extinguished, leaving an atmosphere of 3% CO_2. Inoculated agar is then transported and/or incubated at 37°C in this CO_2 environment. Candle jars are an alternative method to the commercially available systems for the transport of and recovery of *Neisseria* spp.

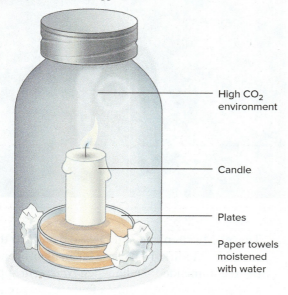

20 to 30 seconds. This color will eventually turn purple.

8. To test for catalase, pipette several milliliters of the hydrogen peroxide reagent over select colonies.

The appearance of gas bubbles indicates a positive test; the absence of gas bubbles is a negative test (*see Figure 25.1*).

9. The REMEL BactiCard™ Neisseria rapid test for the presumptive identification of *Neisseria* spp. can be used per group of students or for demonstration purposes.

Second Period

1. Note the characteristics of any isolated colonies of *N. subflava, N. flavescens,* and *M. nonliquefaciens* on tryptic soy, modified Thayer-Martin agar, chocolate agar, and DNase plates. Describe these colonies.
2. Test and observe the nitrate broth cultures for the reduction of nitrate and/or nitrite (*see Figure 30.2*).
3. Read the results of the carbohydrate degradation tubes.
4. Complete the report for this exercise.

Disposal

When you are finished with all of the tubes, pipettes, and plates, discard them in the designated place for sterilization and disposal.

Laboratory Report 45

Name: _____

Date: _____

Lab Section: _____

Neisseriae

1. *Neisseria* and *Moraxella* culture data (colony size, form, surface, margin, color).

Agar	N. subflava	N. flavescens	M. nonliquefaciens
Tryptic soy agar	_____	_____	_____
Modified Thayer–Martin medium	_____	_____	_____

2. *Neisseria* and *Moraxella* biochemical data.

Test	N. subflava	N. flavescens	M. nonliquefaciens
Oxidase (+ or −)	_____	_____	_____
Catalase (+ or −)	_____	_____	_____
DNase (+ or −)	_____	_____	_____
Acid production			
Glucose	_____	_____	_____
Fructose	_____	_____	_____
Maltose	_____	_____	_____
Sucrose	_____	_____	_____
Nitrate and nitrite reduction			
Nitrate	_____	_____	_____
Nitrite	_____	_____	_____
Gram stain			
Reaction	_____	_____	_____
Morphology	_____	_____	_____
Arrangement	_____	_____	_____

3. In summary, what is the major purpose of this experiment?

ASSESSMENT
Critical Thinking and Learning Outcomes Review

1. From what type of clinical specimen would one expect to isolate *N. gonorrhoeae*? What about *N. subflava* and *M. nonliquefaciens*?

2. Why is the oxidase test useful in the separation of Gram-negative, facultatively anaerobic bacteria?

3. Give one example in which gonococci might give a false-negative test in a CTA carbohydrate test.

4. How does a candle jar function?

5. Since the *Neisseria* are fastidious and sensitive to atmospheric changes, what precautions must be taken in isolating and culturing them?

6. How are pathogenic *Neisseria* identified?

7. What are intracellular Gram-negative diplococci?

EXERCISE 46 — Normal Human Flora

SAFETY CONSIDERATIONS

- Extreme caution should be taken when working with and disposing of microbes isolated from humans because they are potentially pathogens.
- When finished, place all used swabs in disinfectant.
- Keep culture tubes upright in a test-tube rack.

Legal Considerations

Instructors should consult with college or university officials prior to performing this exercise to consider the legal implications of collecting and propagating potentially pathogenic microbes that may be present on human skin.

MATERIALS

- tryptic soy agar (TSA) plate
- mannitol salt agar (MSA) plate
- FTO agar plate
- sterile swabs
- sterile saline
- glass slides
- 3% hydrogen peroxide
- wax pencil or lab marker
- Bunsen burner
- inoculating loop
- 37°C incubator
- Gram-stain reagents
- test-tube rack
- container for biohazard waste
- safety glasses
- disposable gloves
- lab coat

LEARNING OUTCOMES

Upon completion of this exercise, each student will demonstrate the ability to

1. Isolate pure cultures of bacteria from human skin
2. Become better acquainted with the normal microbiota of the human body
3. Use selective and differential media to isolate three unique species of *Micrococcus*

SUGGESTED READING IN TEXTBOOK

1. A Functional Core Microbiome is Required for Host Homeostasis, section 33.3.
2. Identification of Microorganisms from Specimens, section 36.2; see also figure 36.3.

PRINCIPLES

The microorganisms that constitute the **normal microbiota** of the human body are usually harmless, although some are potential pathogens or opportunists. These latter microorganisms may cause disease under certain circumstances (**Figure 46.1**).

Three main reasons for learning about the normal human microbiota are 1) to gain an understanding of the different microorganisms at specific body locations, which provides greater insight into the possible infections that might result from injury to these body sites; 2) to gain a knowledge of indigenous microbiota, which aids in understanding the consequences of overgrowth of those microorganisms normally absent at a specific body site; and 3) to develop an awareness of the role these indigenous microbes play in stimulating the host immune response, which provides protection against microorganisms that might otherwise cause disease.

In this exercise, you will isolate bacteria from your skin. Despite how it might feel to you, human skin is an inhospitable environment for most microbes (**Figure 46.2**). Skin is dry and acidic, and the natural process of shedding dead skin cells makes it difficult for most microbes to colonize. In addition, human skin contains a variety of immune defenses to ward off potential pathogens. Common inhabitants of human skin include members of the genera *Staphylococcus*, *Propionibacterium*, and *Micrococcus* (**Figure 46.3**).

Procedure

First Period

1. Moisten a sterile cotton swab in 0.85% saline. Squeeze out the excess fluid by pressing the swab against the side of the test tube. Using a circular motion, gently rub the swab on the skin site you intend to sample.
2. Use this swab to inoculate the surface of a TSA plate. Rub the swab over the entire surface of the plate.

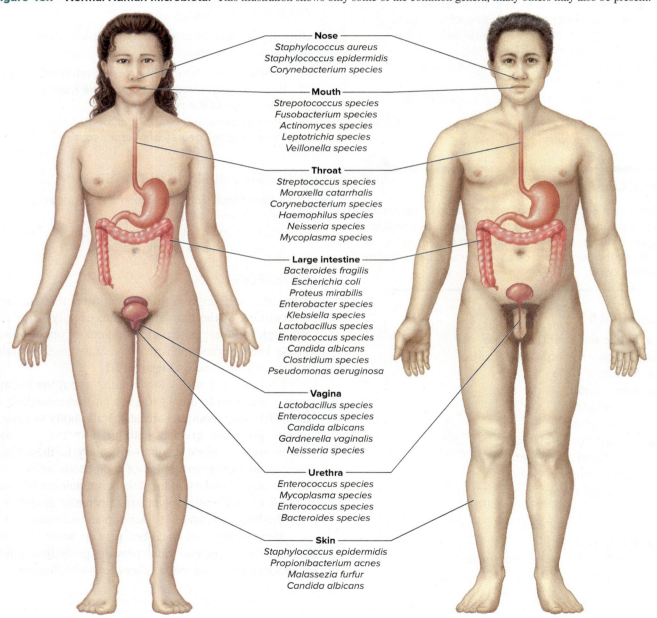

Figure 46.1 Normal Human Microbiota. This illustration shows only some of the common genera; many others may also be present.

330   Medical Microbiology

Figure 46.2 **The architecture of human skin.** On closer inspection, human skin contains many different niches where microbes can proliferate.

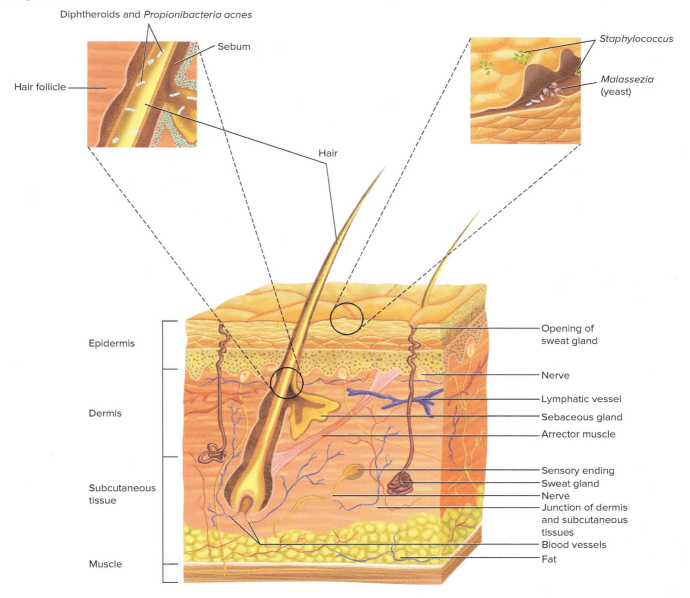

Repeat this process for as many different skin sites as you wish. When labeling your plates, make sure to include this information, in addition to your name and date.

3. Incubate the plates for 24–72 hours at 37°C, checking them each day for growth. Note the appearance of any colonies.

Second Period

1. Streak several well-isolated colonies for isolation on TSA plates. Try to include as many differently pigmented colonies as possible (see **Table 46.1**). Incubate your plates for 24–72 hours at 37°C, checking them daily.

2. Select isolated colonies, describe their appearance, and do a Gram stain. Be sure to note the arrangement of any Gram-positive cocci. Perform a catalase assay on any Gram-positive cocci.

3. Streak any isolates that are Gram-positive cocci on mannitol salt agar (MSA) and furoxone-tween-Oil Red (FTO) agar. MSA is selective for *Staphylococcus* species, while FTO agar is selective for *Micrococcus* species. Incubate the plates at 37°C for 24–72 hours, observing them for growth daily.

4. Record your observations in the lab report for this exercise.

Figure 46.3 The genus *Micrococcus*. There are many different species of *Micrococcus* commonly isolated from human skin including *M. luteus* and *M. roseus*. (**a**) Colonies of *M. luteus* growing on the surface of an agar plate. Note the bright yellow pigment. (**b**) Scanning electron micrograph of *M. luteus*. Members of the genus typically grow in tetrads or small clusters of cocci. *(a: Lisa Burgess/McGraw Hill; b: Source: Janice Carr/CDC)*

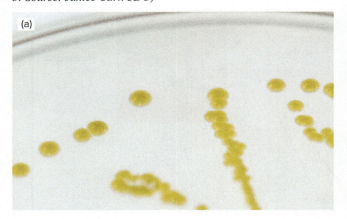
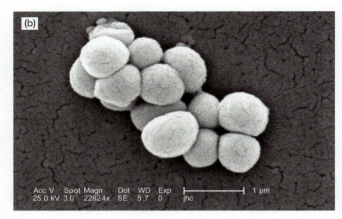

Table 46.1	Some Common Microorganisms Associated with Human Skin	
Organism	**Cellular Morphology**	**Colony Morphology**
Staphylococcus epidermidis	Gram-positive Coccus Clusters	Large, round, white, glistening colony
Staphylococcus aureus	Gram-positive Coccus Clusters	Large, round, glistening, generally yellow colony
Corynebacterium spp. (diphtheroids)	Gram-positive Pleomorphic rods Club shaped, bipolar, or barred	Generally very dry and wrinkled
Yeasts	Large budding cells Single	Large, round, moist colony
Bacillus spp.	Gram-positive Rod Single and chains	Round or irregular; surface dull becoming thick and opaque
Streptococcus viridans	Gram-positive Cocci Chains	Small, dome-shaped colony surrounded by an area of discoloration
Micrococcus luteus	Gram-positive Cocci Single, tetrads	Large, round, yellow, glistening colony

Third Period

Disposal

When you are finished with all of the tubes, pipettes, swabs, and plates, discard them in the designated place for sterilization and disposal.

> **HELPFUL HINTS**
> - When swabbing human skin, try several different sites including the nose, cheek, ears, armpit, or feet.
> - Make sure to check your plates daily as different bacteria grow at different rates. For example, staphylococci will form colonies overnight while micrococci take 3 days for colonies to form.

Laboratory Report 46

Name: _____

Date: _____

Lab Section: _____

Normal Human Flora

1. Complete the following table for the bacteria you isolated from human skin.

	Colony Type 1	Colony Type 2	Colony Type 3
Location of skin sample			
Cell morphology			
Colony morphology			
Catalase test			
Growth on FTO?			
Growth on MSA?			

2. In summary, what is the major purpose of this experiment?

ASSESSMENT
Critical Thinking and Learning Outcomes Review

1. When attempting to culture bacteria from human skin, what types of bacteria do you expect to isolate?

2. How is it possible to isolate a bacterial pathogen from the skin of an individual who is completely healthy?

3. What are some reasons for knowing which microorganisms are associated with different parts of the body?

4. Why is it important to check your isolation plates daily?

5. Based on the culture conditions used in this exercise, what types of microbes are you unlikely to isolate in pure culture? Design an experiment to isolate this group of microbes.

PART 8
Eukaryotic Microbiology

Eukaryotic microorganisms include the fungi (yeasts and molds), the algae, and the protozoa. All of these eukaryotes consist of single cells (or simple aggregates of cells) that contain a membrane-bound nucleus and intracellular membrane-bound organelles. Free-living, eukaryotic microorganisms are widely distributed in the environment. Many eukaryotic microorganisms also participate in various types of symbiotic relationships. While some of them benefit society, others cause human diseases and are of medical importance; others are domestic animal and agricultural plant pathogens.

This section of the manual contains two laboratory exercises that have been designed to better acquaint you with the fungi and the scientific discipline called mycology. You will work with both prepared specimens and living microorganisms in order to better understand the life cycles, salient characteristics, and morphology of selected examples.

After completing one or more of the exercises in Part 8, you will, at the minimum, be able to demonstrate an increased understanding of microbial diversity. This will meet the following American Society for Microbiology Core Curriculum concepts:

- **impact of microorganisms on the biosphere and humans**
- **microbial systems**

Source: James Gathany/CDC

Among the breadth of eukaryotic microorganisms, fungi are considered to be excellent tools for studying important biological processes, particularly from the point of view of eukaryotic cell biology and as model systems for larger eukaryotic organisms like ourselves. Fungi play crucial roles in our microbiomes, our health, the cycling of nutrients in the environment, and many industrial applications that include the production of important industrial enzymes, food additives, biofertilizers, and biofuels. Becoming familiar with the techniques for the study and maintenance of fungal cultures is essential because these techniques are different from those used for handling bacterial cultures.

EXERCISE 47

Fungi I: Yeasts (*Ascomycota*)

SAFETY CONSIDERATIONS

- Be careful with the Bunsen burner flame.
- Keep all culture tubes upright in a test-tube rack or can.

Legal Considerations

Either the course director or the laboratory instructor should check with the appropriate college or university officials (legal department), to see if it is considered an invasion of students' privacy to ask students to provide throat or skin swabs as well as other personal information concerning the microbiota of their bodies.

MATERIALS

- suggested yeast strains:
  7- to 10-day Sabouraud dextrose plate culture of *Saccharomyces cerevisiae* and/or *Rhodotorula rubrum*
- dried commercial baker's yeast
- iodine solution (3 ml water to 1 ml Gram's iodine)
- methylene blue solution
- Bunsen burner
- inoculating loop
- clean glass slides
- coverslips
- wax pencil or lab marker
- Sabouraud dextrose agar plates
- 25°C and 30°C incubators
- 1 glucose fermentation tube with Durham tube (Difco: neopeptone + 1% yeast extract + 1% glucose)
- 1 sucrose fermentation tube with Durham tube (Difco: neopeptone + 1% yeast extract + 1% sucrose)
- sterile cotton swabs
- REMEL IDS RapidID Yeast Plus Panel
- container for biohazard waste
- safety glasses
- disposable gloves
- lab coat

LEARNING OUTCOMES

Upon completion of this exercise, students will demonstrate the ability to

1. Explain the morphological characteristics of yeast cells
2. Culture a typical yeast and study carbohydrate fermentation
3. Stain a typical yeast cell

SUGGESTED READING IN TEXTBOOK

1. Dikarya Are The Most Diverse Fungal Group, section 24.4; see also figure 24.9.
2. Fungal Biology Reflects Vast Diversity, section 24.1; see also figures 24.1–24.3.

Pronunciation Guide

- *Rhodotorula rubrum* (ro-do-TOR-u-lah ROO-broom)
- *Saccharomyces cerevisiae* (sak-ah-ro-MI-seez ser-ah-VEES-ee-eye)
- *Candidia albicans* (kan'-DI-dah al'BA-kans)

Why Are the Following Fungi Used in This Exercise?

The major objectives of this exercise are to learn yeast morphology and culture some typical yeasts. To achieve these objectives, the authors have chosen the common yeast *Saccharomyces cerevisiae* (or *Rhodotorula rubrum*) for this exercise. Because *Candida albicans* is a yeast that is commonly found in the mouth, students should be able to isolate it.

Figure 47.1 The Life Cycle of the Yeast *Saccharomycetes cerevisiae*. When nutrients are abundant, haploid and diploid cells undergo mitosis and grow vegetatively. When nutrients are limited, diploid *S. cerevisiae* undergo meiosis to produce four haploid cells that remain bound within a common cell wall, the ascus. Upon the addition of nutrients, two haploid cells of opposite mating types (*a* and α) fuse to create a diploid cell.

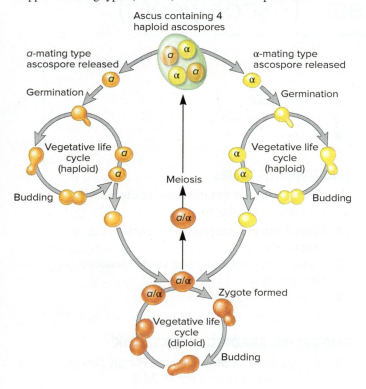

Figure 47.2 Budding in *Saccharomycetes*. Scanning electron micrograph of *Saccharomyces cerevisiae* during budding division. Notice how the cells tend to hang together in chains. (J. Forsdyke/Gene Cox/Science Source)

Figure 47.3 Yeasts. Oval yeast cells (*Saccharomyces cerevisiae*) in a wet mount stained with methylene blue (×1,000). (Javier Izquierdo/McGraw Hill)

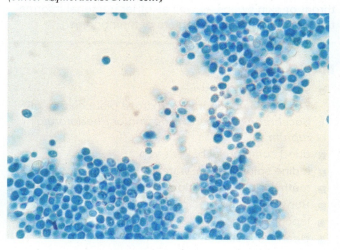

PRINCIPLES

Yeasts are unicellular fungi that are spherical, ellipsoidal, or oval in shape, and usually do not form hyphae (fungal filaments), with the exception of the dimorphic yeasts such as *Candida*. They are about 5 to 10 times larger than bacteria. Yeasts commonly reproduce asexually by **budding,** a process in which a new cell (called a daughter cell) is formed by the parent cell from a protuberance called a **bud (Figures 47.1–47.3).** When yeasts reproduce sexually, they produce several types of **sexual spores** (e.g., ascospores). The type of spore produced is very useful in classifying yeasts. Metabolic activities are also used to identify and classify yeasts. For example, the yeast *Saccharomyces cerevisiae* will ferment glucose but not sucrose. In the laboratory, **Sabouraud dextrose agar** is commonly used to isolate yeasts (**Figure 47.4**). It is a selective medium containing glucose and peptone, and has a low pH, which inhibits the growth of most other microorganisms.

This exercise will allow you to examine the colony and cell morphology of the common yeast, *S. cerevisiae,* and its fermentative capabilities. This yeast is used in making bread and in various alcoholic fermentations. In addition, students will attempt to isolate and culture a yeast from their mouth. A yeast that is most commonly found in the mouth is *C. albicans*. In healthy individuals, *C. albicans* does not produce disease. However, *Candida albicans* is an opportunistic pathogen and may overgrow if the normal microbiota of the mouth (or other areas of the body) is upset or the individual is compromised. This is referred to as the disease **candidiasis.**

Figure 47.4 *Saccharomycetes* cerevisiae **Colonies of Sabouraud Dextrose Agar Plate.** Notice that the appearance of the colonies is similar to that of bacterial colonies and not "fuzzy" like in mold colonies. *(Lisa Burgess/McGraw Hill)*

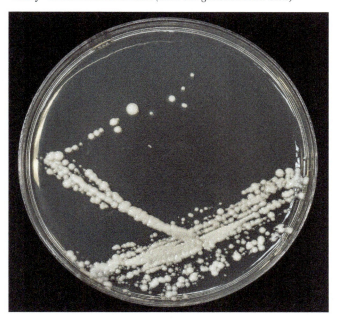

Procedure

First Period

1. With a wax pencil or lab marker, draw two circles on a clean glass slide. Place several drops of the water-iodine solution into one circle and several drops of the methylene blue solution into the other circle.
2. Suspend a loopful of yeast culture (*Saccharomyces* or *Rhodotorula* species) into each circle. Place a coverslip over each.
3. Examine both yeast cell preparations under low and high power. Note the shape and relative size and the presence or absence of budding. Look for the small nucleus and larger vacuole.
4. Draw the representative yeast cells in the report for this exercise.
5. Suspend a pinch of commercial baker's yeast in 2 to 3 ml of warm water. With the inoculating loop, streak onto one half of a Sabouraud dextrose agar plate. Using the inoculating loop, remove some *S. cerevisiae* from the stock plate. Streak the other half of the plate. Label the plate with your name, date, and yeast species. Incubate the plate at 30°C until growth is seen.
6. Inoculate about 1 ml of the commercial baker's yeast suspension into a glucose fermentation tube and a sucrose fermentation tube. Incubate these tubes at 30°C and observe daily until growth is seen. Look for fermentation (gas bubbles in the Durham tubes) and growth as indicated by turbidity.
7. Try to isolate *C. albicans* by swabbing the surface of your tongue and streaking onto a Sabouraud dextrose agar plate. Incubate the plate at room temperature or 25°C until growth has occurred.
8. The REMEL IDS RapidID Plus Panel provides 4-hour identification of yeast. Database includes over 40 taxa. This can be used per group of students or for demonstration purposes.

Second Period

1. Carefully examine the smell of the Sabouraud dextrose agar plate. Observe the appearance of the yeast colonies. Record your results in the report for this exercise.
2. Examine the glucose and sucrose fermentation tubes. Record your results in the report for this exercise.
3. Examine the Sabouraud dextrose agar plate prepared by swabbing your tongue. Smell the plate. Stain several of the colonies with methylene blue as performed in the first period. Record your results in the report for this exercise.

HELPFUL HINTS

- If phase-contrast or ultraviolet microscopes are available, the observation of cellular organelles and inclusions can be particularly revealing when students work with unstained yeast wet mounts.
- Make sure that you avoid incubating these plates at high temperatures, since these fungi tend to prefer cooler temperatures (25–30°C).

NOTES

Laboratory Report 47

Name: _____

Date: _____

Lab Section: _____

Fungi I: Yeasts (*Ascomycota*)

1. Draw typical yeast cells in the following space.

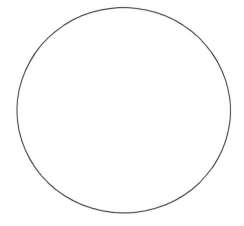

Water-iodine suspension

Magnification: × _____

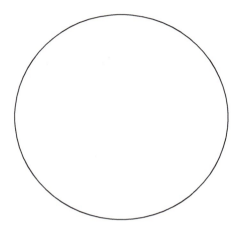

Methylene blue suspension

Magnification: × _____

2. Record the typical features of yeast colonies in the table below.

Feature	Commercial baker's yeast	Lab stock of *S. cerevisiae*
Color		
Smell		

3. Record the fermentation capabilities of *S. cerevisisae*.

   a. What is the color of the glucose broth after _____ days' incubation? _____

   Was gas produced? Explain. _____

   b. What is the color of the sucrose broth after _____ days' incubation? _____

   Was gas produced? Explain. _____

4. Draw the appearance of the yeast colony (potentially *Candida albicans*) from your mouth in the following space. Do the same for the cell morphology.

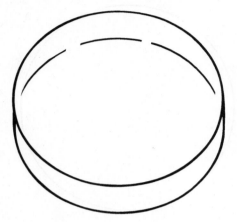
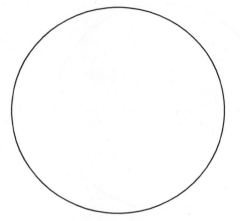

Magnification × _____

342   Eukaryotic Microbiology

ASSESSMENT
Critical Thinking and Learning Outcomes Review

1. Define:

    a. budding

    b. hyphae

    c. yeast

2. Compare and contrast the characteristics of baker's yeast and *Saccharomyces cerevisiae*.

3. Why are yeast colonies larger than bacterial colonies?

4. Why do yeasts generally have to be cultured for longer periods than most bacteria?

5. What are some similarities and differences between the yeasts that were cultured from your mouth and baker's yeast?

6. Why are stains not required for yeast identification?

7. Why is Sabouraud dextrose agar used to cultivate yeasts?

EXERCISE 48

Fungi II: *Zygomycota (Rhizopus), Ascomycota (Penicillium),* and *Basidiomycota (Agaricus)*

SAFETY CONSIDERATIONS

- When working with sporulating fungi, avoid prolonged exposure to the plates.
- Some individuals may be allergic to the fungi, and the laboratory may become excessively contaminated with spores when the plates are left uncovered.
- Although many procedures used in this experiment are unlike those for bacteria, aseptic technique is still required.

MATERIALS

- suggested fungal strains:
  7- to 10-day Sabouraud dextrose plate cultures of *Penicillium notatum* and *Aspergillus niger,* preserved mushrooms (*Agaricus* or *Coprinus*), plus and minus strains of *Rhizopus stolonifer*
- Sabouraud dextrose agar deep
- potato dextrose agar deeps
- 50°C water bath
- sterile Petri plates
- methylene blue in 90% methanol
- inoculating loop
- wax pencil or lab marker
- clean microscope slides
- 22 × 40-mm coverslips
- tweezers
- Bunsen burner
- syringes containing paraffin and oil, or tubes of silicone tub caulk
- wooden swab sticks (or a syringe)
- 1-ml pipettes with pipettor
- glass or wooden support rods
- sterile Pasteur pipettes and bulbs
- prepared slides of Zygomycetes (e.g., *Rhizopus, Saprolegnia*), Ascomycetes (e.g., *Penicillium, Aspergillus, Morchella*), Basidiomycetes (e.g., *Polyporus, Lycoperdon, Coprinus, Puccinia*)
- container for biohazard waste

- safety glasses
- disposable gloves
- lab coat

LEARNING OUTCOMES

Upon completion of this exercise, students will demonstrate the ability to

1. Describe a typical mold
2. Grow colonies of several molds on Sabouraud agar plates
3. Prepare culture slides of several molds
4. Identify the morphology and reproductive structures of representative molds
5. Observe a complete mold life cycle

SUGGESTED READING IN TEXTBOOK

1. The Fungi, chapter 24. It is suggested that students bring their textbook to the laboratory since all of the morphology data, life cycles, and terminology covered in this exercise can be found in chapter 24 and will not be repeated here.

Pronunciation Guide

- *Aspergillus niger* (as-per-JIL-us NI-jer)
- *Penicillium notatum* (pen-a-SIL-ee-um no-TAY-tum)
- *Rhizopus stolonifer* (ri-ZO-pus sto-LON-a-fer)

Why Are the Following Molds Used in This Exercise?

In this exercise, you will become familiar with several aspects of fungus biology (mycology). The authors have chosen three of the most common molds (*Penicillium notatum, Rhizopus stolonifer,* and *Aspergillus niger*) for the student to culture and study. Other fungi are introduced using prepared slides and preserved materials.

345

PRINCIPLES

Molds are multicellular, filamentous fungi in contrast with the unicellular growth observed in yeasts. The techniques of culturing and observing fungi differ from the methods used to study bacteria. Fungi grow at comparatively slow rates, often requiring several days to weeks to form macroscopically visible colonies. Usually, growth will spread over the entire culture plate, producing spores on brightly colored aerial hyphae. Like yeasts, most molds grow best at room temperature (25°C) rather than at 37°C.

The basic medium for the culture of many molds is Sabouraud dextrose agar. The high sugar concentration and low pH (5.6) of this medium make it unsuitable for the growth of most bacteria, thus guarding against contamination. Most fungi grow well at pH 5.6.

Mold colonies may be examined directly in culture with a dissecting microscope. However, it is better to tease away a portion of the growth and place it on a slide with a drop of water or stain (a wet-mount). An advantageous alternative method is a **slide culture,** which allows us to study a growing mold and identify specific structures without disturbing the culture. This approach is extremely useful because fungi are identified primarily by examining their reproductive structures, morphological characteristics, and colony growth.

The macroscopic aggregation (colony) of mold cells is called a **thallus**. A thallus is composed of a mass of strands called a **mycelium;** each strand is a **hypha. Vegetative hyphae** grow on the surface of culture media. They form aerial hyphae, called **reproductive hyphae,** that bear asexual reproductive **spores** or **conidia**. The hyphae that grow below the surface of culture media are called **rhizoidal hyphae.** The hyphal strand of some molds may be separated by a crosswall called a **septum**. Hyphae that contain septa are called **septate hyphae**. Molds with hyphae that lack septa are called **coenocytic hyphae**.

Molds are classified primarily by their stages of sexual reproduction. They are also characterized and classified according to the appearance of the colony (e.g., color, size), organization of the hyphae (e.g., septate or coenocytic), and structure and organization of the spores (e.g., sporangiospores, conidiospores). Molds are important clinically, industrially, and as decomposers (saprophytes) in the environment. Mold spores are a very common source of contamination of laboratory cultures.

In this exercise, the methods used to study molds will include the preparation of colonies on Petri plates, the preparation of special culture slides, the study of the life cycle of *Rhizopus* (the common bread mold), the dissection of a mushroom, and the examination of commercially prepared slides of preserved molds.

Procedure

Preparation and Observation of Colonies

1. Melt one tube of Sabouraud dextrose agar and one of potato dextrose agar.
2. Cool to 50°C in a water bath.
3. Pour into two Petri plates respectively and allow to harden.
4. Using the wax pencil or lab marker, label the Sabouraud dextrose agar plate *Aspergillus* and the potato dextrose agar plate *Penicillium*. Add your name and date to each plate.
5. Using aseptic technique, inoculate the plates as labeled with a single loopful of the mold suspension. Place the loopful of mold inoculum in the center of the plate. Do not spread the inoculum. Handle plates carefully so that they are not jostled.
6. **Do not invert the Petri plates.** Incubate them at room temperature for 2 to 7 days.
7. After the colonies have developed properly (**Figures 48.1** and **48.2**), sketch and describe the macroscopic appearance (e.g., color, texture) of each in the report for this exercise. If dissecting microscopes are available, examine the hyphae and conidia under the microscopes and draw the conidia.

Figure 48.1 **Molds.** *Aspergillus* colonies showing black conidiospores. *(sinhyu/123RF)*

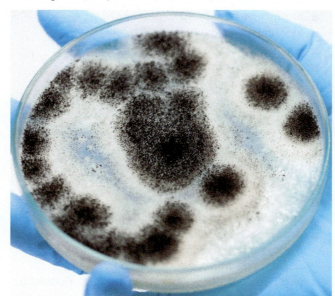

Figure 48.2 Appearance of *Penicillium* Colonies Growing on Sabouraud Dextrose Agar. The green granular surface (at times with radial furrows) and a white apron are typical of this genus. *(Kateryna Kon/123RF)*

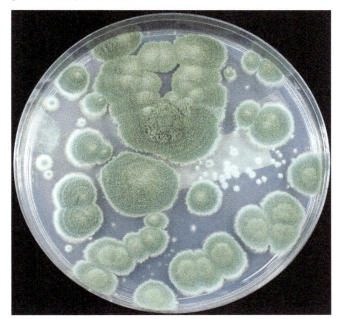

Figure 48.3 Mold Culture Slide. (**a**) Apply hot Parafilm to the surface of a slide to support coverslip. (**b**) Place coverslip on Parafilm about 1mm above the slide. (**c**) Fill half of the chamber with molten agar inoculated with a fungal culture. (**d**) Place the slide inside of a Petri plate on top of moist paper towel while supported by two wooden sticks. Slides can be observed in 2 to 4 days.

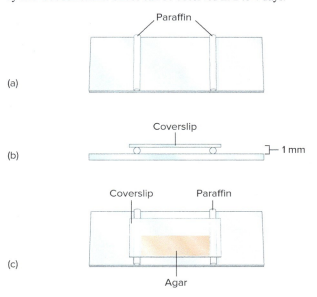

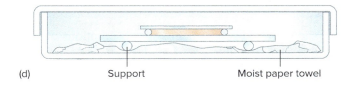

Preparation of Slides for Microscopic Examination of Molds

1. Obtain two clean glass slides and coverslips.
2. Flame the surface of the slides and coverslips in order to sterilize them. Use tweezers to hold the slides and coverslips.
3. Use the wooden sticks (or a heated syringe) to transfer sufficient melted paraffin or silicone caulk to each slide to support the coverslip about 1 mm over the surface of the slide. Let the paraffin or silicone harden. Each slide should appear as in **Figure 48.3a**.
4. Heat the coverslips sufficiently so that they will form a seal when you set them over the hardened paraffin or silicone. Each slide should look like **Figure 48.3b**.
5. Melt and cool to 50°C a tube of Sabouraud dextrose agar (for *Aspergillus*) and a tube of potato dextrose agar (for *Penicillium*). Label the tubes accordingly.
6. Using a sterile pipette, transfer and mix 0.5 ml of the proper mold suspension with the proper agar tube.
7. For each mold, use a sterile Pasteur pipette and quickly let sufficient inoculated agar run under the coverslip of the prepared slide to half fill the chamber. Each slide should now appear as shown in **Figure 48.3c**. This procedure must be completed before the agar hardens.
8. Moisten two circles of paper towel that just fit the bottom of the Petri plates, and place one in each plate. Each slide should then be placed in its own Petri plate. Label the plates accordingly. The slide should be supported above the moist paper by two wooden sticks or a piece of a V-shaped glass rod (**Figure 48.3d**). Place the lid on the Petri plate and incubate at room temperature for 2 to 4 days.
9. After incubation, observe the slides with a microscope using the low-power objective. Add a few drops of methylene blue in methanol to stain the various structures. Methanol is necessary to soften the cell wall and allow the stain to enter.

Commercially Prepared Slides

1. Obtain commercially prepared slides illustrating the various classes of fungi, their morphology, and their reproduction. Carefully study each of these slides. Draw and label the indicated

Figure 48.4 **The Basidiomycota.** (a) Close-up of *Agaricus* displaying structural details of the cap, stipe, and annulus. (b) A close-up of a wild mushroom's gills. (c) Life cycle of a typical soil basidiomycetes, with a close look at the structure of the basidium, and primary and secondary mycelia. *(a: Ingram Publishing/SuperStock; b: IT Stock/age fotostock)*

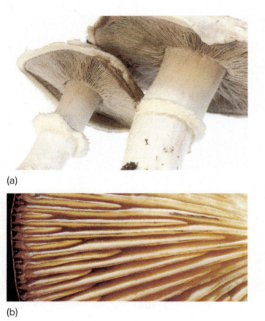
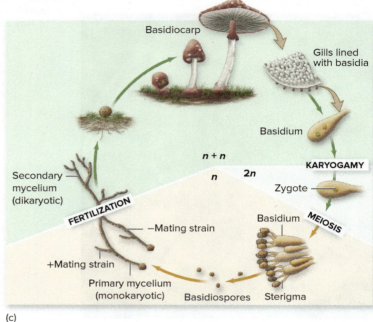

structures in the report for this exercise. These slides should supplement your observations of live fungi and give you a good picture of fungus morphology and the various forms of fungal reproduction.

Zygomycetes:
   Rhizopus (sporangia and zygotes)
   Saprolegnia (fruiting bodies, sporangia)
Ascomycetes:
   Penicillium (sections, conidia)
   Aspergillus (conidia)
   Morchella (sections with asci)
Basidiomycetes:
   Polyporus (sections with basidia)
   Lycoperdon (longitudinal section)
   Coprinus (medium longitudinal section, cross section)
   Puccinia (specialized reproductive structures such as aecia, telia, uredia)

Dissection of Preserved Mushrooms

1. Examine a specimen of *Agaricus* (**Figure 48.4**) or *Coprinus*. Note and draw its major anatomical features (cap, stipe, annulus, and mycelia at the base of the stipe).

2. Carefully dissect out a gill section, mount it in water, cover it with a coverslip, and gently crush it. Examine your specimen under the microscope. At the gill surface, you should be able to see the basidia, sterigmata, and basidiospores.
3. Make your drawings in the report for this exercise.

Rhizopus Morphology and Reproduction

1. The life cycle of *Rhizopus stolonifer* can be easily studied.
2. Inoculate a potato dextrose agar plate with the plus and minus culture strains provided.
3. A spot of each spore type should be placed at opposite sides of the plate, about 4 cm or more apart.
4. Within 4 to 7 days, the two strains will have grown together and a line of zygospores will be formed in the center of the plate. Carefully remove the cover of the Petri plate and observe the morphology of the zygomycete with a dissecting scope. You should be able to see the hyphae and sporangia (**Figure 48.5**). Zygospores can be studied by carefully teasing

Figure 48.5 *Rhizopus stolonifer*, A Zygomycete That Grows on Simple Sugars, Such as Those Found in Moist Bread or Fruit. **(a)** A microscopic view of *Rhizopus*. The sporangium is found at the end of a long, unbranched, nonseptate sporangiophore. **(b)** A micrograph (×430) of a zygosporangium. **(c)** The life cycle of *Rhizopus stolonifer* involves sexual and asexual phases. The subclass *Zygomycota* is named for the zygosporangia characteristic of *R. stolonifer*. *(a: Rattiya Thongdumhyu/Shutterstock; b: Richard Gross/McGraw Hill)*

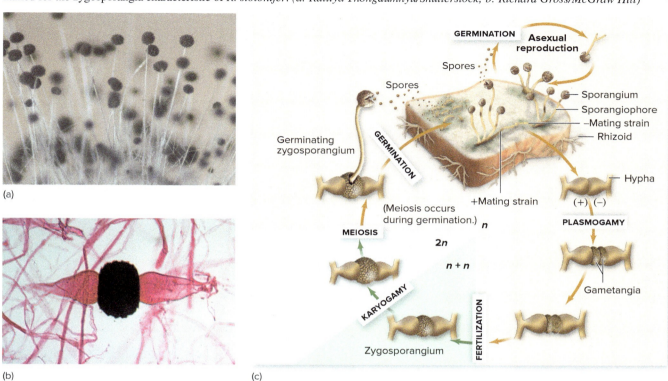

apart the mycelium in the center of the plate. Zygospores in various stages of development will be seen beneath the surface. You should also be able to observe gametangia (Figure 48.5c) as well. Portions of the thallus may also be removed and pressed under a coverslip for observation at higher magnifications.

Disposal

When you are finished with all of the plates and tubes, place them in the designated place for sterilization and disposal.

HELPFUL HINTS

- When preparing slides for fungal culture and observation, it is essential to heat the coverslip properly (step 4). If the coverslip is heated too little, it will not stick to the paraffin. However, be careful not to overheat the coverslip, or the paraffin will liquify and run over the slide surface.
- After the fungi are incubated for several days, desiccation of the agar and cultures can be avoided if the Petri plates are taped for a tighter seal.

Fungi II: *Zygomycota* (*Rhizopus*), *Ascomycota* (*Penicillium*), and *Basidiomycota* (*Agaricus*)

NOTES

Laboratory Report 48

Name: _____

Date: _____

Lab Section: _____

Fungi II: *Zygomycota (Rhizopus)*, *Ascomycota (Penicillium)*, and *Basidiomycota (Agaricus)*

1. Record your observations of the mold plate colonies.

Feature	*Aspergillus*	*Penicillium*
Extent of growth		
Pigmentation		
Aerial hyphae present		

2. Draw the appearance of the mold slide cultures.

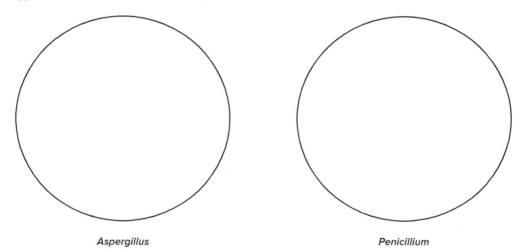

Aspergillus   *Penicillium*

3. Commercially prepared slides.

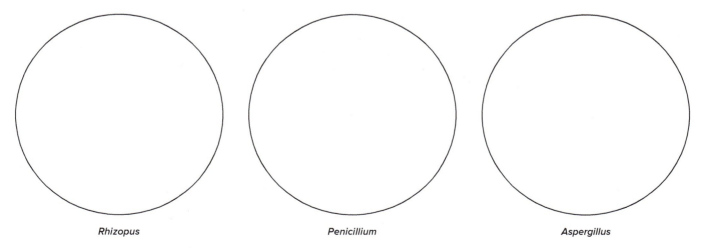

Rhizopus   *Penicillium*   *Aspergillus*

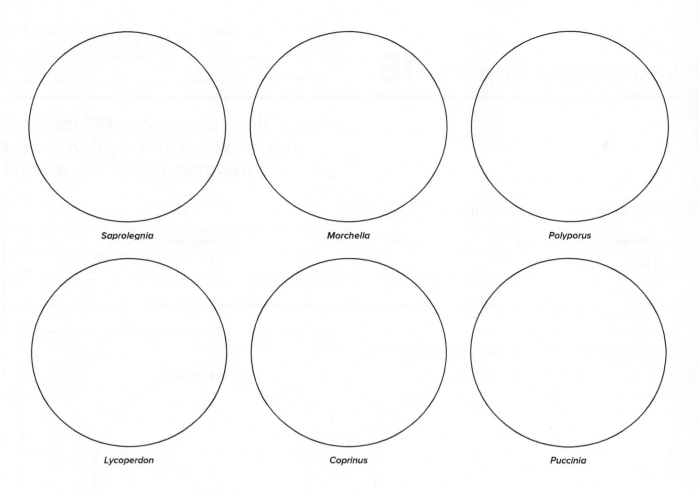

4. Draw the following features for the dissection of preserved mushrooms.

   a. Major anatomical features of a mushroom. b. Mushroom gill section.

5. In summary, what is the major purpose of this experiment?

ASSESSMENT
Critical Thinking and Learning Outcomes Review

1. What is the difference between vegetative and aerial mycelia?

2. Can bacteriological media be used for the cultivation of molds? Explain your answer.

3. In the common bread-mold life cycle, what do the plus and minus signs mean?

4. Why are molds not streaked for isolation like previous exercises have done for bacteria?

5. What is the difference between molds and yeasts?

6. Are *Rhizopus* hyphae coenocytic or septate?

7. How would you describe the fruiting bodies of *Aspergillus*? What about *Penicillium*?

PART 9
Microbial Genetics and Genomics

All of the characteristics of microorganisms (e.g., their chemical composition, growth patterns, metabolic pathways, and pathogenicity) are inherited. These characteristics are transmitted from parent cell to daughter cells through genes. The science of **genetics** is concerned with the study of what genes are, how they carry genetic information, how they replicate, and how their information is expressed within a microorganism to determine its particular characteristics.

Bacteria are especially useful in the study of genetics because large numbers of them can usually be cultured in a short period, and their relatively simple genetic composition facilitates studying the structure and function of specific genes.

DNA is the genetic material that gives a microorganism its inherited characteristics. In the first two exercises, DNA will be extracted and used for the identification of its source organism using polymerase chain reaction. **Genetic engineering** involves combining DNA (genes) from different sources and the introduction of new genetic material into a suitable cell in which it will be both replicated and expressed. Several steps are usually part of the process: (1) provision of a suitable carrier such as a **plasmid**; (2) the formation of composite DNA molecules, or **chimeras**; (3) introduction of the composite DNA into a functional **recipient cell**; and (4) an appropriate **selection method** to isolate the desired recombinant bacterium. The last four exercises in this part of the manual use some of these genetic techniques.

Genomics is the study of the molecular organization of genomes (all of the genetic material in an organism), their information content, and the gene products they encode.

After completing one or more of the exercises in Part 9, you will be able to demonstrate an increased understanding of microbial genetics. This will meet the following American Society for Microbiology Core Curriculum concept:

- **information flow and genetics**

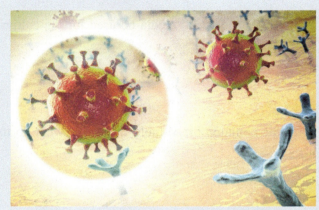

Kateryna Kon/Shutterstock

Molecular techniques, particularly those based on our understanding of nucleic acids, are essential in the world that we live in today. For example, they enhance our ability to detect emerging diseases like COVID-19 by way of a test based on the polymerase chain reaction (PCR) that targets specific regions of the SARS-CoV-2 virus. Rapid genome sequencing of this virus and expression of its genes using modern genetics techniques have been crucial in the development of the proper detection methods as well as in developing a better understanding of how this virus works and how it can be treated.

EXERCISE 49

Isolation and Purification of Genomic DNA from Yeast and Bacteria

SAFETY CONSIDERATIONS

- The alcohols used in this experiment are volatile and flammable liquids. Do not use near an open flame.
- Prolonged exposure to ethanol can cause central nervous system depression.
- The cell lysate solution contains SDS, which can be harmful if inhaled.
- Disposable exam gloves, lab coats, and chemical splash goggles are required.
- Discard cultures and contaminated wastes only in the appropriate biohazard container indicated by your instructor.

MATERIALS

- suggested yeast strain: baker's yeast, dehydrated or *Saccharomyces cerevisiae*
- suggested bacterial strains: 10–12 hour (mid-log growth phase) tryptic soy broth cultures of *Escherichia coli*, *Serratia marcescens*, and *Enterobacter aerogenes*. Cultures can be maintained in the log phase by immersion in an ice-water bath.
- 0.5- to 10-μl micropipettes and sterile tips
- 10- to 100-μl micropipettes and sterile tips
- 100- to 1,000-μl micropipettes and sterile tips
- microcentrifuge tubes
- microcentrifuge
- 1-ml pipettes
- Pasteur pipettes
- 250-ml beaker
- glass stirring rod
- disinfectant
- 80°C modular dry bath incubator
- 37°C modular dry bath incubator
- heating block modules
- 65°C water bath
- 50–60°C water
- Zymolyase 20T (2.5 mg/ml)
- Adolph's meat tenderizer
- 20-mg/ml RNase A solution (DNase free; dissolve 1 mg/ml in 1 × TE [10 mM Tris, pH 7.4 + 1 mM EDTA, pH 8.0])
- cell lysate solution (Tris-EDTA-SDS) from Puregene
- protein precipitation solution (2.5M ammonium acetate)
- 1.5-ml tubes
- 100% isopropanol
- cold 70% ethanol
- cold 95% ethanol
- cold 100% absolute ethanol
- vortex mixer
- hydration solution (1× Tris-EDTA hydration solution)
- wax pencil or lab marker
- container for biohazard waste
- safety glasses
- disposable gloves
- lab coat

LEARNING OUTCOMES

Upon completion of this exercise, students will demonstrate the ability to

1. Demonstrate how to isolate genomic DNA from a Gram-negative bacterium
2. Explain why different protocols are needed for DNA isolation from various taxa

SUGGESTED READING IN TEXTBOOK

1. All of chapter 18 on Microbial Genomics.
2. Experiments Using Bacteria and Viruses Demonstrate That DNA Is the Genetic Material, section 13.1; see also figures 13.1 to 13.3.

Pronunciation Guide

- *Escherichia coli* (esh-er-l-ke-a KOH-lie)
- *Enterobacter aerogenes* (en-ter-oh-BAK-ter a-RAH-jen-eez)
- *Saccharomyces cerevisiae* (sak-ah-ro-MI-seez SERah-VEES-ee-eye)
- *Serratia marcescens* (se-RA-she-ah mar-SES-sens)

Why Are the Following Microorganisms Used in This Exercise?

Isolation and purification of DNA are often the first steps in starting genetic analysis. Chemically, DNA is identical in structure from one organism to another. It is the sequence of base pairs that distinguishes DNA from various species. The composition of an organism (whether it is unicellular or multicellular, the exact makeup of the individual cells and tissues, and whether or not a cell wall is present) affects protocols used to isolate DNA. Also the use for which DNA is required may dictate the specific method employed in its isolation. We will carry out a simple protocol to extract DNA from a eukaryotic cell (*Saccharomyces cerevisiae*). Since *Escherichia coli* is one of the most studied prokaryotic cells, it will be used in this exercise as a source of bacterial DNA. We will also use *Serratia marcescens* and *Enterobacter aerogenes*, two other Gram-negative organisms. Once DNA has been obtained and purified, it can be saved and used in more complex exercises, such as quantitation by gel electrophoresis, PCR amplification, digestion with restriction endonucleases, and membrane hybridizations such as Southern blots. Although this experiment stops with pure DNA from a Gram-negative bacterium depending on time and equipment, instructors can supplement it with additional genomic protocols.

PRINCIPLES

Isolation and purification of DNA are essential procedures in molecular biology. Extraction and purification of genomic DNA from prokaryotic and eukaryotic micro organisms have facilitated the study of complex genomes. This has also led to the construction of genomic DNA libraries for many species.

In the first part of this exercise, you will be spooling DNA onto a glass stirring rod or Pasteur pipette, which is a common way to isolate DNA once it has been precipitated out of an aqueous solution using ethanol. One of the reasons the yield of yeast DNA is so large is that the size of the yeast genome averages 1.3×10^7 base pairs, compared to bacteria that average 4.7×10^6 base pairs.

In the second part of this exercise you will isolate DNA from a group of Gram-negative bacteria. The genomic DNA isolation and purification technique that will be used in this exercise uses a four-step process:

1. Bacterial cells must be lysed to release genome DNA. The most common solution contains the detergent sodium dodecyl sulfate and sodium hydroxide.
2. Ribonuclease is added to the cell bacterial lysate to remove RNA.
3. The bacterial proteins are removed next by the addition of ammonium or potassium acetate. The acetate precipitates the proteins but leaves the large genomic DNA in solution.
4. The genomic DNA is concentrated and desalted by using cold isopropanol precipitation followed by an ethanol wash.

Procedure

DNA Isolation from Baker's Yeast

1. Measure 100 ml of 50–60°C tap water into a 250-ml beaker.
2. Add one-half package of dry yeast to the tap water and stir until completely suspended.
3. Incubate for 5–15 minutes at 50–60°C. The warm water activates the yeast.
4. Add 10 ml of Zymolyase; stir at 37°C for 1 hour on a hot plate. Zymolyase lyses the yeast cell wall, releasing the contents of the cell spheroplasts, including the DNA, into the solution.
5. Add 3 g of meat tenderizer and stir. The meat tenderizer contains the proteolytic enzyme papain that digests the cellular proteins.
6. Incubate at room temperature for 20 minutes.
7. Using a plastic 10-ml transfer pipette, carefully layer 10–15 ml of cold 95% ethanol on top of the suspension of lysed yeast cells by allowing the alcohol to run down the side of the 250-ml beaker. The alcohol should form a distinct layer on the top of the detergent and cell mixture (**Figure 49.1a**). The 95% ethanol dehydrates the DNA so it precipitates out of solution.
8. Insert a glass rod through the 95% ethanol into the cell suspension. Gently stir and roll the rod between your fingers, in one direction, but do not mix the two layers (**Figure 49.1b**).
9. Observe the glass rod. DNA will stick to the glass rod and appear as slimy white threads (**Figure 49.2**). Note that in this simple isolation procedure, DNA will remain associated with some cellular protein.

Figure 49.1 DNA Isolation from Baker's Yeast. (a) Layering of ethanol on homogenate. (b) DNA spooled onto a glass stirring rod.

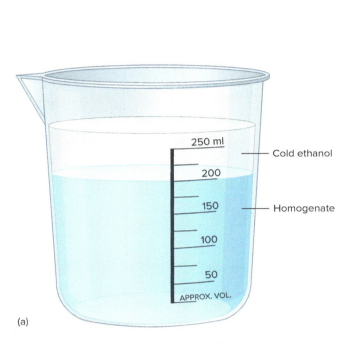

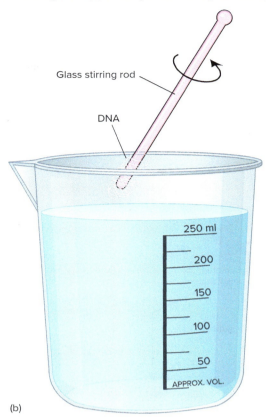

Figure 49.2 DNA Being Spooled onto a Glass Rod. Alcohol is used to precipitate the yeast DNA from the solution and the DNA spun onto the tip of this glass rod. *(Lisa Burgess/McGraw Hill)*

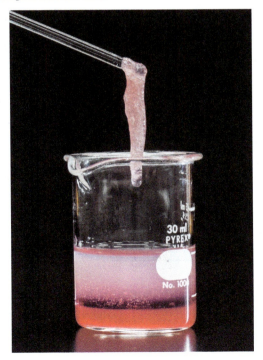

10. Scrape the DNA off the glass rod into a plastic weighing boat and determine its wet weight in micrograms.
11. If the DNA isolation goes well, it is possible to store and use this DNA for subsequent experiments. The DNA can be resuspended in water, but for long-term storage a low-salt buffer-like Tris-EDTA (TE) is recommended.

DNA Extraction from a Gram-Negative Bacterium

1. Obtain a 10- to 12-hour tryptic soy broth culture of *E. coli, E. aerogenes,* or *S. marcescens* as assigned by your instructor to carry out the DNA extraction protocol below (**Figure 49.3**).
2. Label a microcentrifuge tube. Using a pipette, transfer 1 ml of the bacterial culture into the microcentrifuge tube.
3. Pulse the tube at top speed in the microcentrifuge for 5 to 10 seconds to cause the bacteria to form a pellet.
4. Carefully remove most of the supernatant by pipetting it into a container of disinfectant.

Figure 49.3 Summary of Genomic DNA Isolation.

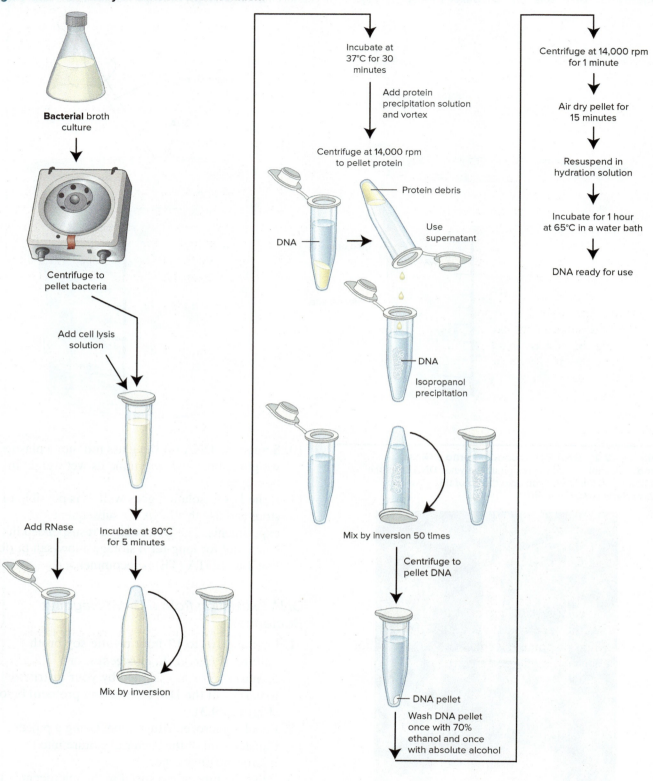

358 Microbial Genetics and Genomics

5. Add 600 μl of cell lysate solution and gently pipette up and down to resuspend the bacterial pellet. The SDS in the lysis solution is a detergent that disrupts the plasma membrane of the cells.
6. Incubate at 80°C for 5 minutes.
7. Slowly cool the sample in the microcentrifuge tube at room temperature. The sample must reach room temperature before adding RNase since heat will denature the enzyme. Once the sample is at room temperature, add 3 μl of RNase solution to the bacterial cell lysate.
8. Mix by inverting the microcentrifuge tube 25 times and incubate at 37°C for 30 minutes.
9. Cool the sample to room temperature. Add 200 μl of protein precipitation solution and vortex **very gently** for 20 seconds. The ammonium acetate in the solution precipitates the protein, leaving the genomic DNA in solution.
10. Microcentrifuge the sample for 3 minutes at 14,000 rpm to pellet the protein.
11. Pour the supernatant into a clean 1.5-ml tube, leaving the protein pellet behind.
12. Add 600 μl of 100% isopropanol, cap the tube, and mix **very gently** by inverting the tube at least 50 times. Vigorous mixing will break the DNA into small fragments.
13. Centrifuge in the microcentrifuge at 14,000 rpm for 1 minute to pellet the DNA.
14. Pour off the supernatant and drain the liquid onto an absorbent towel. The pellet will appear off-white to yellow.
15. Add 600 μl of 70% ethanol and invert the tube several times to wash the pelleted DNA. Decant the 70% ethanol and add 600 μl of absolute alcohol. Invert the tube several times to wash the pelleted DNA. The alcohol washes remove any salts from the DNA.
16. Centrifuge at 14,000 rpm for 1 minute.
17. Pour off the supernatant **very slowly and carefully,** watching the pellet so that it does not come out of the microcentrifuge tube.
18. Air dry the DNA pellet for at least 15 minutes. Air drying removes any residual alcohol, which will interfere with any subsequent analyses.
19. Add 100 μl of hydration solution to the pellet and place the microcentrifuge tube in a water bath at 65°C for 1 hour or incubate overnight at room temperature.
20. The isolated bacterial DNA is now ready to use. At this point, depending on the available equipment and time, your instructor may have you perform other experiments on your samples of DNA.
21. Pure DNA samples that are free of contaminating nucleases can be stored at 4°C for several years. However, since slow degradation may occur at this temperature, −80°C is recommended for prolonged storage. If a non-frost-free freezer is available, DNA can be stored at −20°C.

Disposal

When you are finished with all of the pipettes and tubes, place them in the designated place for disposal.

> **HELPFUL HINTS**
>
> - Remember that heat denatures the RNase enzyme and vigorous mixing will shear DNA into small fragments.
> - To locate the tiny DNA pellet, orient the hinge on the microcentrifuge tube outward in the microcentrifuge. After centrifugation, the DNA pellet will be located toward the bottom of the tube on the same side as the hinge.
> - Any residual alcohol will interfere with subsequent DNA analysis.
> - Do not pipette the DNA to mix, as this will cause shearing.

NOTES

Laboratory Report 49

Name: _____

Date: _____

Lab Section: _____

Isolation and Purification of Genomic DNA from Yeast and Bacteria

1. How much DNA, in micrograms wet weight, did you isolate from the yeast?

2. For each of the steps or reagents in the isolation and purification of genomic DNA from your assigned bacterial culture, indicate what each accomplishes.

Step/Reagent	Accomplishes
Cell lysis solution	
Incubation	
RNase	
Protein precipitation solution	
Centrifugation	
Isopropanol precipitation	
Alcohol washes	
Hydration solution	

ASSESSMENT
Critical Thinking and Learning Outcomes Review

1. Where is DNA located in a yeast cell?

2. What is the role of Zymolyase in the extraction of DNA from yeast and why is it not used in the Gram-negative extraction?

3. Using your knowledge of the cell structure of various prokaryotic organisms, hypothesize why specific conditions are used in the cell lysis step.

4. You have just isolated and identified a new bacterium and you need to design a protocol for isolating its genomic DNA. Explain your rationale for what information you will need in your design.

5. How might you optimize the purity of DNA you extracted from your new organism?

6. Why must you handle DNA gently in this extraction procedure?

EXERCISE 50: 16S rRNA Gene PCR and Sequencing

SAFETY CONSIDERATIONS

- Ethidium bromide is a potent mutagen. Wear gloves and dispose of all materials that come in contact with ethidium bromide in an allocated waste container.
- UV light can cause damage to the eyes if looked at directly. It should only be used to indirectly observe the gel (e.g., through a gel viewing station).

MATERIALS

- DNA samples (prepared during Exercise 49, or from other sources as assigned by the instructor)
- molecular-grade water (nuclease free)
- 16S rRNA PCR primers: 27F (5′- AGAGTTTGATCMTGGCTCAG-3′) and 1492R (5′- GGTTACCTTGTTACGACTT-3′)
- RedTaq ReadyMix (Sigma-Aldrich)
- PCR tubes
- thermocycler
- agarose
- 1x TAE Buffer
- 500-ml flask
- 100-ml graduated cylinder
- balance
- spatulas
- weighing boats
- electrophoresis gel rig
- power supply
- ethidium bromide
- imaging station for gel visualization
- laptop
- container for waste
- container for ethidium bromide waste
- safety glasses
- disposable gloves
- lab coat

LEARNING OUTCOMES

Upon completion of this exercise, students will demonstrate the ability to

1. Describe how 16S rRNA gene sequencing techniques can aid the identification of microorganisms
2. Perform PCR and gel electrophoresis
3. Carry out a BLAST search of a DNA sequence to identify an unknown bacterium

SUGGESTED READING IN TEXTBOOK

1. Polymerase Chain Reaction Amplifies Targeted DNA, section 17.2; see also figure 17.7.
2. DNA Sequencing Methods, section 18.1; see also figures 18.1 to 18.3.
3. Microbes Have Evolved and Diversified for Billions of Years, section 1.2; see also figures 1.8 to 1.10.

Pronunciation Guide

- *Thermus aquaticus* (THER-mus ah-kw-A-ti-kuhs)

PRINCIPLES

Molecular methods can be extremely useful for the identification of an unknown microorganism. The **16S rRNA gene** is a conserved gene universally present in all *Bacteria* and *Archaea* that has become the molecule of choice for determining the taxonomic assignment of these microorganisms. Since this gene produces an RNA molecule that has an essential function in all ribosomes, it is (a) present in all microbes and (b) not prone to changing too much over time. In addition, this gene tends to have highly conserved regions interspersed with variable regions, where most of the sequence diversity is

Figure 50.1 The 16S rRNA Gene. Diagram of the 16S rRNA gene including the approximate size and location of conserved regions (red) and variable regions (orange), as well as the location of two universal primers (27F and 1492R).

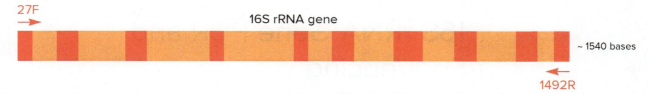

observed (**Figure 50.1**). The conserved regions serve as reference points when comparing these sequences side by side, but also for the design of small molecules of DNA known as primers that can bind to these regions in every known 16S rRNA gene.

In order to verify the identity of an organism, the 16S rRNA gene sequence needs to be amplified and sequenced. Amplification is achieved by way of the **polymerase chain reaction (PCR)**. This is one of the most important techniques in molecular labs since it allows you to generate billions of copies of a specific DNA sequence of your choice. This is achieved by repeating three steps (**Figure 50.2**). First, a denaturation step briefly heats the DNA sample to separate double-stranded DNA into individual strands. Next, an annealing step cools down the reaction so that short pieces of DNA, called primers, are allowed to bind to the complementary regions surrounding the target DNA. Finally, in an extension step, a thermostable DNA polymerase from the organism *Thermus aquaticus* known as *Taq* polymerase will synthesize new DNA by adding deoxyribonucleotide triphosphates (dNTPs) to the 3′-end of the annealed primers. As these three steps are repeated, they increase the concentration of the target DNA sequence exponentially from a single copy to billions of copies.

Agarose gel electrophoresis allows us to separate and view DNA based on size. For this technique, an agarose gel is submerged in a buffer solution and DNA is loaded at one end of the gel (**Figure 50.3**). As an electric current is applied, DNA travels through the gel polysaccharide matrix. Because DNA is negatively charged it will travel toward the positive end of the gel. Using a **DNA size ladder** (a mixture of DNA fragments of known size), it is possible to determine the approximate size of the DNA samples loaded onto the gel. For their visualization, the gel needs to be stained with ethidium bromide or a chemical of similar properties. Ethidium bromide binds itself strongly to DNA and fluoresces in the presence of UV light.

Once the fragments of DNA are confirmed to be of the correct size they can be sent to a commercial service for sequencing. Sequencing services will usually provide you with a text file in FASTA format and a **chromatograph of the sequence** (**Figure 50.4**). A file in FASTA format uses text in a convention that is universal in the field of bioinformatics where the name of the sequence is identified with a ">" symbol, followed by the sequencing information immediately below. The chromatograph provides you with the quality and the fluorescent intensity of each dNTP. This information can be used for deciding how much of the sequence is of sufficient quality to be used in further analysis. A clean sequence can then be submitted to a Basic Local Alignment Search Tool (BLAST) search online that will look for the best match, comparing your sequence to billions of sequences on the GenBank databases.

In this lab, we will be using a pair of universal primers that target regions shared by all *Bacteria* and *Archaea* for PCR amplification of a portion of the 16S rRNA gene and perform basic analysis of the sequencing results.

Procedure

First Period

1. Obtain an unknown DNA sample (from Exercise 49 or provided by your instructor). Label your PCR reaction tube accordingly.
2. Keep your reagents and your final PCR reaction mix on ice throughout its preparation.
3. You will need to prepare a 30-μl reaction consisting of the following:

   - 15 μl of 2× RedTaq mix
   - 3 μl of primer 27F (5 μM)
   - 3 μl of primer 1492R (5 μM)
   - 6 μl of molecular-grade water
   - 3 μl of DNA sample

4. Carefully mix and pipette these volumes in the order listed above into a PCR tube (or PCR tube strip you may be sharing with others). Extreme caution should be used to not reuse micropipette tips. Dispose of tips in between every reagent.
5. In a thermocycler, run a PCR program for these reactions that consists of the following steps: one initial denaturation step of 95°C (5 minutes); 30 cycles of denaturation at 94°C (30 seconds), annealing at 56°C (30 seconds), and extension at 72°C (30 seconds); and a final extension step at

Figure 50.2 **The Polymerase Chain Reaction (PCR).** DNA is synthesized through repeated cycles of denaturation, annealing, and extension. After denaturation, DNA primers (light orange) bind to the template DNA (blue) and serve as starting points for the *Taq* polymerase to synthesize a new strand during the extension step. Subsequent cycles 2 and 3 demonstrate how synthesis will enrich for the sequence flanked by the primers.

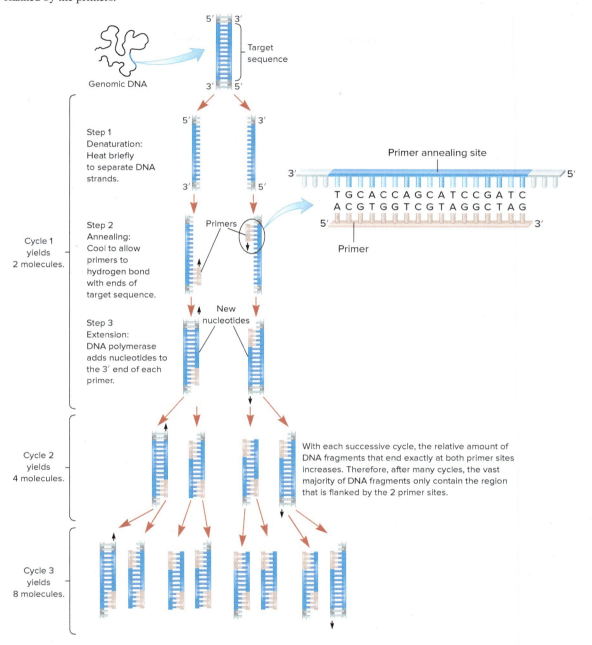

72°C for 5 minutes. Your instructor will show you how to load the reactions on a thermocycler with the appropriate program and we will evaluate success in the next period through gel electrophoresis.

Second Period

1. Prepare a 1% agarose gel. Measure 100-ml of 1× TAE buffer and transfer to a 500-ml flask. Using a spatula and weighing boat, weigh out 1 g of agarose and dissolve in the TAE buffer. You will need to melt the agarose by heating this solution in a microwave until it is completely dissolved. Allow the agarose solution onto cool down at room temperature for 30 seconds.
2. Before pouring the agarose solution onto the gel casting system, make sure to add 2 μl of ethidium bromide (final concentration of 0.2 μg/ml) and mix well.

Figure 50.3 Gel Electrophoresis Equipment. Equipment for running gel electrophoresis consists of a main buffer chamber, a gel tray for the agarose gel, a gel casting comb, and the power supply. *(Javier Izquierdo/McGraw Hill)*

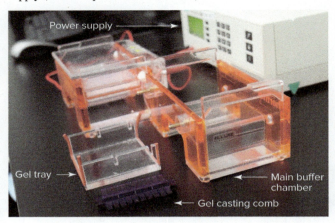

Figure 50.4 DNA Sequencing Chromatograph. Example of chromatograph data, generated between bases 538 and 580. Each nucleotide is labeled with a different colored dye as depicted in the chromatograph.

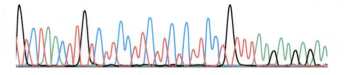

3. Carefully pour the gel into the gel casting tray and immediately add the appropriate comb to make wells. Allow it to solidify for 30 minutes.
4. Once it is solidified, position the gel so that the wells are close to the negative side of the gel running system. The gel must then be submerged in 1× TAE buffer inside the main buffer chamber and the comb can then be removed.
5. A DNA ladder should be loaded into the first well. Then proceed to load each PCR reaction into the wells in the gel. Your instructor will let you know the volumes to add to the gel, since they will depend on the comb you used.
6. Connect the gel box to a power source and run at 100V for 30–45 minutes.
7. Observe the gel under UV light to confirm that you obtained the right size of PCR product prior to sending it out for sequencing. The primer names (27F and 1492R) indicate their position in the 16S rRNA gene.
8. Once confirmed, your instructor will send the PCR product out to be sequenced by a commercial DNA sequencing company. Dispose of the agarose gel in the appropriate container for ethidium bromide waste.

Third Period

1. Your instructor will provide you with the sequencing files for each of your reactions. In general, this will include a text file in FASTA format and/or a DNA chromatograph.
2. If a chromatograph file is available, use it to determine where the sequencing reaction starts and ends. With the assistance of your instructor, examine the chromatograph as a guide to determine which portion of the sequence in the FASTA file you must use. Make sure to only use the high-quality portion of the sequence (i.e., regions where the chromatograph peaks are clearly demarcated).
3. Copy and paste only the appropriate portion of the sequence onto the Query Sequence box of NCBI's nucleotide BLAST (BLASTN) tool. Your instructor will provide the appropriate link and will discuss how to adjust the parameters of the search. For example, you may want to limit your search to sequences from type material or may want to exclude uncultured/environmental sample sequences.
4. Click the BLAST button at the bottom of the page.
5. When you get your results, record your best match. This is usually the sequence at the very top of the list. For the lab report of this exercise, write down the names of the most closely related organisms. You will also want to record the percentage of sequence similarity for the sequences that you choose.

Disposal

Make sure to discard all the tips and tubes in the designated place. Materials with ethidium bromide will have a separate designated place for their disposal.

> **HELPFUL HINTS**
>
> - Be very careful in your preparation of the PCR reaction to minimize exposure of reagents and to dispose of every tip after use. Remember that we are amplifying a gene present in all microorganisms, so any contamination could lead to the amplification of other 16S rRNA genes other than your sample.
> - After fully dissolving the agarose in the 1× TAE buffer it can be cooled down in a waterbath at 55–60°C.

Laboratory Report 50

Name: _____

Date: _____

Lab Section: _____

16S rRNA Gene PCR and Sequencing

1. Draw the appearance of the gel including the ladder that you used and your PCR reactions.

2. List the five closest BLAST result matches to your sequence with their percent of sequence similarity.

3. What is the purpose of this experiment?

ASSESSMENT
Critical Thinking and Learning Outcomes Review

1. What is the approximate size in base pairs of the PCR product you obtained?

2. Are all the 16S rRNA gene PCR products that you ran on the gel the same length? Propose why they could be of different sizes.

3. How would you write the FASTA format file for the 27F primer?

4. What are some possible reasons for a negative PCR reaction?

5. With universal primers that target the 16S rRNA gene, contamination of reagents can be a major problem. How can you test whether the molecular-grade water you used is contaminated?

EXERCISE 51 Mutations

SAFETY CONSIDERATIONS

- Handle rifampicin with care because it is toxic to humans at high doses.
- Keep all culture tubes upright in a test-tube rack.
- Ethidium bromide is a potent mutagen. Wear gloves and dispose of all materials that come in contact with ethidium bromide in an allocated waste container.
- UV light can cause damage to the eyes if looked at directly. It should only be used to indirectly observe the gel or through a gel viewing station.

MATERIALS

- suggested bacterial strain: overnight broth culture of *Escherichia coli*
- LB agar plates
- LB agar plates with rifampicin (10 μg/ml to 100 μg/ml)
- dilution tubes containing 0.9-ml LB broth
- cell spreaders
- microfuge tubes
- micropipettes and tips
- ice and ice bucket
- *rpoB* primers; forward (5′-TGTTCCATTTTCCGGTCAAC-3′) and reverse (5′-TAATAAATCTTTCACGGATT-3′); sequencing primers: 1 (5′-CGCCGTAAACTGCCTGCGAC-3′), 2 (5′-GTCTGCGGTTGGTCGTATGA-3′), 3 (5′-TCCAACTTGGATGAAGAAGG-3′), and 4 (5′-CGTGACACCAAGCTGGGTCC-3′).
- RedTaq ReadyMix (Sigma-Aldrich)
- thermocycler
- agarose
- 1× TAE Buffer
- 500-ml flask
- 100-ml graduated cylinder
- balance
- spatulas
- weighing boats
- electrophoresis gel rig
- power supply
- ethidium bromide
- imaging station for gel visualization
- PCR tubes with pre-aliquoted PCR master mix solution (30 μl per reaction)
- container for biohazard waste
- container for ethidium bromide waste
- safety glasses
- disposable gloves
- lab coat

LEARNING OUTCOMES

Upon completion of this exercise, students will demonstrate the ability to

1. Describe the mechanisms by which mutations can arise in bacterial cultures
2. Isolate spontaneous rifampicin-resistant mutants of *E. coli*
3. Map the location of the rifampicin-resistant mutations onto the sequence of a gene
4. Explain how mutations contribute to the growing problem of antibiotic-resistant bacterial infections

SUGGESTED READING IN TEXTBOOK

1. Antibacterial Drugs, section 9.4; see also table 9.1.
2. Mutations: Heritable Changes in a Genome, section 16.1; see also figure 16.1 and table 16.1.
3. Detection and Isolation of Mutants; section 16.2.
4. Polymerase Chain Reaction Amplifies Targeted DNA, section 17.2; see also figure 17.7.
5. DNA Sequencing Methods, section 18.1; see also figures 18.1 to 18.3.

Pronunciation Guide

- *Escherichia coli* (esh-er-I-ke-a KOH-lie)

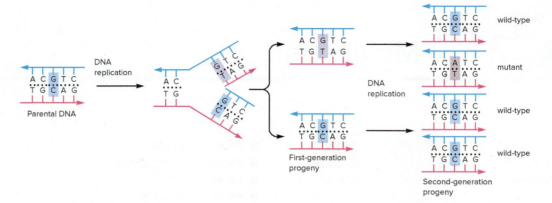

Figure 51.1 Mutations Occur During DNA Replication. Formation of incorrect base pairs (purple box) during DNA replication results in one progeny DNA molecule that contains a mutation. If left uncorrected, this mutation will persist in subsequent rounds of DNA replication.

PRINCIPLES

Broadly defined, **mutations** are stable, heritable changes in the nucleotide sequence of DNA as compared to a "normal" reference sequence that is naturally occurring (**Figure 51.1**). By definition, this normal sequence is referred to as the **wild-type**. In this exercise, you will attempt to isolate spontaneous chromosomal mutations that confer resistance to the antibiotic rifampicin.

By manipulating the growth conditions of a bacterial culture, you can increase the frequency in which mutations are isolated. If a mutant bacterium, by virtue of the change that it has undergone, is better suited to the environment in which it is formed than its parent, growth of the mutant will be favored. In such cases, the mutant (and its progeny) will quickly become the dominant form in a culture. This simple example perfectly illustrates the concept of natural selection, the cornerstone of evolutionary theory.

Mutations occur in one of two ways. **Spontaneous mutations** arise occasionally in all bacteria and develop in the absence of any added pressure from chemical or physical agents. Uncorrected errors in DNA replication, insertion of mobile genetic elements, and damage to DNA bases as a side-effect of normal metabolism can all lead to spontaneous mutations. In contrast, **induced mutations** are the result of exposure to a mutagen, which can be either a chemical or physical agent. Chemical mutagens can directly damage DNA bases or intercalate between bases of DNA (**Figure 51.2a**). Physical mutagens like radiation can also damage DNA bases, preventing normal base-pairing interactions (**Figure 51.2b**).

Regardless of the cause of the lesion, mutations in protein-coding genes can be further classified as missense, nonsense, or silent. **Missense mutations** alter the codon for one amino acid, changing it into a codon for a different amino acid. **Nonsense mutations** also result in a codon change. However, in this case, the new codon is one of the stop codons. **Silent mutations** are alterations in the DNA sequence of a codon that do not change the amino acid for that position. Depending on the change, missense mutations can have a relatively minor or absolutely devastating impact on the structure and function of a protein. Nonsense mutations are also tricky; occurrence of a nonsense mutation early in the coding sequence of a gene will likely result in an extremely truncated protein devoid of function. Nonsense mutations that only truncate a protein by a few amino acids are likely to retain at least some function.

Spontaneous mutations that become resistant to antibiotics (such as rifampicin) can be readily detected in the laboratory. The reason for this easy detection is that the spontaneous mutations will grow in the presence of high antibiotic concentrations that inhibit the growth of normal bacteria (i.e., those bacteria that have not mutated and thus remain sensitive to antibiotics). Thus, this exercise in microbial genetics employs a simple method to isolate and select rifampicin-resistant mutant *Escherichia coli* strains. Rifampicin prevents transcription by binding to the beta subunit of RNA polymerase and interfering with its function. In this exercise, you will attempt to select for mutations in *rpoB*, the gene that codes for the beta subunit of RNA polymerase. In consultation with your instructor, this same methodology can be applied to mutant hunts using other bacteria or different antibiotics.

Figure 51.2 DNA Damaging Agents. The structure of DNA can be damaged by both chemical agents like methylnitrosoguanidine (**a**) and physical agents like UV radiation (**b**).

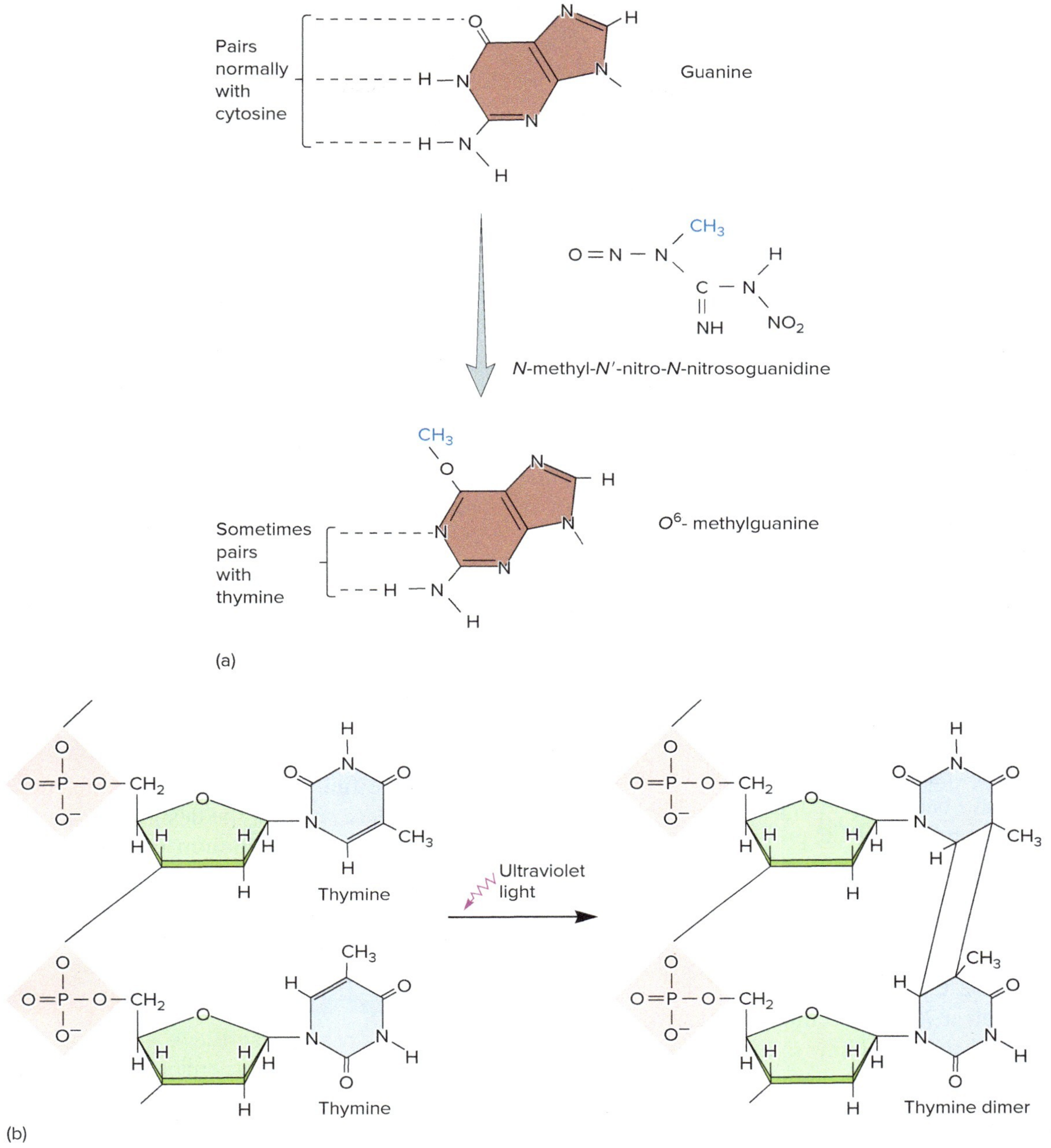

Procedure

First Period

1. Obtain 7 dilution tubes containing 0.9 ml of LB broth. Label the tubes with the appropriate dilution (10^{-1} through 10^{-7}).
2. Create a dilution series of the *E. coli* overnight culture. Pipet 0.1 ml of overnight culture into the 10^{-1} tube and mix by vortexing.
3. Transfer 0.1 ml from the 10^{-1} dilution tube to the 10^{-2} dilution tube. Mix by vortexing. Continue making dilutions to 10^{-7}.
4. Label the LB plates without antibiotic 10^{-5}, 10^{-6}, and 10^{-7}. Label the LB plates supplemented with rifampicin "undiluted," 10^{-1}, 10^{-2}, and 10^{-3}.
5. Plate 0.1 ml of the appropriate *E. coli* dilution onto the corresponding plates. For the undiluted plate, transfer 0.1 ml directly from the overnight

Mutations

culture. Spread the plates using the cell spreaders.

6. Incubate the plates at 37°C overnight.

Second Period

1. After incubation, count the colonies on the LB plates to determine the number of CFUs in the starting overnight cultures.
2. Now observe the LB plates containing rifampicin. Count any visible colonies. If no colonies are visible, allow the plates to incubate for an additional 24–48 hours.
3. Re-streak any rifampicin-resistant colonies onto fresh LB rifampicin plates. Incubate them overnight at 37°C.

Third Period

1. Put on clean gloves. Wipe your hands, work area, and pipettes with ethanol-soaked paper towels.
2. Obtain a PCR tube for each strain you want to sequence. Keep these tubes on ice at all times.
3. Your instructor will pre-aliquot 30 µl of the PCR master mix into each tube. Using a sterile pipette tip, transfer a small amount of colony material from the rifampicin-resistant isolate into a PCR tube. Do this for each isolate to be analyzed. Reserve one tube as a negative control (i.e., no colony material is added).
4. Place the tubes back on ice. After all samples in your class have been prepared, place the tubes in a thermocycler and run a PCR program for these reactions, which consists of the following steps: one initial denaturation cycle at 94°C for 5 minutes; 30 cycles of denaturation at 94°C (30 seconds), annealing at 56°C (30 seconds), and extension at 72°C (3 minutes); and a final extension step at 72°C for 5 minutes. Your instructor will show you how to load the reactions on a thermocycler with the appropriate program and you will evaluate success in the next period through gel electrophoresis.

Fourth Period

1. Analyze the reactions for product as follows. Run out 5 µl of each reaction on a 0.7% agarose gel. Image the gel and look for production of a single band.
2. Successful reactions can then be processed for DNA sequence analysis. Your instructor will describe this process as it will likely involve sending your PCR products to an outside vendor for sequencing.

Fifth Period

1. Download and open your raw FASTA sequence file and copy the sequence data as plain text. Make sure to only use high-quality sequence (i.e., regions where the chromatograph peaks are clearly demarcated).
2. Paste your sequence file into the Query Sequence box using NCBI's nucleotide BLAST (BLASTN) tool.
3. Paste the sequence for *E. coli rpoB* into the Subject Sequence box. (*Note:* You can download the sequence by visiting ecocyc.org and searching for *rpoB*.) Click the BLAST button on the lower left.
4. Look at the output. Are there any regions of difference between the two sequences? If so, translate the DNA sequence into protein to determine if your mutation caused an amino acid change. (*Note:* If your school does not have an institutional license for bioinformatics analysis software, freely available programs to view and align DNA sequences are readily available online.)

Disposal

When you are finished with all of the pipettes, tubes, and plates, place them in the designated place for disposal. Materials with ethidium bromide will have a separate designated place for their disposal.

> **HELPFUL HINTS**
>
> - Rifampicin is light sensitive, so keep any media containing this antibiotic in the dark as much as possible.
> - Consider using several different concentrations of rifampicin. Different lab strains of *E. coli* may have slightly different intrinsic levels of resistance, which will impact the outcome of the experiment.
> - Use a high-fidelity DNA polymerase for the colony PCR so that you can be confident that any mutations uncovered during the sequence analysis are legitimate.

Laboratory Report 51

Name: _____

Date: _____

Lab Section: _____

Mutations

1. Determine how many total bacteria were in the starting *E. coli* culture provided by your instructor. Calculate this value in colony forming units (CFU)/ml. Show your work.

2. Determine the frequency of rifampicin-resistant mutations in your starting *E. coli* culture.

3. List any mutations in *rpoB* that you uncovered as part of your analysis. Include the nucleotide position. Classify the mutations as being silent, missense, or nonsense.

ASSESSMENT
Critical Thinking and Learning Outcomes Review

1. What is the difference between a spontaneous and an induced mutation?

2. What is the difference between a mutation rate and mutation frequency? Consult your textbook or the primary literature to answer this question in full.

3. What is the mechanism of action of rifampicin? What is the mechanism of resistance to rifampicin?

4. Do you think resistance to rifampicin is likely to spread via horizontal gene transfer? Why or why not?

5. What bacterial infections is rifampicin often used to treat?

6. Do you think mutations that confer resistance to rifampicin could be detrimental to the bacterial cell? Explain your answer.

EXERCISE 52 Transformation

SAFETY CONSIDERATIONS

- Make sure the centrifuge tubes are balanced.
- Keep all culture tubes upright in a test-tube rack.

MATERIALS

- suggested bacterial strains:
  Escherichia coli
  Acinetobacter baylyi ADP1 (ATCC 33305)
- miniprep of plasmid pMMB66 (ATCC 37621)
- sterile microfuge tubes
- sterile 10% glycerol
- sterile centrifuge bottles
- centrifuges
- ice-water bath
- 2 sterile 250-ml flasks
- LB broth
- LB agar, LB agar supplemented with 100 mg/l ampicillin, and LB agar supplemented with 150 mg/l ampicillin
- pipettes and pipettor
- sterile micropipet tips and micro pipettor
- wax pencil or lab marker
- cell spreaders
- Bunsen burner
- container for biohazard waste
- safety glasses
- disposable gloves
- lab coat

LEARNING OUTCOMES

Upon completion of this exercise, students will demonstrate the ability to

1. Describe the process of transformation
2. Prepare electrocompetent bacterial cells
3. Demonstrate the successful transformation of plasmid DNA into two different bacterial hosts
4. Differentiate between natural and artificial competence

SUGGESTED READING IN TEXTBOOK

1. Transformation Is the Uptake of Free DNA, section 16.7; see also figures 16.21 and 16.22.

Pronunciation Guide

- *Escherichia coli* (esh-er-I-ke-a KOH-lie)
- *Acinetobacter baylyi* (as-i-net-o-BAK-ter bay-LEE-eye)

Why Is the Following Bacterium Used in This Exercise?

In this exercise, you will learn how to form recombinant bacterial cells by transformation. To accomplish this objective, the authors have selected *Escherichia coli* and *Acinetobacter baylyi*, two Gram-negative bacteria that are easy to manipulate in the laboratory. One of these organisms is naturally competent while the other is not. At the end of this exercise, you will use your experimental results to determine which is which!

PRINCIPLES

Transformation is the uptake by a recipient bacterium of a naked DNA molecule or a DNA fragment from the environment and the maintenance of this molecule (or fragment) in a heritable form (**Figure 52.1**). In natural transformation, the DNA comes from a donor bacterium. The process is random, and any portion of the genome may be transferred between bacteria. **Plasmids,** extrachromosomal DNA elements commonly found in bacteria, can also be transferred via transformation. When bacteria lyse, they release considerable amounts of DNA into the surrounding environment. These fragments may be relatively large and contain several genes. If a fragment contacts a **competent bacterium,** one able to take up DNA and be transformed, it can be bound by the cell and taken inside. Many bacteria are naturally

Figure 52.1 Bacterial Transformation. Transformation occurring with (**a**) DNA fragments or (**b**) plasmid DNA producing a successful (stable) and an unsuccessful (unstable) transformation. The transforming DNA is purple, and integration is at a homologous region of the genome.

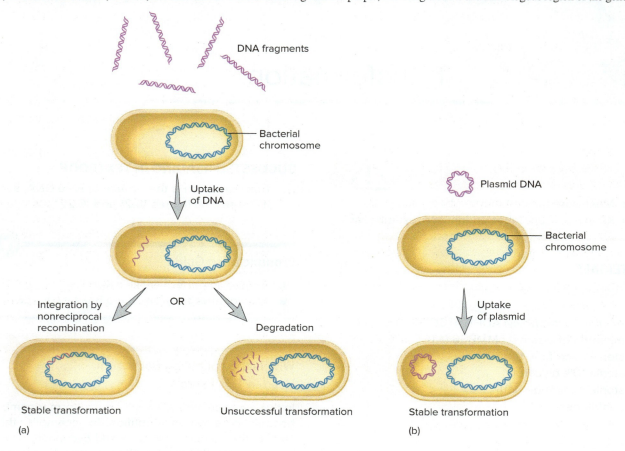

competent, meaning that they possess the cellular machinery to bring exogenous DNA inside the cell. Therefore, transformation is an important route of genetic exchange in nature. Many well-studied bacteria cannot take up DNA on their own, but can be made artificially competent by manipulation in the laboratory. One method for making cells artificially competent is called **electroporation.** In this technique, an electrical pulse is used to permeabilize the bacterial cell envelope, allowing uptake of exogenous DNA. This technique is fast and efficient, but it requires the use of specialized equipment (**Figure 52.2**).

In this introductory transformation exercise, you will transform a broad-host range plasmid pMMB66 into both *Acinetobacter baylyi* and *Escherichia coli*. The wild-type version of both organisms is sensitive to the antibiotic ampicillin. Plasmid pMMB66 encodes resistance to ampicillin. Thus, you will use the pressure of antibiotic resistance to select for only those bacteria that were able to take up pMMB66 from the environment.

Figure 52.2 Electroporation Apparatus. This is an example of a commercially available electroporation apparatus. *(Javier Izquierdo/McGraw Hill)*

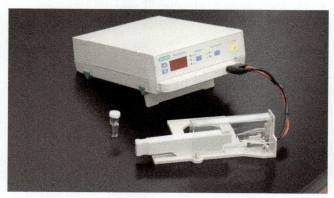

Procedure

Preparation and Transformation of Naturally Competent Cells

1. Obtain cultures of *A. baylyi* and *E. coli* that have been grown overnight in LB broth at 30°C.

2. In a sterile test tube, combine 1 ml fresh LB broth, ~100 ng of transforming pMMB66 plasmid DNA, and 70 µl of the *A. baylyi* culture. Label this tube "nAb + DNA."
3. In another tube, combine 1 ml of fresh LB broth and 70 µl of *A. baylyi*. Label this tube "nAb − DNA."
4. Repeat steps 2 and 3 for the *E. coli* culture, labeling these tubes "nEc + DNA" and "nEc − DNA."
5. Incubate the tubes at 30°C for at least 3 hours (preferably overnight).
6. Make and plate dilutions (10^{-1} to 10^{-3}) of the transformed culture onto selective and non selective LB agar plates. Plate 100 µl of each dilution. (*Note:* For *A. baylyi*, plate on LB supplemented with ampicillin at 150 mg/l. For *E. coli*, plate on LB supplemented with ampicillin at 100 mg/l.) Incubate the plates at 30°C overnight.
7. Observe your plates and note where you observe growth. Record your observations in the lab report for this exercise.

Preparation and Transformation of Electrocompetent Cells

1. Obtain cultures of *A. baylyi* and *E. coli* that have been grown overnight in LB broth at 30°C.
2. Separately dilute each overnight culture 1:100 in 50 ml of fresh LB broth in 250-ml flasks. Grow the cultures at 30°C for 2–3 hours while shaking until the cells reach an $OD_{600\,nm}$ of 0.3–0.5.
3. Place the cells on ice for 20 minutes. (*Note:* For the rest of this experiment, keep the cells on ice at all times unless otherwise directed).
4. Harvest the cells by centrifugation at $5,000 \times g$ for 10 minutes at 4°C. Wash the cell pellets 3× in an equal volume of sterile, ice-cold 10% glycerol. Pellet the cells between each wash with another round of centrifugation at $5,000 \times g$ for 10 minutes at 4°C.
5. After the final wash, resuspend the cell pellet in 1/10 the starting volume of 10% glycerol.
6. Label four microfuge tubes as follows: "eAb + DNA, eAb − DNA, eEc + DNA, and eEc − DNA".
7. Gently add 50 µl of the cells to the appropriately labeled tube.
8. In the tubes labeled "+ DNA," add 1 µl of the transforming plasmid pMMB66 DNA and mix gently. Allow the cells to rest on ice for 5 minutes.
9. Transfer the cells from the microfuge tube into a pre-chilled electroporation cuvette. (*Note:* Use cuvettes with a 0.1-cm gap.) Pulse the cuvettes in the electroporation apparatus according to manufacturer's protocols.
10. Immediately recover the cells in 700 µl of LB pre-warmed to 30°C. Transfer the recovery mixture to a sterile culture tube. Incubate the cells at 30°C while shaking for 1 hour.
11. Spread the cells over several selective and non-selective agar plates using a cell spreader as follows. Plate 20 µl and 200 µl of the *E. coli* electroporations on LB and LB supplemented with 100 mg/l ampicillin. Plate 20 µl and 200 µl of the *A. baylyi* electroporations on LB and LB supplemented with 150 mg/l ampicillin. Incubate the plates overnight at 30°C.
12. Observe your plates and note where you observe growth. Record your observations in the lab report for this exercise.

Disposal

When you are finished with all of the pipettes, plates, and tubes, place them in the designated place for disposal.

NOTES

Laboratory Report 52

Name: _____

Date: _____

Lab Section: _____

Transformation

1.

Growth of colonies after overnight incubation								
	nAb +DNA	nAb −DNA	eAb +DNA	eAb −DNA	nEc +DNA	nEc −DNA	eEc +DNA	eEc −DNA
LB − ampicillin								
LB + ampicillin								

2. Based on your results, which organism do you think is naturally competent? Explain your answer.

ASSESSMENT
Critical Thinking and Learning Outcomes Review

1. How would you define bacterial transformation?

2. What is the difference between natural and artificial competence?

3. What is electroporation?

4. How does transformation of plasmid DNA differ from transformation of DNA fragments? Explain your answer.

5. From an evolutionary point of view, why is transformation important?

6. What are some important uses of transformation in the laboratory?

EXERCISE 53 Conjugation

SAFETY CONSIDERATIONS

- Sterilize all loops before use; cool slightly before inoculum pick-up to avoid killing the bacteria on contact.
- When finished, all cultures and tubes will be autoclaved at 121°C for 15 minutes.
- In the event that a culture is accidentally spilled during handling, it is imperative to flood the immediate area with disinfectant solution.

MATERIALS

- suggested bacterial strains: *Escherichia coli* ATCC 23740 (Hfr donor) and *Escherichia coli* ATCC 23724 (F^- recipient). Both strains should be grown in a rich medium like LB or tryptic soy broth
- M9 agar plates supplemented with thiamine (100 mg/l)
- M9 agar plates supplemented with thiamine (100 mg/l) and streptomycin (25 mg/l)
- M9 agar plates supplemented with thiamine, leucine, and threonine (all at 100 mg/l), and streptomycin (25 mg/l)
- M9 broth (without additives)
- dilution tubes containing 0.9 ml M9 broth
- Bunsen burner
- inoculating loop
- vortex mixer
- 37°C incubator
- wax pencil or lab marker
- container for biohazard waste
- safety glasses
- disposable gloves
- lab coat

LEARNING OUTCOMES

Upon completion of this exercise, students will demonstrate the ability to

1. Perform conjugal transfer of DNA between donor and recipient strains of *E. coli*
2. Describe how the process of conjugation allows for horizontal gene transfer
3. Explain how prototrophs and auxotrophs are used in conjugation experiments

SUGGESTED READING IN TEXTBOOK

1. Antibacterial Drugs, section 9.4; see also tables 9.1 and 9.2.
2. Antimicrobial Drug Resistance Is a Public Health Threat, section 9.8.
3. Conjugation Requires Cell-Cell Contact; section 16.6.

Pronunciation Guide

- *Escherichia coli* (esh-er-I-ke-a KOH-lie)

Why Is the Following Bacterium Used in This Exercise?

In this exercise, you will perform a bacterial conjugation experiment. The authors have selected two strains of *Escherichia coli* for this purpose. The ATCC 23740 donor strain of *E. coli* has an F factor integrated into its chromosome, making it an Hfr strain. The ATCC 23724 recipient carries a mutation that confers resistance to the antibiotic streptomycin. It also contains several mutations that prevent biosynthesis of some amino acids. Recombinants of these two strains will have different combinations of either the donor or the recipient strains' characteristics and can be readily detected by using selective plating media.

PRINCIPLES

Conjugation describes the mating between two bacteria, when DNA is transferred from one cell to the other. In order for conjugation to occur, there must be direct physical contact between bacteria. The **donor** bacterium produces a special structure called the sex pilus which extends from the cell surface and makes contact with a **recipient** cell (**Figure 53.1**). The genes for sex pilus formation are encoded on conjugative plasmids, the most famous of which is the **F factor.**

Plasmids are small, circular, double-stranded DNA molecules that can act as extrachromosomal elements (i.e., they can carry genetic information and exist apart from the chromosome). These small pieces of DNA reproduce independently of the bacterial chromosome. In some cases, plasmids can become integrated into the chromosome. When this happens in the case of the F factor, the resulting strain is known as an **Hfr** (for high frequency of recombination). Because Hfr strains still contain the genes necessary to construct sex pili and initiate DNA transfer, these strains will attempt to transfer their entire chromosome when used to mate with a recipient cell (**Figure 53.2**). This fact was critical for mapping the *E. coli* chromosome prior to the advent of modern DNA sequencing technologies.

In this experiment, you will perform a mating between two *E. coli* strains: *E. coli* ATCC 23724 (recipient F$^-$ cell) and *E. coli* ATCC 23740 (donor Hfr cell). ATCC 23724 is an **auxotrophic** strain (a mutant that has a growth factor requirement) for the amino acids leucine and threonine. It also carries a chromosomal gene for streptomycin resistance. ATCC 23740 is a derivative of the commonly used *E. coli* lab strain C600 that is sensitive to streptomycin. It is a **prototroph** for the amino acids leucine and threonine, meaning these nutrients do not need to be provided in the media for this strain to grow. By the end of this exercise, you will have harnessed a naturally occurring mechanism of **horizontal gene transfer** to construct a hybrid strain of *E. coli*.

Procedure

1. Prepare the mating mixture by washing the donor (**ATCC 23740**) and recipient (**ATCC 23724**) cells in M9 broth. Pellet 1 ml of cells from an overnight culture of donor and recipient by centrifugation for 5 minutes at $5,000 \times g$.
2. Pour off the supernatant in the biohazard waste container. Gently resuspend the cell pellets in 1 ml of M9 broth. Repeat the 5-minute centrifugation as before.
3. Pour off the supernatant and gently resuspend the cell pellets in 0.5 ml of M9 broth. Using a sterile inoculating loop, streak each strain for isolated colonies on all three types of M9 agar: + thiamine, + thiamine/streptomycin, and + thiamine/leucine/threonine/streptomycin. Incubate these plates at 37°C overnight.
4. Set up the mating mixture. Combine all of the washed donor and recipient cells together into a clean, sterile culture tube. Mix the cells gently and incubate them at 37°C for at least 3 hours. (*Note:* Longer incubations of up to 24 hours are encouraged.) Do not shake this tube as it will disrupt the mating pairs.
5. After incubation, vortex the tube to disrupt the mating pairs. Make serial dilutions of the mating reaction using the 0.9-ml M9 dilution tubes.
6. Plate 0.1 ml of each dilution onto all three types of M9 agar. Incubate the plates overnight at 37°C.
7. Observe your plates and record any growth in the lab report for this exercise.

Disposal

When you are finished with all of the pipettes, plates, and tubes, place them in the designated place for disposal.

Figure 53.1 Bacterial Conjugation. The F$^+$ cell to the left is covered with fimbriae (short, dark-staining appendages) and a sex pilus (long, light-staining appendage) connects to the F$^-$ recipient cell on the right. *(Dennis Kunkel Microscopy, Inc./Phototake)*

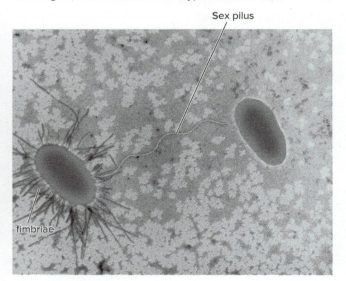

Figure 53.2 Conjugation Using an Hfr Strain. In an Hfr X F⁻ mating, the F factor is integrated into the chromosome of the Hfr donor strain. Upon donor-recipient contact via the sex pilus, DNA transfer begins from the origin of transfer *oriT*. Transfer of the chromosome from the donor to the recipient will continue until the mating pair is broken. Any incoming DNA must recombine into the recipient chromosome for it to be maintained.

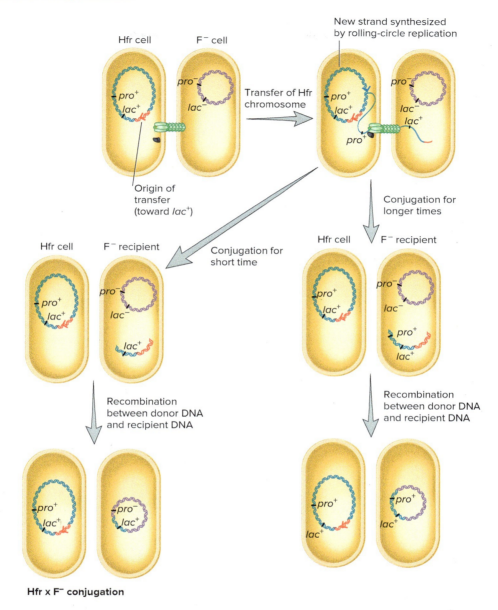

Hfr x F⁻ conjugation

Conjugation 383

NOTES

Laboratory Report 53

Name: _____

Date: _____

Lab Section: _____

Conjugation

1. What does growth (or lack of it) tell you about the genotype of the donor and the recipient on each of the following plates?

    a. M9 + thiamine

    b. M9 + thiamine + streptomycin

    c. M9 + thiamine + leucine + threonine + streptomycin

2. Where did you observe growth of the plated mating mixture?

3. Describe how cells from the mating mixture could grow on M9 plates supplemented with only thiamine and streptomycin.

4. In summary, what is the major purpose of this experiment?

ASSESSMENT
Critical Thinking and Learning Outcomes Review

1. What is the difference between a prototroph and an auxotroph?

2. Why do you think thiamine was included in the M9 broth used to propagate both the donor and the recipient strain?

3. In this experiment, which strain is the prototroph and which strain is the auxotroph? Explain your answer.

4. How is an Hfr strain different from a standard F^+ strain?

5. Do you think the results of your experiment would have been different if the mating occurred for 3 minutes instead of 3 hours? Explain your answer.

6. Do bacteria reproduce sexually? Explain.

EXERCISE 54 Generalized Transduction

SAFETY CONSIDERATIONS

- Chloroform should be handled carefully and used inside a chemical fume hood.
- Exercise caution when handling phage. Any spills should be cleaned up immediately to prevent contamination of other bacterial cultures in your lab.

MATERIALS

- sterile culture tubes
- sterile glass vials with screwcap lid
- fume hood
- vortex mixer
- micropipettes and sterile tips
- sterile cell spreaders
- LB plates
- LB plates supplemented with kanamycin (25 µg/ml)
- LB broth supplemented with calcium chloride (5 mM) and glucose (0.2%)
- LB broth supplemented with sodium citrate (0.1 M)
- transducing solution (5 mM calcium chloride and 10 mM magnesium sulfate)
- sodium citrate (1 M)
- chloroform
- broth cultures of *Escherichia coli*
  - wild-type (e.g., *E. coli* B)
  - *trxA732(del)::kan^R* (Yale University *E. coli* Genetic Stock Center)
- Phage P1
- container for biohazard waste
- safety glasses
- disposable gloves
- lab coat

LEARNING OUTCOMES

Upon completion of this exercise, students will demonstrate the ability to

1. Describe how phage-mediated cell lysis can be visualized in bacterial cultures
2. Explain how phages can be used to move genetic material between bacteria
3. Understand why phages are useful tools for molecular biologists
4. Understand the mechanism of phage infection
5. Use different phages to transfer antibiotic resistance to another bacterial strain

Why Is The Following Phage Used in This Exercise?

Phage P1 has been a mainstay of bacterial genetics for decades. Even with the explosion of modern molecular biology techniques, routine genetic manipulation (i.e., strain building) in *E. coli* is still accomplished using generalized transduction with phage P1. Phage particles are robust and efficient reagents that allow creative geneticists to conduct exquisitely complicated experiments.

SUGGESTED READING IN TEXTBOOK

Transduction Is Virus-Mediated DNA Transfer, section 16.8.

Pronunciation Guide

- *Escherichia coli* (esh-er-I-ke-a KOH-lie)

PRINCIPLES

Just like humans, bacteria are susceptible to viral infections. **Bacteriophages,** or phages for short, are viruses that infect bacterial cells. To initiate infection, phages first bind to specific structures on the bacterial cell surface, often proteins. Each phage particle must then inject its genetic material so that it reaches the bacterial cytoplasm. The phage genome is then replicated and expressed, leading to synthesis and assembly of new progeny phage particles. Phage-encoded proteins degrade the bacterial cell envelope, which will eventually kill the cell. This scenario describes a **lytic** infection (**Figure 54.1**). During the final stages of a lytic phage infection, the components of the bacterial cell begin to fragment. The newly synthesized phage are finally released from the cytoplasm when the host bacterium bursts, allowing the infection to spread to other cells.

Bacteria that grow in the environment are constantly exposed to bacteriophages. While you might think that a phage infection is detrimental to bacterial health, this is not always the case. Sometimes phage infections are beneficial to a bacterium given that phages can move genetic information between bacterial cells. In fact, this is an important mechanism by which genetic diversity increases in bacterial populations. For example, phages can carry genes that allow growth on metabolites that the host bacterium was previously unable to use. Phages can also move genes that confer resistance to antibiotics, or encode toxins that are harmful to human cells.

Studies of bacteriophages provided much of the framework for understanding the central dogma of molecular biology. Many of these fundamentals were established using phage like **lambda** and the so-called **T phages.** Other phages, like **P1** (**Figure 54.2**) and **P22,** gained widespread use for their ability to transfer genetic material between bacterial cells. Sometimes during the end stages of a lytic phage infection, pieces of the bacterial chromosome get packaged into the P1 phage head by mistake. The resulting particles are not capable of sustaining further lytic infections, yet they retain some key functions. Importantly, when such particles encounter another bacterial cell, they are still able to attach to their cognate receptor and inject their genetic material just like a normal phage. However, since the injected genetic material is derived from a bacterial chromosome, no phage genetic material will be delivered into the host cell. Without this blueprint, no additional infectious phage particles will be produced. It is this unique set of circumstance that allows these "one-hit wonder" phages to move DNA between bacterial cells in a technique called **generalized transduction** (**Figure 54.3**).

Figure 54.1 **Lytic Phage Infection of *E. coli*.** Bacterial cells were mixed with viral particles and then overlayed on an agar plate. Following overnight incubation, a lawn of bacterial growth appears. The circular zones of clearing, called plaques, are due to lysis of infected cells. *(Lisa Burgess/McGraw Hill)*

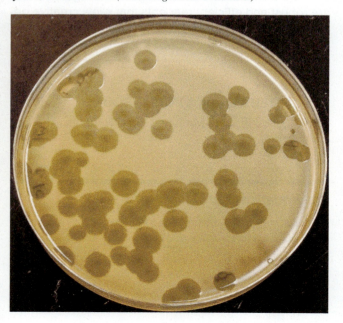

Figure 54.2 **Transmission Electron Micrograph of Bacteriophage P1.** *(B. Heggeler/Biozentrum, University of Basel/Science Source)*

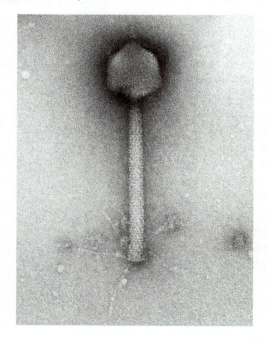

Figure 54.3 Overview of Generalized Transduction. In this example, a P1 phage lysate is raised on a *his⁺ lys⁺* strain of bacteria. Infrequently, a fragment of the bacterial chromosome containing the *his⁺* gene (blue) is mistakenly packed into a newly synthesized P1 particle. This particle can then be used to infect a *his⁻ lys⁻* strain of bacteria. Since this fragment of DNA has no way of replicating, the only way it will survive is to integrate into the host cell chromosome by recombination. The resulting cell will thus become *his⁺ lys⁻*.

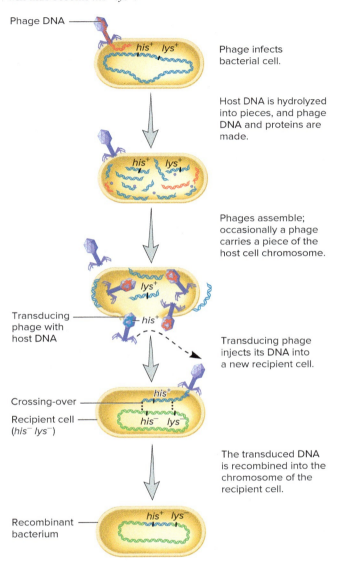

During generalized transduction, any piece of a bacterial chromosome can become packed inside a phage particle. Since the packaging of bacterial DNA is random, it is possible for any piece of the bacterial chromosome to be moved in this way. This differs from **specialized transduction,** where only a few particular genes can be moved between bacterial chromosomes via phage. Once the bacterial DNA is released into the cytoplasm by a transducing phage, this DNA must integrate into the chromosome or risk being degraded by host cell nucleases. As a result, a bacterium can gain new genetic material—and whatever properties this material encodes—after the transduction is complete. In this exercise, you will use phage P1 to move a drug-resistant cassette into a kanamycin-sensitive strain of *E. coli*.

Procedure

First Period

Preparation of Phage Lysate

1. Obtain two tubes containing 5 ml of LB supplemented with $CaCl_2$ and glucose. Label one tube "lysate" and the other tube "control."
2. Inoculate each tube by diluting an overnight culture of *E. coli trxA732(del)::kan* 1:100.
3. Grow the culture with aeration in a shaker or roller drum at 37°C for 30 minutes or until the culture starts to appear turbid.
4. Add 0.05 ml of P1 to the tube labeled "lysate." Do not add P1 to the "control" tube. Continue incubation for at least 2 hours.
5. Remove the two tubes from the incubator and note any differences in appearance. Discard the control tube.
6. In a fume hood, carefully add 0.2 ml of chloroform to the remaining tube and vortex. This will destroy any remaining viable bacterial cells. P1 phage particles are resistant to chloroform treatment.
7. Centrifuge the tube at maximum speed for 10 minutes.
8. Without disturbing the chloroform layer/cellular debris at the bottom of the tube, transfer the supernatant to a sterile glass vial and secure the screwcap.
9. For storage, add 0.1 ml of chloroform to the vial and store at 4°C.

Preparation of Transducing Cells

1. Centrifuge 5 ml of an overnight broth culture of wild-type *E. coli* for 5 minutes at $5,000 \times g$.
2. Pour off the culture supernatant in the biohazard waste container.
3. Resuspend the cell pellet in 2.5 ml of the transducing solution.

Generalized Transduction of trxA732(del)::kan

1. Obtain three new sterile culture tubes. Label the tubes "phage only," "cells only," or "transduction."
2. Carefully pipet 0.1 ml of the transducing cells into the "cells only" and "transduction" tubes. Add 0.05 ml of the P1 lysate prepared above to the "phage only" and "transduction" tubes.
3. Gently mix the tubes and incubate them without shaking at 30°C for 30 minutes. This will allow the phage time to adsorb to the cells.
4. Add 1 ml of LB + sodium citrate to each tube. Incubate the tubes at 37°C in a shaker or on a roller drum for 1 hour. Sodium citrate will chelate the excess divalent cations which are required for P1 to bind the bacterial cells.
5. Remove the tubes from the incubator and centrifuge the tubes for 10 minutes at $5,000 \times g$. Pour off the supernatant in the biohazard waste container.
6. Resuspend the cell pellet in 0.1 ml of 1 M sodium citrate. Vortex the tubes vigorously to make sure the cell pellets go into solution.
7. Obtain three LB plates supplemented with kanamycin, and three LB plates without antibiotics. Label one of each plate as follows: "phage only," "cells only," or "transduction."
8. From each tube, pipet half the solution (about 0.05 ml) onto an LB plate and the other half onto an LB kanamycin plate. Spread the solution across the surface of each plate using a sterile cell spreader.
9. Invert the plates and incubate them overnight at 37°C.

Second Period

1. The next day, observe the plates for signs of growth.
2. Determine if your transduction was successful and record observations for each of the plates in the report for this exercise.

Disposal

When you are finished with the plates tubes, and pipette tips, dispose of them in the biohazard waste container.

> **HELPFUL HINTS**
>
> - Carefully monitor the turbidity of the tube when preparing a P1 lysate. If the tube starts becoming more turbid after 40 minutes, add more P1 to the tube.
> - Vortex tubes vigorously to ensure mixing of chloroform.

Laboratory Report 54

Name: _____

Date: _____

Lab Section: _____

Generalized Transduction

1. Describe the appearance of the "lysate" and "control" tubes during the preparation of the P1 lysate. How did these two tubes look compared to a tube of sterile LB broth?

2. Complete the following table after observation of your transduction plates.

	"Phage only"	"Cells only"	"Transduction"
LB			
LB kanamycin			

ASSESSMENT
Critical Thinking and Learning Outcomes Review

1. What does an increase in turbidity of a bacterial culture indicate?

2. How does phage lysis appear in a broth culture of growing bacterial cells?

3. Name two well-studied bacteriophages.

4. How is it possible that any gene can be transferred via generalized transduction?

5. How does generalized transduction differ from specialized transduction?

6. Imagine your lab partner forgot to supplement the LB with 0.1 M sodium citrate. What consequence would this error have for the transduction experiment?

7. For the "phage only" and "cells only" controls, did you expect to see growth on either LB or LB kanamycin? Why or why not? In each case, what does growth or absence of growth indicate?

APPENDIX A
Dilutions with Sample Problems

Dilutions can be divided into five categories:

1. Diluting a solution to an unspecified final volume in just one step.
2. Diluting a solution to a specified final volume in just one step.
3. Diluting a solution to an unspecified final volume in several steps.
4. Diluting a solution to a specified final volume in several steps.
5. Serial dilutions.

Diluting a Solution to an Unspecified Volume in Just One Step

One must first calculate how many times to dilute (this is called the **dilution factor**) the initial material (**stock solution**) to obtain the final concentration. To accomplish this type of dilution, use the following formula:

$$\frac{\text{Initial Concentration (IC)}}{\text{Final Concentration (FC)}} = \text{Dilution Factor (DF)}$$

For example, if you want to dilute a solution with an initial concentration of solute of 5% down to 1%, using the above formula gives

$$\frac{\text{IC}}{\text{FC}} = \text{DF} = \frac{5\%}{1\%} = 5$$

Thus, in order to obtain a 1% solution from a 5% solution, the latter must be diluted five times. This can be accomplished by taking one volume (e.g., cc, ml, liter, gallon) of the initial concentration (5%) and adding four volumes (e.g., cc, ml, liter, gallon) of solvent for a total of five volumes. Stated another way, 1 ml of a 5% solution + 4 ml of diluent will give a total of 5 ml, and each ml contains 1% instead of 5%.

Diluting a Solution to a Specified Volume in Just One Step

First, calculate the number of times the initial concentration must be diluted by dividing the final concentration (FC) into the initial concentration (IC).

Second, divide the number of times the initial concentration must be diluted into the final volume specified to determine the aliquot (or portion) of the initial concentration to be diluted.

Third, dilute the aliquot of the initial concentration calculated in step 2 by the volume specified.

For example, you have a 10% solution and want a 2% solution. However, you need 100 ml of this 2% solution.

$$\frac{\text{IC}}{\text{FC}} = \text{DF} = \frac{10\%}{2\%} = 5$$

Divide the number of times the 10% solution must be diluted (DF) into the final volume specified:

$$\frac{100 \text{ ml}}{5} = 20 \text{ ml}$$

Dilute the portion of 10 to the volume specified:

20 ml of a 10% solution + 80 ml of diluent = 100 ml (each milliliter = 2%)

Another method for performing this type of dilution is to use the following formula:

$$\frac{C_1}{C_2} = \frac{V_2}{V_1} \text{ or } C_1V_1 = C_2V_2$$

C_1 = standard concentration available
C_2 = standard concentration desired
V_2 = final volume of new concentration
V_1 = volume of C_1 required to make the new concentration

For example, if you want to prepare 100 ml of 10% ethyl alcohol from 95% ethyl alcohol, then

$$C_1 = 95\%, C_2 = 10\%,$$
$$V_2 = 100 \text{ ml},$$
$$V_1 = x \text{ and}$$
$$\frac{95}{10} = \frac{100}{x}$$
$$x = 1{,}000/95$$
$$x = 10.5 \text{ ml}$$

Thus 10.5 ml of 95% ethyl alcohol + 89.5 ml of H_2O = 100 ml of a 10% ethyl alcohol solution.

Diluting a Solution to an Unspecified Volume in Several Steps

Frequently in the microbiology laboratory, large dilutions must be employed. They cannot be done in one step because they are too large. As a result, they must be done in several

A-1

steps to conserve not only amounts of diluent to be used but also space. For example, a 0.5 g/ml solution diluted to 1 µg/ml is a 500,000-fold dilution.

$$0.5 \text{ g} = 0.5 \text{ g} \times 10^6 \text{ µg/g}$$
$$= 500,000 \text{ µg}$$

To obtain a solution containing 500,000 µg/ml in one step would require taking 1 ml of 0.5 gm/ml stock solution and adding 499,999 ml of diluent. As you can see, it would be almost impossible to work with such a large fluid volume.

A 500,000 times dilution can be easily performed in two steps by (1) taking 1 ml of the initial concentration and diluting it to 500 ml, and, (2) diluting 1 ml of the first dilution to 1,000 ml.

1 ml of 500,000 µg/ml + 499 ml of diluent = 1,000 µg/ml
1 ml of 1,000 µg/ml + 999 ml of diluent = 1 µg/ml

Thus by this two-step procedure, we have cut down the volume of diluent used from 499,999 ml to 1,498 ml (499 ml + 999 ml).

Dilution Ratios Used in This Manual

According to the *ASM* (*American Society for Microbiology*) *Style Manual*, dilution ratios may be reported with either colons (:) or shills (/), but note there is a difference between them. A shill indicates the ratio of a part to a whole: for example, 1/2 means one of two parts, with a total of two parts. A colon indicates the ratio of one part to two parts, with a total of three parts. Thus 1/2 equals 1:1, but 1:2 equals 1/3.

Diluting a Solution to a Specified Volume in Several Steps

This type of dilution is identical to all previous dilutions with the exception that the specified final volume must be one factor of the total dilution ratio.

For example, you want a 1/10,000 dilution of whole serum (undiluted) and you need 50 ml.

Divide dilution needed by the volume:

$$\frac{10,000}{50} = 200$$

200 (1/200 dilution) = the first step in the dilution factor; the second is 1/50, obtained as follows:

1 ml of serum + 199 ml of diluent = 1/200 dilution
1 ml of 1/200 dilution + 49 ml of diluent = 1/50
To check: 50 × 200 = 10,000

Serial Dilutions

The usefulness of dilutions becomes most apparent when small volumes of a material are required in serological procedures. For example, if 0.01 ml of serum were required in a certain test, instead of measuring out this small volume with a consequent sacrifice of accuracy, it would be more advantageous to dilute the serum 100 times. One ml of this 1/100 dilution would then contain 0.01 ml of the serum. Each ml of this dilution would be equivalent to the required 0.01 ml of undiluted serum.

Dilutions represent fractional amounts of a material and are generally expressed as the ratio of one volume of material to the final volume of the dilution. Thus a 1/10 dilution of serum represents 1 volume of serum in 10 volumes of dilution (1 volume of serum + 9 volumes of diluent). Undiluted serum may be expressed as 1/1.

1 ml of serum + 1 ml of saline may be expressed as 1/2. Each milliliter of this dilution is equivalent to 0.5 ml of undiluted serum.

1 ml of serum + 2 ml of saline may be expressed as 1/3. Each milliliter of this dilution is equivalent to 0.33 ml of undiluted serum.

1 ml of serum + 99 ml of saline may be expressed as 1/100. Each milliliter of this dilution is equivalent to 0.01 ml of undiluted serum.

From the above, one can see that dilution expressions are fractions written as ratios where the numerator is unity and the denominator is the dilution value.

Division of the numerator by the denominator will give the amount of material per milliliter of the dilution. For example, in a 1/25 dilution, 1/25 = 0.04. Therefore, each milliliter of this dilution contains 1/25 or 0.04 ml of the original material (e.g., serum).

To convert a ratio into a dilution expression, divide both the numerator and denominator by the value of the numerator. For example, in a mixture consisting of 4 ml of serum and 6 ml of saline,

4 ml of serum + 6 ml of saline
= 10 ml of serum dilution
Ratio of serum dilution = 4/10

Dividing both numerator and denominator by the numerator value (4),

$$\frac{4 \div 4}{10 \div 4} = 1/2.5 = \text{serum dilution}$$

In the preparation of dilutions, any multiple or submultiple of the constituent volumes may be used. For example,

1/30 dilution = 1 ml serum + 29 ml saline
= 0.5 ml serum + 14.5 ml saline
= 2 ml serum + 58.0 ml saline

Serial dilutions indicate that an identical volume of material is being transferred from one vessel to another. The purpose of this procedure is to increase the dilutions of a substance by certain increments. For example, in a twofold dilution, the dilution factor is doubled each time (e.g., 1/2, 1/4, 1/8, etc.). See table on page A-3 for further examples.

Tube	Amount of Saline Added to Each Tube	Serum	Protocol	Final Dilution of Serum
1	0	2 ml	Remove 1 ml to tube 2	1/1
2	1 ml	1 ml of 1/1	Mix, remove 1 ml to tube 3	1/2
3	1 ml	1 ml of 1/2	Mix, remove 1 ml to tube 4	1/4
4	1 ml	1 ml of 1/4	Mix, remove 1 ml to tube 5	1/8
5	1 ml	1 ml of 1/8	Mix, remove 1 ml to tube 6	1/16
6	1 ml	1 ml of 1/16	Mix, discard 1 ml	1/32

Sample Problems

Diluting a Solution to an Unspecified Volume in Just One Step

Problem 1
Dilute a solution that has an initial concentration of solute of 10% down to 2%.

Problem 2
Dilute a solution that has an initial concentration of 0.01% to 0.0001%.

Diluting a Solution to a Specified Volume in Just One Step

Problem 3
You have a 0.01% solution. You want a 0.001% solution and you need 25 ml.

Problem 4
Prepare 50 ml of a 3% solution from a 4% solution.

Problem 5
Prepare 100 ml of 10% alcohol from 95% alcohol.

Problem 6
You need 45 ml of 50% alcohol. How can you prepare this from 70% alcohol?

Problem 7
What volume of a 20% dextrose broth solution should be used to prepare 250 ml of 1% dextrose broth?

Problem 8
Prepare 45 ml of a 2% suspension of RBCs from a 5% suspension.

Problem 9
What volume of a 0.02% solution can be prepared from 25 ml of a 0.1% solution?

Problem 10
What percent concentration of alcohol is prepared when 10 ml of 95% alcohol is diluted with H_2O to make a final volume of 38 ml?

Problem 11
What percent of dextrose is prepared when 3 ml of a 10% dextrose solution is mixed with 12 ml of broth?

Diluting a Solution to an Unspecified Volume in Several Steps
Problem 12
You have a stock solution of protein containing 10 g/ml. You want a concentration of 2 mg/ml. How would you perform this dilution in just three steps?

Problem 13
You have a stock solution of 10 mg/ml of vitamins and want to obtain a solution of 0.5 µg/ml. How would you perform this dilution in just three steps?

Diluting a Solution to a Specified Volume in Several Steps
Problem 14
You want a 1:128 dilution of serum and you need 4 ml. How would you perform this dilution in several steps?

Problem 15
You want a 1:3,000 dilution of serum and you need 2 ml. How would you perform this dilution in several steps?

Problem 16
How would you prepare 1 ml of a 1:5 dilution of serum?

Problem 17
How would you prepare 8 ml of a 1:20 dilution of serum?

Problem 18
Based on the following dilutions, how many bacteria were present in the original sample? _____

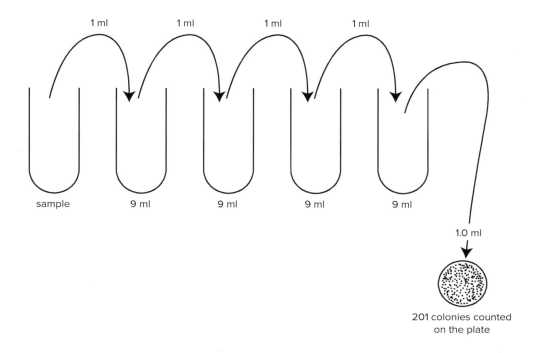

201 colonies counted on the plate

Problem 19
How many bacteria were present in the following sample? _____

Dilution	Amount Plated	Number of Colonies Counted
10^{-4}	1.0 ml	256
10^{-5}	1.0 ml	28
10^{-6}	1.0 ml	7

Problem 20
If 0.1 ml of a urine culture from a 10^{-6} dilution yielded 38 colonies, how many bacteria were there per milliliter in the original sample? _____

Problem 21
How many bacteria were present in the following sample? _____

Dilution	Volume Plated	Number of Colonies Counted
10^{-7}	0.1 ml	26

Problem 22

How many bacteria were present in the following sample? _____

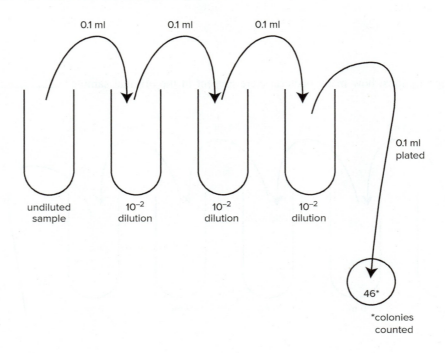

Problem 23

How many bacteria were there in the original sample? _____

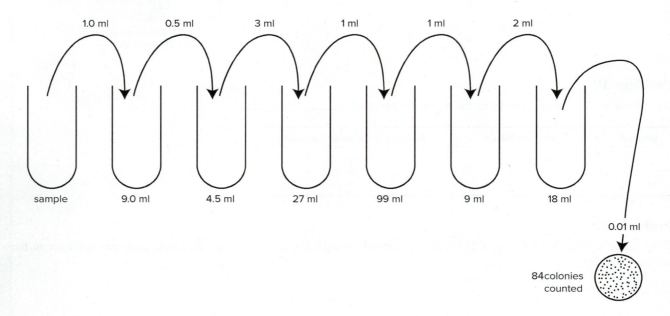

Answers to Sample Problems

Problem 1

Take 1 volume of the initial concentration (10%) and 4 volumes of solvent for a total of 5 volumes: 1 ml of a 10% solution + 4 ml of diluent will give a total of 5 ml and each ml will now contain 2% instead of 10%.

Problem 2

In order to obtain a 0.0001% solution from a 0.01% solution, the latter must be diluted 100 times. Therefore, take 1 volume of the initial concentration (0.01%) and add 99 volumes of solvent for a total of 100 volumes. 1 ml of 0.01% solution + 99 ml of diluent will give a total of 100 ml and each ml now contains 0.0001% instead of 0.01%.

Problem 3

A 0.01% solution is 10 times more concentrated than the 0.001% solution. Divide the number of times the 10 must be diluted into the final volume specified: 25/10 = 2.5. 2.5 ml of a 0.01% solution + 22.5 ml of diluent will give a total of 25 ml and each ml now contains 0.001%.

Problem 4

$C_1 = 4\%$
$C_2 = 3\%$
$V_2 = 50$ ml
$V_1 = x$

$$\frac{4\%}{3\%} = \frac{50}{x} \qquad 4x = 37.5 \text{ ml}$$

Thus 37.5 ml of a 4% solution + 12.5 ml of solvent = 50 ml of a 3% solution.

Problem 5

$C_1 = 95\%$
$C_2 = 10\%$
$V_2 = 100$ ml
$V_1 = x$

$$\frac{95\%}{10\%} = \frac{100}{x} \qquad x = \frac{1{,}000}{95} \qquad x = 10.5 \text{ ml}$$

Thus 10.5 ml of 95% alcohol + 89.5 ml of H_2O = 100 ml of a 10% alcohol solution.

Problem 6

$C_1 = 70\%$
$C_2 = 50\%$
$V_2 = 45$ ml
$V_1 = x$

$$\frac{70\%}{50\%} = \frac{45}{x} \qquad x = \frac{50 \times 45}{95} \qquad x = 32.1 \text{ ml}$$

Thus 32.1 ml of 70% alcohol + 12.9 ml of H_2O = 45 ml of a 50% alcohol solution.

Problem 7

$C_1 = 20\%$
$C_2 = 1\%$
$V_2 = 250$ ml
$V_1 = x$

$$\frac{20\%}{1\%} = \frac{250}{x} \qquad 20x = 250 \qquad x = 12.5 \text{ ml}$$

Thus 12.5 ml of a 20% solution + 237.5 ml of broth = 250 ml of a 1% dextrose broth solution.

Problem 8

$C_1 = 5\%$
$C_2 = 2\%$
$V_2 = 45$ ml
$V_1 = x$

$$\frac{5\%}{2\%} = \frac{45 \text{ ml}}{x} \qquad x = \frac{90}{5} \qquad x = 18 \text{ ml}$$

Thus 18 ml of 5% + 27 ml of saline = 45 ml of a 2% red blood cell suspension.

Problem 9

$C_1 = 0.1\%$
$C_2 = 0.02\%$
$V_2 = x$
$V_1 = 25$ ml

$$\frac{0.1\%}{0.02\%} = \frac{x}{25 \text{ ml}} \qquad 0.02x = 2.5 \qquad x = 125 \text{ ml}$$

Thus prepare a total volume of 125 ml of a 0.02% solution from 25 ml of 0.01% solution by adding 100 ml of diluent to the latter.

Problem 10

$C_1 = 95\%$
$C_2 = x$
$V_2 = 38$ ml
$V_1 = 10$ ml

$$\frac{95\%}{x} = \frac{38 \text{ ml}}{10 \text{ ml}} \quad 38x = 950 \quad x = 25\%$$

Problem 11

$C_1 = 10\%$
$C_2 = x$
$V_2 = 15$ ml
$V_1 = 3$ ml

$$\frac{10\%}{x} = \frac{15 \text{ ml*}}{3 \text{ ml}}$$

*3 + 12 = 15 ml (total volume)

$15x = 30 \quad x = 2\%$

Problem 12

In order to make all units equal, you have to convert 10 grams to milligrams.

Stock solution = 10,000 mg/ml
Final concentration = 2 mg/ml

$$\frac{IC}{FC} = \frac{10{,}000 \text{ mg/ml}}{2 \text{ mg/ml}} = 5{,}000$$

Thus the dilution factor is 5,000.

To perform a 1:5,000 dilution:
Step #1   1 ml of 10,000 mg/ml stock solution protein + 49 ml of diluent = 1:50 dilution = 200 mg/ml
Step #2   1 ml of 200 mg/ml + 9 ml of diluent = 1:10 dilution = 20 mg/ml
Step #3   1 ml of 20 mg/ml + 9 ml of diluent = 1:10 dilution = 2 mg/ml
To check to make sure the correct dilution was made: 50 × 10 × 10 = 5,000

Problem 13

Make all units the same.

10 mg × 1,000 µg/mg = 10,000 µg
Stock solution = 10,000 µg/ml
Final concentration = 0.5 µg/ml

$$\frac{IC}{FC} = DF = \frac{10{,}000 \text{ mg/ml}}{0.5 \text{ mg/ml}} = 20{,}000$$

Thus the dilution factor is 20,000.

Step #1   1 ml of 10,000 µg/ml + 19 ml of diluent = 1:20 dilution = 500 µg/ml (10,000/20 = 500)
Step #2   1 ml of 500 µg/ml + 99 ml of diluent = 1:100 dilution = 5 µg/ml (500/100 = 5)
Step #3   1 ml of 5 µg/ml + 9 ml of diluent = 1:10 dilution = 0.5 µg/ml (5/10 = 0.5)
To check to make sure the correct dilution was made:
20 × 100 × 10 = 20,000

To help you with these types of problems, always remember that volumes may change depending on the protocol, but the dilution factor will always remain the same. For example:

1 + 99			1 + 9	
10 + 990	1:100 dilution		10 + 90	1:10 dilution
0.1 + 9.9			0.5 + 4.5	
1 + 4			1 + 7	
10 + 40	1:5 dilution		5 + 35	1:8 dilution
0.1 + 0.4			0.5 + 3.5	
1 + 19			1 + 14	
10 + 190	1:20 dilution		5 + 70	1:15 dilution
0.1 + 1.9			0.5 + 7.0	
1 + 249				
10 + 2,490	1:250 dilution			
0.1 + 24.9				

Problem 14

The first step is to establish the initial dilution as follows:

$$\frac{128}{4} = 32$$

1:32 is the first step dilution, and the second is 1:4.
1 ml of serum + 31 ml of diluent = 1:32 (individual dilution)
1 ml of the 1:32 dilution = 3 ml of diluent = 1:4 (individual dilution)
To check to make sure the dilution was correctly made: $32 \times 4 = 128$

Problem 15

We can obtain a 1:3,000 dilution in 3 steps by using 1:30 and 1:10 dilutions.

$$\frac{3,000}{2} = 1,500$$

1 ml of serum + 29 ml of diluent = 1:30 (individual dilution)
1 ml of 1:30 dilution + 9 ml of diluent = 1:10 (individual dilution)
1 ml of 1:10 + 9 ml of diluent = 1:10 (individual dilution)
To check to make sure the dilution was correctly made: $30 \times 10 \times 10 = 3,000$

Problem 16

$\frac{D_1}{D_2} = \frac{V_1}{V_2}$

$D_1 = 1$
$D_2 = 5$
$V_1 = x$
$V_2 = 1$

$\frac{1}{5} = \frac{x}{1}$ $5x = 1$ $x = 0.2$ ml

Thus 0.2 ml (undiluted serum) + 0.08 ml of saline = 1.0.

Problem 17

$D_1 = 1$
$D_2 = 20$
$V_1 = x$
$V_2 = 8$

$\frac{1}{20} = \frac{x}{8}$ $20x = 8$ $x = 0.4$ ml

Thus 0.4 ml of undiluted serum + 7.6 ml of saline = 1:20.

Problem 18 2.01×10^6

Problem 19 2.8×10^6

Problem 20 3.8×10^8

Problem 21 2.6×10^9

Problem 22 4.6×10^8

Problem 23 8.4×10^{10}

APPENDIX B
Reagents, Solutions, Stains, and Tests

Reagents and stains appear in this appendix as the authors have presented the material in the individual laboratory exercises and are listed in alphabetical order. When necessary, methodology is given with the reagents, stains, or tests. The detailed procedures, however, are presented in the exercise in which their use is discussed.

Acid-Alcohol (for Ziehl-Neelsen stain)
- Concentrated hydrochloric acid .... 3 ml
- 95% ethyl alcohol .... 97 ml

Alcohol, 90%, 500 ml (from 95%)
- 95% alcohol .... 473 ml
- Distilled water .... 27 ml

Alcohol, 80%, 500 ml (from 95%)
- 95% alcohol .... 421 ml
- Distilled water .... 79 ml

Alcohol, 75%, 500 ml (from 95%)
- 95% alcohol .... 395 ml
- Distilled water .... 105 ml

Alcohol, 70%, 500 ml (from 95%)
- 95% alcohol .... 368 ml
- Distilled water .... 132 ml

Barritt's Reagent (for Voges-Proskauer test)
Solution A: 6 g of α-naphthol in 100 ml of 95% ethyl alcohol. Dissolve the α-naphthol in the ethanol with constant stirring.
Solution B: 40 g of potassium hydroxide in 100 ml of water. Store in the refrigerator.

Bile Solubility Test (10% bile)
- Sodium deoxycholate .... 1 g
- Sterile distilled water .... 9 ml

To test for bile solubility, prepare two tubes, each containing a sample of fresh culture (a light suspension of the bacterium in buffered broth, pH 7.4). To one tube add a few drops of a 10% solution of sodium deoxycholate. The same volume of sterile physiological saline is added to the second tube. If the bacterial cells are bile soluble, the tube containing the bile salt will lose its turbidity in 5 to 15 minutes and show an increase in viscosity.

Cleaning Solution for Glassware Strong:
- Potassium dichromate .... 20 g
- Distilled water .... 200 ml
- Dissolve dichromate in water; when cool, add very slowly:
- Concentrated sulfuric acid .... 9 parts
- 2% aqueous potassium dichromate .... 1 part

Copper Sulfate Solution (20%)
- Copper sulfate ($CuSO_4 \cdot 5H_2O$) .... 20 g
- Distilled water .... 80 ml

Crystal Violet Capsule Stain (1%)
- Crystal violet (85% dye content) .... 1 g
- Distilled water .... 100 ml

Decolorizers (for Gram stain)
1. Intermediate agent, 95% ethyl alcohol.
2. Fastest agent, acetone.
3. Slowest agent, acetone-isopropyl alcohol (isopropyl alcohol, 300 ml; acetone, 100 ml). For the experienced microbiologist, any one of the three decolorizing agents will yield good results.

Diphenylamine Reagent (for the nitrate test)
Working in a fume hood, dissolve 0.7 g of diphenylamine in a mixture of 60 ml of concentrated sulfuric acid and 28.8 ml of distilled water. Allow to cool. Slowly add 11.3 ml of concentrated hydrochloric acid. After the solution has stood for 12 to 24 hours, some of the base will separate. This indicates that the reagent is saturated.

Eosin Blue
- Eosin blue stain .... 1 g
- Distilled water .... 99 ml

Ferric Chloride Reagent
- $FeCl_3 \cdot 6H_2O$ .... 12 g
- 2% aqueous HCl .... 100 ml

The 2% aqueous HCl is made by adding 5.4 ml of concentrated HCl (37%) to 94.6 ml of distilled H_2O.

Gram's Iodine (Lugol's)
According to the *ASM Manual for Clinical Microbiology,* dissolve 2 g of potassium iodide in 300 ml of

distilled water and then add 1 g of iodine crystals. Rinse the solution into an amber bottle with the remainder of the distilled water. Discard when the color begins to fade.

Gram Stain

(A) **Crystal violet** (Hucker modification)

1. Crystal violet 85% dye............................ 2 g
   95% ethyl alcohol................................... 20 ml
   Mix and dissolve.
2. Ammonium oxalate .... 0.8 g
   Distilled water... 80.0 ml

Add solution A to solution B. Let stand for a day, then filter. If the crystal violet is too concentrated, solution A may be diluted as much as 10 times.

(B) **Gram's Iodine Solution** (mordant)

Iodine crystals... 1 g
Potassium iodide.. 2 g
Distilled water.. 300 ml

Store in an amber bottle; discard when the color begins to fade.

(C) **Safranin** (counterstain) **Solution**

Safranin... 2.5 g
95% ethyl alcohol....................................... 100.0 ml

For a working solution, dilute stock solution 1/10 (10 ml of stock safranin to 90 ml of distilled water).

India Ink (for capsule stain)

Mix the specimen with a small drop of India ink on a clean slide. If the India ink is too dark, dilute it to 50% with distilled water.

Kinyoun Acid-Fast Stain

(A) **Kinyoun Carbolfuchsin**

Basic fuchsin... 4 g
95% alcohol .... 20 g
Phenol crystals.. 8 g
Distilled water.. 100 ml

(B) **Acid-alcohol**

Concentrated hydrochloric acid.................. 3 ml
95% ethyl alcohol.. 97 ml

Methylene Blue Counterstain

Methylene blue .... 0.3 g
Distilled water... 100 ml

Kovacs' Reagent (for the indole test)

N-amyl or isoamyl alcohol .... 150 ml
Concentrated hydrochloric acid.................. 0 ml
p-dimethylaminobenzaldehyde................... 10 g

Working in a fume hood, dissolve the aldehyde in alcohol and then slowly add the acid. The dry aldehyde should be light in color. Alcohols that result in indole reagents that become deep brown should not be used. Store in a dark bottle with a glass stopper in a refrigerator when not in use.

Malachite Green Solution (for endospore stain)

Malachite green oxalate.............................. 5 g
Distilled water.. 100 ml

Methylene Blue (Löffler's alkaline)

Solution A: Dissolve 0.3 g of methylene blue (90% dye content) in 30 ml of 95% ethyl alcohol.

Solution B: Dissolve 0.01 g of potassium hydroxide in 100 ml of distilled water.
Mix solutions A and B. Filter with Whatman No. 1 filter paper before use.

Methylene Blue Stain (simple staining)

Methylene blue .... 0.3 g
Distilled water.. 100.0 ml

Methyl Red Reagent (for detection of acid)

Methyl red... 0.1 g
95% ethyl alcohol.. 300.0 ml

Dissolve the dye in alcohol and add sufficient distilled water to make 500 ml. Positive tests are red-orange, and negative tests are yellow.

Naphthol, Alpha (for the Enterotube II System)

5% α-naphthol in 95% ethyl alcohol.

Nessler's Reagent (for the ammonia test)

Working in a fume hood, dissolve 50 g of potassium iodide in 35 ml of cold (ammonia-free) distilled water. Add mercuric chloride drop by drop until a slight precipitate forms. Add 400 ml of a 50% solution of potassium hydroxide. Dilute to 1 liter, allow to settle, and decant the supernatant for use. Store in a tightly closed dark bottle.

Alternate procedure:

Solution A:

Mercuric chloride .... 1 g
Distilled water.. 6 ml
Dissolve completely.

Solution B:

Potassium iodide... 2.5 g
Distilled water.. 6.0 ml

Solution C:

Potassium hydroxide................................... 6 g
Distilled water.. 6 ml

Dissolve solution C completely and add to the mixture of solutions A and B. Add 13 ml of distilled water. Mix well and filter through Whatman No. 1 filter paper before use. Store in a dark, stoppered bottle.

Nigrosin Solution (Dorner's, for negative staining)

Water-soluble nigrosin................................ 10.0 g
Distilled water.. 100.0 ml
Formalin (40% formaldehyde).................... 0.5 ml

Gently boil the nigrosin and water approximately 30 minutes. Add 0.5 ml of 40% formaldehyde as a preservative. Filter twice through Whatman No. 1 filter paper and store in a dark bottle in the refrigerator.

Nitrate Test Reagent (see under diphenylamine)

Nitrite Test Reagents (Caution—solution B may be carcinogenic. Use safety precautions such as the avoidance of aerosols, mouth pipetting, and contact with skin.)

(A) **Solution A:** Dissolve 8 g of sulfanilic acid in 1 liter of 5 N acetic acid (1 part glacial acetic acid to 2.5 parts distilled water).

(B) **Solution B:** Dissolve 6 ml of N, N,-dimethyl-1-naphthylamine in 1 L of 5 N acetic acid.

DO NOT MIX SOLUTIONS.

Oxidase Test Reagent

Mix 1 g of dimethyl-p-phenylenediamine hydrochloride in 100 ml of distilled water. This reagent should be made fresh daily and stored in a dark bottle in the refrigerator.

***O*-nitrophenyl-β-D-Galactoside (ONPG)**

0.1 M sodium phosphate buffer .... 50.0 ml
ONPG (8×10^{-4} M) .... 12.5 mg

Phosphate Buffers

Stock buffers:

Alkaline buffer, 0.067 M Na_2HPO_4 solution. Dissolve 9.5 g of Na_2HPO_4 in 1 liter of distilled water.

Acid buffer, 0.067 M NaH_2PO_4 solution. Dissolve 9.2 g of $NaH_2PO_4 \cdot H_2O$ in 1 liter of distilled water.

Buffered water (pH 7.0 to 7.2)

Acid buffer (NaH_2PO_4) .... 39 ml
Alkaline buffer (Na_2HPO_4) .... 61 ml
Distilled water .... 900 ml

BE SURE GLASSWARE IS CLEAN. Buffered water, if sealed, is stable for several weeks.

Physiological Saline

Dissolve 8.5 g of sodium chloride in 1 liter of distilled water (0.85%) or 9 g in 1 liter of distilled water (0.9%).

Physiological Saline (Buffered)

Sodium chloride (0.85%; 8.5 g in 1 liter of distilled water) is buffered to pH 7.2 with 0.067 M potassium phosphate mixture.

Phosphate-Buffered Saline

10× stock solution, 1 liter
80 g NaCl
2 g KCl
11 g $Na_2HPO_4 \cdot 7H_2O$
2 g KH_2PO_4

Stock Solutions

1 M $CaCl_2$	1 M HCl
147 g	Mix in the following order:
$CaCl_2 \cdot 2H_2O$	913.8 ml H_2O
H_2O to 1 liter	86.2 ml concentrated HCl

1 M KCL	1 M $MgCl_2$
74.6 g KCl	20.3 g
H_2O to 1 liter	$MgCl_2 \cdot 6H_2O$
	H_2O to 100 ml

5 M NaCl	10 M NaOH
292 g NaCl	Dissolve 400 g
H_2O to 1 liter	NaOH in 450 ml of distilled water. Add water to 1 liter.

Triton X-100 Stock Solution (10%)

Triton X-100 .... 10 ml
Distilled water.. 90 ml

Mix and store in a tightly stoppered bottle at room temperature; the solution will keep indefinitely.

Trommsdorf's Reagent (for the nitrite test)

Working in a fume hood with a beaker on a hot plate, slowly add, with constant stirring, 100 ml of a 20% aqueous zinc chloride solution to a mixture of 4 g of starch in water. Continue heating until the starch is completely dissolved and the solution is clear. Dilute with water and add 2 g of potassium iodide. Dilute to 1 liter with distilled water, filter once through Whatman No. 1 filter paper, and store in a capped, dark bottle.

Vaspar

Melt 1 pound of Vaseline and 1 pound of paraffin. Store in small student-use bottles.

Voges-Proskauer Reagent (see Barritt's reagent)

West Stain (flagella)

Solution A:

Mordant: 50 ml of saturated aqueous aluminum potassium sulfate + 100 ml of 10% tannic acid solution + 10 ml of 5% ferric chloride. This solution should be stored in an aluminum foil-covered bottle at 5°C until used.

Solution B:

Stain: 7.5 g of silver nitrate ($AgNO_3$) in 150 ml of distilled water. While working in a fume hood, add concentrated NH_4OH dropwise to 140 ml of the silver nitrate solution while it is being stirred on a magnetic mixer. A brown precipitate will form at the start of NH_4OH addition. Enough NH_4OH should be added so that the brown precipitate just dissolves. Finally, add 5% silver nitrate dropwise until a faint cloudiness persists. This solution should be stored at 5°C in an aluminum foil-covered bottle until used.

Ziehl-Neelsen Acid-Fast Stain

(A) **Solution A:** Dissolve 0.3 g of basic fuchsin (90% dye content) in 10 ml of 95% ethyl alcohol.

(B) **Solution B:** Dissolve 5 g of phenol in 95 ml of distilled water.

Mix solutions A and B. Note: Add either 1 drop of Tergitol No. 4 per 30 ml of carbolfuchsin or 2 drops of Triton X-100 per 100 ml of stain for use in the heatless method. Tergitol No. 4 and Triton X act as detergents, emulsifiers, and wetting agents.

Acid-alcohol, 3%

Concentrated hydrochloric acid	3 ml
95% alcohol	97 ml

APPENDIX C
Culture Media

Sterilization of all tubed media is accomplished at 15 lb pressure (121°C) for 30 minutes unless otherwise specified. Longer sterilization times will be required for large volumes of media. Most of the media are available commercially in powdered form, with specific instructions for their preparation and sterilization.

Actidione (Cycloheximide) Agar (pH 5.5)

Glucose	50.0 g
Agar	15.0 g
Pancreatic digest of casein	5.0 g
Yeast extract	4.0 g
Potassium dihydrogen phosphate	0.5 g
Potassium chloride	0.42 g
Calcium chloride	0.12 g
Magnesium sulfate	0.12 g
Bromcresol green	22.0 mg
Actidione (cycloheximide)	10.0 mg
Ferric chloride	2.5 mg
Distilled water	1,000.0 ml

Ammonium Sulfate API Broth (pH 7.5)

Bacto yeast extract	1.0 g
Ascorbic acid	0.1 g
Sodium lactate	5.2 g
Magnesium sulfate	0.2 g
Dipotassium phosphate	0.01 g
Ferrous ammonium sulfate	0.1 g
Sodium chloride	10.0 g
Distilled water	1,000.0 ml

Azotobacter Nitrogen-Free Broth (pH 7.2)

Dipotassium phosphate	1.0 g
Magnesium sulfate	0.2 g
Sodium chloride	0.2 g
Ferrous sulfate	5.0 mg
Distilled water	1,000.0 ml

Beggiatoa Medium (pH 7.2)

Yeast extract	2.0 g
Sodium acetate	0.5 g
$CaCl_2$	0.1 g
Distilled water	1,000.0 ml

Add 10 sigma units of filter-sterilized catalase per milliliter of medium after autoclaving.

Bile Esculin Agar (pH 6.8)

Oxgall	20.0 g
Agar	15.0 g
Pancreatic digest of gelatin	5.0 g
Beef extract	3.0 g
Esculin	1.0 g
Ferric citrate	0.5 g
Distilled water	1,000.0 ml

Blood Agar (pH 7.3)

Infusion from beef heart	500.0 g
Tryptose	10.0 g
Sodium chloride	5.0 g
Agar	15.0 g
Distilled water	1,000.0 ml

Note: Dissolve the above ingredients and autoclave. Cool the sterile blood agar base to 45° to 50°C and aseptically add 50 ml of sterile, defibrinated blood. Mix thoroughly and then dispense into plates while a liquid. Blood agar base for use in making blood agar also can be purchased. A combination of hemoglobin and a commercial nutrient supplement can be used in place of defibrinated blood.

Bottom Agar (pH 7.0)

Use 12-ml sterile nutrient agar pours to prepare plates.

Brain-Heart Infusion Agar (pH 7.4)

Calf brains, infusion from	200.0 g
Beef hearts, infusion from	250.0 g
Proteose peptone	10.0 g
Dextrose	2.0 g
Sodium chloride	5.0 g
Disodium phosphate	2.5 g
Agar	15.0 g
Distilled water	1,000.0 ml

Brewer's Anaerobic Agar (pH 7.2)

Bacto tryptone	5.0 g
Proteose peptone	10.0 g
Bacto yeast extract	5.0 g
Bacto dextrose	10.0 g
Sodium chloride	5.0 g
Agar	20.0 g
Sodium thioglycollate	2.0 g
Sodium formaldehyde sulfoxylate	1.0 g
Resazurin	0.002 g
Distilled water	1,000.0 ml

Brilliant Green Bile Lactose (2%) Broth

Peptone	10.0 g
Oxgall	20.0 g
Lactose	10.0 g
Brilliant green	0.0133 g
Distilled water	1,000.0 ml

***Chlorobium* Medium (pH 7.0)**

$NaHCO_3$	2.0 g
$Na_2S \cdot 9H_2O$	2.0 g
KH_2PO_4	1.0 g
NH_4Cl	1.0 g
$MgCl_2 \cdot 6H_2O$	0.5 g
$FeCl_2 \cdot 6H_2O$	0.0005 g
Distilled water	1,000.0 ml

Chocolate Agar (pH 7.0)

Proteose peptone	20.0 g
Dextrose	0.5 g
Sodium chloride	5.0 g
Disodium phosphate	5.0 g
Agar	15.0 g
Distilled water	1,000.0 ml

Note: Aseptically add 5% sterile, defibrinated sheep blood to the sterile and molten agar. Heat at 80°C for 15 minutes or until a chocolate color develops.

***Chromatium* Medium (pH 7.0)**

$NaHCO_3$	2.0 g
$Na_2S \cdot 9H_2O$	1.0 g
Sodium pyruvate	0.5 g
KH_2PO_4	0.5 g
NH_4Cl	0.4 g
$MgSO_4 \cdot 7H_2O$	0.2 g
$CaCl_2 \cdot 2H_2O$	0.05 g
Distilled water	1,000.0 ml

Cystine Tryptic Agar (pH 7.3)

Tryptose	20.0 g
L-Cystine	0.5 g
Sodium chloride	5.0 g
Sodium sulfite	0.5 g
Agar	2.5 g
Phenol red	0.017 g
Distilled water	1,000.0 ml

After autoclaving and cooling to 50°C, add appropriate Bacto Differentiation Disk Carbohydrate (e.g., dextrose, fructose, maltose, sucrose). Allow to cool unslanted in an upright position.

Deoxyribonuclease (DNase) Test Agar (pH 7.3)

Deoxyribonucleic acid	2.0 g
Phytone peptone	5.0 g
Sodium chloride	5.0 g
Trypticase	15.0 g
Agar	15.0 g
Distilled water	1,000.0 ml

DNase Test Agar with Methyl Green

Bacto tryptose	20.0 g
Deoxyribonucleic acid	2.0 g
Sodium chloride	5.0 g
Bacto agar	15.0 g
Methyl green	0.05 g

Endo Agar (pH 7.5)

Peptone	10.0 g
Lactose	10.0 g
Dipotassium phosphate	3.5 g
Sodium sulfite	2.5 g
Basic fuchsin	0.4 g
Agar	15.0 g
Distilled water	1,000.0 ml

Enriched Nitrate Broth

See Nitrate Broth.

Eosin-Methylene Blue (EMB) Agar (pH 7.2)

Peptone	10.0 g
Lactose	5.0 g
Sucrose	5.0 g
Dipotassium phosphate	2.0 g
Agar	13.5 g
Eosin Y	0.4 g
Methylene blue	0.06 g
Distilled water	1,000.0 ml

Eugon Agar (pH 7.0)

Tryptose	15.0 g
Soytone	5.0 g
Dextrose	5.0 g
L-Cystine	0.2 g
Sodium chloride	4.0 g
Sodium sulfite	0.2 g
Agar	15.0 g
Distilled water	1,000.0 ml

Eugon Broth (pH 7.0)

Tryptose	15.0 g
Soytone	5.0 g
Dextrose	5.0 g
L-Cystine	0.2 g
Sodium chloride	4.0 g
Sodium sulfite	0.2 g
Distilled water	1,000.0 ml

Furoxone-Tween-Oil Red (FTO) Agar

TSA base

Trypticase (tryptone)	15.0 g
Phytone (soytone)	5.0 g
Sodium chloride	5.0 g
Yeast extract	1.0 g
Agar	15.0 g
Distilled water	1,000.0 ml

Mix these components together and sterilize using an autoclave.

Separately, prepare the following stock solutions:

A. 0.1% Furoxone

Nitrofuran (Furoxone)	1.0 g
Acetone	10 ml

Do not autoclave this solution or use near open flame.

B. 20% Tween 80

Tween 80	20 ml
Distilled water	80 ml

Slowly mix the solution on a magnetic stir plate. When the Tween 80 is fully in solution, filter sterilize using a 0.2 micron filter and store the solution at room temperature in the dark for no more than one month.

C. 0.1% Oil Red O

Oil Red O	0.01 g
Acetone	10 ml

Do not autoclave this solution or use near open flame.

Allow the TSA base to cool to 55°C. Then, carefully add 10 ml of 0.1% Furoxone, 25 ml of 20% Tween 80, and 5 ml of 0.1% Oil Red O. Mix on a magnetic stir plate prior to pouring the plates.

Gel Diffusion Agar

Noble agar	10.0 g
Distilled water	1,000.0 ml

A 1/1,000 dilution of merthiolate can be added as a preservative. Dispense in appropriate dishes.

Glucose–Minimal Salts

Prepare solutions A–D.

A. Minimal salts

$(NH_4)_2SO_4$	20.0 g
K_2HPO_4	140.0 g
KH_2PO_4	60.0 g
Sodium citrate · $2H_2O$	10.0 g
$MgSO_4 \cdot 7H_2O$	2.0 g
Distilled water	1,000.0 ml

Autoclave under standard conditions.

B. 50% glucose

Glucose	50.0 g
Distilled water	100.0 ml

Autoclave under standard conditions.

C. Amino acids

Prepare solutions of individual amino acids to give 2 mg/ml and filter sterilize.

D. Agar solution

Agar	15.0 g
Distilled water	880.0 ml

To prepare final glucose–minimal salts mix aseptically:

Agar solution (50°C)— solution D	880.0 ml
Minimal salts—solution A	100.0 ml
50% glucose—solution B	10.0 ml
Amino acids—solution C	10.0 ml

KF Streptococcus Agar (pH 7.2)

Proteose peptone, No. 3 Difco	10.0 g
Yeast extract	10.0 g
Sodium chloride	5.0 g
Sodium glycerophosphate	10.0 g
Maltose	20.0 g
Lactose	1.0 g
Sodium azide	0.4 g
Bromcresol purple	0.015 g
Agar	20.0 g
Distilled water	1,000.0 ml

Kligler Iron Agar (pH 7.4)
Beef extract	3.0 g
Yeast extract	3.0 g
Peptone	15.0 g
Proteose peptone	5.0 g
Lactose	10.0 g
Dextrose	1.0 g
Ferrous sulfate	0.2 g
Sodium chloride	5.0 g
Sodium thiosulfate	0.3 g
Agar	12.0 g
Phenol red	0.024 g
Distilled water	1,000.0 ml

Lactose Fermentation Broth (1× and 2×, pH 6.9)
Beef extract	3.0 g
Peptone	5.0 g
Lactose	5.0 g
Distilled water	1,000.0 ml

Note: For the 2×, use twice the ingredients.

Lauryl Tryptose Broth (pH 6.8)
Tryptose	20.0 g
Lactose	5.0 g
Potassium phosphate, dibasic	2.75 g
Potassium phosphate, monobasic	2.75 g
Sodium chloride	5.0 g
Sodium lauryl sulfate	0.1 g
Distilled water	1,000.0 ml

Levine EMB Agar (pH 7.1)
Peptone	10.0 g
Lactose	10.0 g
Dipotassium phosphate	2.0 g
Agar	15.0 g
Eosin Y	0.4 g
Methylene blue	0.065 g
Distilled water	1,000.0 ml

Litmus Milk
Skim milk powder	100.0 g
Litmus	0.75 g
Distilled water	1,000.0 ml

Note: Autoclave at 12 lb pressure for 15 minutes.

Löwenstein–Jensen Medium
Asparagine	3.6 g
Monopotassium phosphate	2.4 g
Magnesium sulfate	0.24 g
Magnesium citrate	0.6 g
Potato flour	30.0 g
Malachite green	0.4 g
Distilled water	600.0 ml

Lysine Iron Agar (pH 6.7)
Peptone	5.0 g
Yeast extract	3.0 g
Dextrose	1.0 g
L-Lysine hydrochloride	10.0 g
Ferric ammonium citrate	0.5 g
Sodium thiosulfate	0.04 g
Bromcresol purple	0.02 g
Agar	15.0 g
Distilled water	1,000.0 ml

Mannitol Salt Agar (pH 7.4)
Beef extract	1.0 g
Peptone	10.0 g
Sodium chloride	75.0 g
D-Mannitol	10.0 g
Agar	15.0 g
Phenol red	0.025 g
Distilled water	1,000.0 ml

M-Endo Broth (pH 7.5)
Yeast extract	6.0 g
Thiotone peptone	20.0 g
Lactose	25.0 g
Dipotassium phosphate	7.0 g
Sodium sulfite	2.5 g
Basic fuchsin	1.0 g
Distilled water	1,000.0 ml

Note: Heat until boiling but do not autoclave.

M-FC Broth (pH 7.4)
Biosate peptone or tryptose	10.0 g
Polypeptone peptone or proteose peptone	5.0 g
Yeast extract	3.0 g
Sodium chloride	5.0 g
Lactose	12.5 g
Bile salts	1.5 g
Aniline blue	0.1 g
Distilled water	1,000.0 ml

Note: Add 10 ml of rosolic acid (1% in 0.2 N sodium hydroxide). Heat to boiling with gentle agitation. Do not autoclave.

MM 1 Medium (pH 7.4)
Spizizen's salts supplemented with the following:
Vitamin-free casein hydrolysate	0.2 g
L-Tryptophan	0.05 g
Dextrose	5.0 g
Distilled water	1,000.0 ml

MM 2 Medium (pH 7.4)
Spizizen's salts supplemented with the following:
Vitamin-free casein hydrolysate	0.2 g
L-Tryptophan	0.005 g
Dextrose	5.0 g
Magnesium sulfate to a final concentration of	5.0 mM
Distilled water	1,000.0 ml

MacConkey's Agar (pH 7.1)
Bacto peptone	17.0 g
Proteose peptone	3.0 g
Lactose	10.0 g
Bile salts mixture	1.5 g
Sodium chloride	5.0 g
Agar	13.5 g
Neutral red	0.03 g
Crystal violet	0.001 g
Distilled water	1,000.0 ml

Moeller's Decarboxylase Broth with Ornithine (pH 7.2)
Peptone	5.0 g
Beef extract	5.0 g
Dextrose	0.5 g
Bromcresol purple	0.01 g
Cresol red	0.005 g
Pyridoxal	0.005 g
L-Ornithine	10.0 g
Distilled water	1,000.0 ml

In the place of 10 g (1%) of L-ornithine, 10 g (1%) of L-lysine or L-arginine can be used.

Motility Test Media (pH 7.2)
Tryptose	10.0 g
Sodium chloride	5.0 g
Agar	5.0 g
Distilled water	1,000.0 ml

MR-VP Broth (pH 6.9)
Peptone	7.0 g
Dextrose	5.0 g
Potassium phosphate	5.0 g
Distilled water	1,000.0 ml

Mueller-Hinton Agar (pH 7.4)
Beef, infusion	300.0 g
Casamino acids	17.5 g
Starch	1.5 g
Agar	17.0 g
Distilled water	1,000.0 ml

Nitrate Agar Slants (pH 6.8)
Peptone	5.0 g
Beef extract	3.0 g
Potassium nitrate	1.0 g
Agar	12.0 g
Distilled water	1,000.0 ml

Nitrate Broth (pH 7.2)
Peptone	5.0 g
Beef extract	3.0 g
Potassium nitrate	1.0 g
Distilled water	1,000.0 ml

Nitrate-Free Broth (pH 7.2)
Peptone	5.0 g
Beef extract	3.0 g
Distilled water	1,000.0 ml

Nutrient Agar (pH 7.0)
Peptone	5.0 g
Beef extract	3.0 g
Agar	15.0 g
Distilled water	1,000.0 ml

Note: Autoclave at 121 lb pressure for 15 minutes.

Nutrient Broth (pH 7.0)
Peptone	5.0 g
Beef extract	3.0 g
Distilled water	1,000.0 ml

Nutrient Gelatin (pH 6.8)
Peptone	5.0 g
Beef extract	3.0 g
Gelatin	120.0 g
Distilled water	1,000.0 ml

Peptone Broth (pH 7.2)
Peptone	10.0 g
Sodium chloride	5.0 g
Distilled water	1,000.0 ml

Phenol Red Dextrose Broth (pH 7.4)
Trypticase (proteose peptone)	10.0 g
Beef extract	1.0 g
Dextrose	5.0 g
Sodium chloride	5.0 g
Phenol red	0.025 g
Distilled water	1,000.0 ml

Note: Autoclave at 15 lb pressure for 15 minutes. Be careful not to autoclave longer.

Phenol Red Lactose Broth (pH 7.4)

Trypticase (proteose peptone)	10.0 g
Beef extract	1.0 g
Lactose	10.0 g
Sodium chloride	5.0 g
Phenol red	0.025 g
Distilled water	1,000.0 ml

Note: Autoclave at 15 lb pressure for 15 minutes. Be careful not to autoclave longer.

Phenol Red Sucrose (Saccharose) Broth (pH 7.4)

Trypticase (proteose peptone)	10.0 g
Beef extract	1.0 g
Sucrose (saccharose)	10.0 g
Sodium chloride	5.0 g
Phenol red	0.025 g
Distilled water	1,000.0 ml

Note: Autoclave at 15 lb pressure for 15 minutes. Be careful not to autoclave longer.

Phenylalanine Deaminase (Phenylalanine) Agar (pH 7.3)

Yeast extract	3.0 g
Dipotassium phosphate	1.0 g
Sodium chloride	5.0 g
DL-phenylalanine	2.0 g
Agar	12.0 g
Distilled water	1,000.0 ml

Plate Count Agar (Standard Methods Agar, Tryptone Glucose Yeast Agar; pH 7.0)

Tryptone	5.0 g
Yeast extract	2.5 g
Dextrose (glucose)	1.0 g
Agar	15.0 g
Distilled water	1,000.0 ml

Presence-Absence Broth (P-A Broth)

Pancreatic digest of casein	10.0 g
Lactose	7.5 g
Pancreatic digest of gelatin	5.0 g
Beef extract	3.0 g
Sodium chloride	2.5 g
Dipotassium phosphate (K_2HPO_4)	1.375 g
Potassium dihydrogen phosphate (KH_2PO_4)	1.375 g
Sodium lauryl sulfate	0.05 g
Bromcresol purple	8.5 mg
Distilled water	1,000.0 ml

Add components to distilled/deionized water and bring volume to 333.0 ml. Mix thoroughly. Distribute into screw-capped 250.0-ml milk dilution bottles in 50.0-ml volumes. Autoclave for 15 min at 15 psi pressure—121°C.

Potato Dextrose Agar (pH 5.6)

Potatoes, infusion from	200.0 g
Dextrose	20.0 g
Agar	15.0 g
Distilled water	1,000.0 ml

Sabouraud (Dextrose) Agar (pH 5.6)

Peptone	10.0 g
Dextrose	40.0 g
Agar	15.0 g
Distilled water	1,000.0 ml

Salt Medium–Halobacterium

Sodium chloride	250.0 g
Magnesium sulfate · $7H_2O$	10.0 g
Potassium chloride	5.0 g
Copper chloride	0.2 g
Yeast extract	10.0 g
Tryptone	2.5 g
Agar	20.0 g
Distilled water	1,000.0 ml

Sea Water Agar (SWA) (pH 7.5)

Agar	15.0 g
Peptone	5.0 g
Yeast extract	3.0 g
Beef extract	3.0 g
Sea water, synthetic	1.0 L

Sea Water Synthetic

NaCl	24.0 g
$MgSO_4(7H_2O)$	7.0 g
$MgCl_2(6H_2O)$	5.3 g
KCl	0.7 g
$CaCl_2$	0.1 g

Preparation of synthetic sea water.

Add components to distilled/deionized water and bring volume to 1.0 L. Mix thoroughly. Adjust pH to 7.5.

Preparation of medium.

Combine components. Mix thoroughly. Gently heat and bring to a boil. Distribute into tubes or flasks. Autoclave for 15 min at 15 psi—121°C. Pour into sterile petri dishes or leave in tubes.

SIM Agar (pH 7.3)

Peptone	30.0 g
Beef extract	3.0 g
Ferrous ammonium sulfate	0.2 g
Sodium thiosulfate	0.025 g
Agar	3.0 g
Distilled water	1,000.0 ml

Simmons Citrate Agar (pH 6.9)

Ammonium dihydrogen phosphate	1.0 g
Dipotassium phosphate	1.0 g
Sodium chloride	5.0 g
Sodium citrate	2.0 g
Magnesium sulfate	0.2 g
Agar	15.0 g
Bromothymol blue	0.08 g
Distilled water	1,000.0 ml

Spirit Blue Agar with 3% Lipase (pH 6.8)

Tryptone	10.0 g
Yeast extract	5.0 g
Agar	20.0 g
Spirit blue	0.15 g
Distilled water	1,000.0 ml

After the above has been autoclaved and allowed to cool to 50° to 55°C, add 30 ml of Bacto lipase reagent (Difco) slowly while agitating the medium in the flask to obtain an even distribution.

Spizizen's Salts

Sodium sulfate	2.0 g
Dipotassium phosphate	14.0 g
Monopotassium phosphate	6.0 g
Sodium citrate	1.0 g
Magnesium sulfate	0.2 g
Distilled water	1,000.0 ml

SS Agar (pH 7.0)

Beef extract	5.0 g
Peptone	5.0 g
Lactose	10.0 g
Bile salts no. 3	8.5 g
Sodium citrate	8.5 g
Sodium thiosulfate	8.5 g
Ferric citrate	1.0 g
Agar	13.5 g
Brilliant green	0.33 mg
Neutral red	0.025 g
Distilled water	1,000.0 ml

Standard Methods Agar

See Plate Count Agar.

Starch Agar (pH 7.5)

Beef extract	3.0 g
Soluble starch	10.0 g
Agar	12.0 g
Distilled water	1,000.0 ml

T-Broth (pH 7.3)

Nutrient broth	8.0 g
Peptone	5.0 g
Sodium chloride	5.0 g
Glucose	1.0 g
Distilled water	1,000.0 ml

To make 2× concentrated T-broth, double all of the above ingredients except the distilled water. Adjust the pH to 7.3 with 0.1 M NaOH.

Thayer-Martin Modified Medium (pH 7.0)

Bacto GC medium base	36.0 g
Hemoglobin	10.0 g
Bacto supplement B or VX	10.0 ml
Bacto antimicrobic vial CNVT	10.0 ml
Distilled water	1,000.0 ml

Thioglycollate Broth (pH 7.1)

Peptone	15.0 g
Yeast extract	5.0 g
Dextrose	5.0 g
L-Cystine	0.75 g
Thioglycollic acid	0.5 g
Agar	0.75 g
Sodium chloride	2.5 g
Resazurin	0.001 g
Distilled water	1,000.0 ml

Todd-Hewitt Broth (pH 7.8)

Beef heart, infusion	500.0 g
Neopeptone	20.0 g
Dextrose	2.0 g
Sodium chloride	2.0 g
Disodium phosphate	0.4 g
Sodium carbonate	2.5 g
Distilled water	1,000.0 ml

Top Agar (pH 7.0)

Nutrient broth plus 0.75% agar.
Prepare 4.5-ml pours.

Triple Sugar Iron Agar (pH 7.4)
Beef extract	3.0 g
Yeast extract	3.0 g
Peptone	15.0 g
Peptose-peptone	5.0 g
Lactose	10.0 g
Saccharose	10.0 g
Dextrose	1.0 g
Ferrous sulfate	0.2 g
Sodium chloride	5.0 g
Sodium thiosulfate	0.3 g
Phenol red	0.024 g
Agar	12.0 g
Distilled water	1,000.0 ml

Tryptic Nitrate Broth (pH 7.2)
Tryptose	20.0 g
Dextrose	1.0 g
Disodium phosphate	2.0 g
Potassium nitrate	1.0 g
Distilled water	1,000.0 ml

Trypticase (Tryptic) Soy Agar (pH 7.3)
Trypticase (tryptone)	15.0 g
Phytone (soytone)	5.0 g
Sodium chloride	5.0 g
Agar	15.0 g
Distilled water	1,000.0 ml

Trypticase Soy Agar with Lecithin and Polysorbate 80 (pH 7.3)
Tryptone	15.0 g
Soy peptone	5.0 g
Sodium chloride	5.0 g
Lecithin	0.7 g
Sorbitan monooleate complex	5.0 g
Agar	15.0 g
Distilled water	1,000.0 ml

Trypticase (Tryptic) Soy Broth (pH 7.3)
Tryptone	17.0 g
Soytone	3.0 g
Dextrose	2.5 g
Sodium chloride	5.0 g
Dipotassium phosphate	2.5 g
Distilled water	1,000.0 ml

Tryptone Agar
Tryptone	10.0 g
Calcium chloride (reagent)	0.01 M
Sodium chloride	5.0 g
Agar	11.0 g
Distilled water	1,000.0 ml

Tryptone Broth
Tryptone	10.0 g
Calcium chloride (reagent)	0.01 M
Sodium chloride	5.0 g
Distilled water	1,000.0 ml

Tryptone Glucose Yeast Agar
See Plate Count Agar.

Urea Broth (pH 6.9)
Yeast extract	0.1 g
Monopotassium phosphate	0.091 g
Disodium phosphate	0.095 g
Urea	20.0 g
Phenol red	0.01 g
Distilled water (sterile)	1,000.0 ml

Violet Red Bile Agar (pH 7.4)
Yeast extract	3.0 g
Peptone	7.0 g
Bile salts no. 3	1.5 g
Lactose	10.0 g
Sodium chloride	5.0 g
Agar	15.0 g
Neutral red	0.03 g
Crystal violet	0.002 g
Distilled water	1,000.0 ml

Vogel-Johnson Agar (pH 7.2)
Tryptone	10.0 g
Yeast extract	5.0 g
Mannitol	10.0 g
Dipotassium phosphate	5.0 g
Lithium chloride	5.0 g
Glycine	10.0 g
Agar	15.0 g
Phenol red	0.025 g
Distilled water	1,000.0 ml

YM Agar
Yeast extract	3.0 g
Malt extract	3.0 g
Peptone	5.0 g
Dextrose	10.0 g
Agar	20.0 g
Distilled water	1,000.0 ml

YM Broth
Yeast extract	3.0 g
Malt extract	3.0 g
Peptone	5.0 g
Dextrose	10.0 g
Distilled water	1,000.0 ml

APPENDIX D
Sources and Maintenance of Microbiological Stock Cultures

Sources of Microbiological Cultures

In addition to the major biological supply houses such as Carolina, Fisher, and Wards, the strains used in this lab manual can be obtained from the following organization:

American Type Culture Collection
10801 University Boulevard
Manassas, VA 20110
USA/Canada 1-703-365-2700
Outside USA/Canada 1-301-881-2600
FAX 1-703-365-2701
www.atcc.org

Hawaii Restriction Notice

Shipment of living materials (viruses, algae, bacteria, protozoa) into Hawaii is restricted. Please contact the Hawaii Department of Agriculture (808-586-0844) to obtain the necessary import permits and other pertinent information.

Maintenance of Microbiological Stock Cultures

In microbiology, a stock culture is a standard strain that conforms to typical morphological, biochemical, physiological, and serological characteristics of the species it represents. Over time, the culture will possess sufficient stability to display or retain these characteristics.

Stock cultures can be maintained at little expense (compared to buying new ones each time a strain is needed) and require small amounts of time with respect to upkeep. The two following general procedures are commonly used:

I. **Freeze-drying** (lyophilization) or **quick-freeze** methods requiring specific equipment
   A. **Freeze-drying** (lyophilization): In this procedure, microorganisms in a liquid medium are quick-frozen in dry ice with a solvent and dried under high vacuum from the frozen state. Many microorganisms can be preserved by freeze-drying almost indefinitely.
   B. **Quick-freeze method:** This method is recommended for both anaerobic and aerobic microorganisms. Most will survive for 6 months or longer.
   1. Grow the microbiological culture in a rich broth until the culture reaches stationary phase.
   2. Label small, sterile, 2 ml screw-cap vials with the name of the microorganism and date.
   3. In each vial, combine 1 ml of the stationary phase culture with 1 ml of sterile 20% glycerol.
   4. Gently mix each vial.
   5. Prepare a dry-ice bath by using crushed dry ice and ethanol.
   6. Place the vials in the dry-ice bath for about 15 seconds. This rapidly freezes the culture in the vial.
   7. Store at −80°C or colder.
   8. To revive the culture, scratch the surface of the frozen glycerol stock with a sterile micropipette tip or inoculating loop.
   9. Using aseptic technique, remove a small amount of frozen culture and subculture onto an appropriate growth medium.

II. **Special media maintenance** methods involve holding at room temperature, holding at incubator temperature, or refrigeration
   A. **CTA (cysteine-trypticase agar)** without carbohydrates is available commercially. This medium will support growth for a long period of time at room temperature.
   1. Aseptically inoculate a loopful of culture into a tube of tryptic soy broth. Incubate at 35°C for 18 to 24 hours.
   2. Using a sterile 1-ml pipette, inoculate a few drops of the tryptic soy broth into the CTA tube.
   3. Usually, this stock culture can be maintained up to 6 months at room temperature.
   4. Every 2 to 3 months check for viability by subculturing onto appropriate growth media.

5. If the subculture growth is scanty but displays typical characteristics of the microorganism, make a transfer from the new culture to a new CTA tube.
6. If the subculture is not in good condition, the stock culture should be discarded and a new one made.

B. **More fastidious microorganisms:** Pneumococcus, α- and β-hemolytic streptococci, enterococcus, and others can be maintained on blood agar or tryptose agar slants in screw-cap tubes streaked for heavy growth. Keep the tubes in a refrigerator and transfer the cultures every 2 weeks to a fresh slant.

C. **Cooked meat medium:** Commercially available cooked meat medium is an excellent medium for both aerobes and anaerobes. Many gram-negative bacteria (e.g., *Salmonella, Shigella, Proteus*) can remain viable for years. Coryne bacteria and staphylococci can remain viable up to 6 months and must then be subcultured.

The microbiological stock culture is inoculated into the meat layer of a tube of cooked meat medium. To maintain anaerobes, first boil the medium to drive off the dissolved oxygen. After inoculation and incubation for 24 to 48 hours, add a layer of 1 ml of sterile paraffin oil. Check the culture for viability every 2 to 3 months.

D. **TSA slants:** TSA slants with screw caps can be inoculated with a specific microorganism and incubated for 24 to 48 hours. It is then covered with a small amount of sterile sheep or horse serum and frozen at −50°C. Most microorganisms can be maintained this way for 6 or more months.

Overall considerations. Regardless of the procedure used to maintain specific microbiological cultures, the following special considerations should always be followed:

1. Do not use media that contain fermentable carbohydrates if you plan to keep the cultures for long periods.
2. Never use selective media.
3. Never allow a culture to dry out. Always use tightly closed screw-cap tubes for storage.
4. Do not refrigerate temperature-sensitive microorganisms (e.g., *Neisseria* and *Vibrio species*), although they survive well by quick-freezing.

APPENDIX E
Laboratory Technical Skills and Laboratory Thinking Skills

The American Society for Microbiology, through its Office of Education and Training, adopted a Laboratory Core Curriculum representing themes and topics considered essential to teach in every introductory microbiology laboratory, regardless of its emphasis.

The core themes and topics are meant to frame objectives to be met somewhere within the introductory microbiology laboratory. Depending on the specific emphasis of the course, a single lab session could meet multiple core objectives, focus on one objective, or emphasize a topic that is not in the lab core but is important to that particular course.

Laboratory Technical Skills

A student successfully completing basic microbiology will demonstrate the ability to

1. **Use a bright-field light microscope** to view and interpret slides, including
   a. correctly setting up and focusing the microscope
   b. proper handling, cleaning, and storage of the microscope
   c. correct use of all lenses
   d. recording microscopic observations

2. **Properly prepare slides** for microbiological examination, including
   a. cleaning and disposal of slides
   b. preparing smears from solid and liquid cultures
   c. performing wet-mount and/or hanging drop preparations
   d. performing Gram stains

3. **Properly use aseptic techniques** for the transfer and handling of microorganisms and instruments, including
   a. sterilizing and maintaining sterility of transfer instruments
   b. performing aseptic transfer
   c. obtaining microbial samples

4. **Use appropriate microbiological media and test systems,** including
   a. isolating colonies and/or plaques
   b. maintaining pure cultures
   c. using biochemical test media
   d. accurately recording macroscopic observations

5. **Estimate the number of microorganisms** in a sample using serial dilution techniques, including
   a. correctly choosing and using pipettes and pipetting devices
   b. correctly spreading diluted samples for counting
   c. estimating appropriate dilutions
   d. extrapolating plate counts to obtain correct CFU or PFU in the starting sample

6. **Use standard microbiology laboratory equipment correctly,** including
   a. using the standard metric system for weights, lengths, diameters, and volumes
   b. lighting and adjusting a laboratory burner
   c. using an incubator

Laboratory Thinking Skills

A student successfully completing basic microbiology will demonstrate an increased skill level in

1. **Cognitive processes,** including
   a. formulating a clear, answerable question
   b. developing a testable hypothesis
   c. predicting expected results
   d. following an experimental protocol

2. **Analysis skills,** including
   a. collecting and organizing data in a systematic fashion
   b. presenting data in an appropriate form (graphs, tables, figures, or descriptive paragraphs)
   c. assessing the validity of the data (including integrity and significance)
   d. drawing appropriate conclusions based on the results

3. **Communications skills,** including
   a. discussing and presenting laboratory results or findings in the laboratory

4. **Interpersonal and citizenry skills,** including
   a. working effectively in groups or teams so that the task, results, and analysis are shared
   b. effectively managing time and tasks to be done simultaneously, by individuals and within a group
   c. integrating knowledge and making informed judgments about microbiology in everyday life

Laboratories typically supplement and integrate closely with the lecture content in ways that are unique to each instructor. Consequently, the laboratory content that is considered essential for laboratory work by one instructor may be covered in the lecture portion of the course by another instructor, making it difficult to define specific topics that should be integral in all microbiology laboratories. As a result, ASM has developed these six core concepts crucial for understanding the central principles of microbiology:

- Evolution
- Cell Structure and Function
- Metabolic Pathways
- Information Flow and Genetics
- Microbial Systems
- The Impact of Microorganisms

In order to meet the above curricular goals, this manual consists of 54 exercises arranged into nine parts covering the following basic topics:

PART ONE: Microscopic Techniques introduces the students to the proper use and care of the different types of microscopes used in the microbiology laboratory for the study of microorganisms.

PART TWO: Bacterial Cell Biology presents the basic procedures for visualization and differentiation of microorganisms based on cell form and various structures.

PART THREE: Basic Culture Techniques acquaints students with proper laboratory procedures in preparing microbiological media and in culture techniques that are used in isolating microorganisms.

PART FOUR: Microbial Biochemistry introduces some of the biochemical activities that may be used in characterizing and identifying bacteria.

PART FIVE: Environmental Factors Affecting Growth of Microbiology acquaints students with some of the various physical and chemical agents that affect microbial growth.

PART SIX: Environmental and Food Microbiology is concerned with the environmental aspects of water, milk, and food.

PART SEVEN: Medical Microbiology presents an overview of some pathogenic microorganisms, and acquaints students with basic procedures used in isolation and identification of pathogens from infected hosts, including those from the student's own body.

PART EIGHT: Eukaryotic Microbiology presents an overview that is intended to help students appreciate the morphology, taxonomy, and biology of the fungi.

PART NINE: Microbial Genetics and Genomics presents six experiments designed to illustrate the general principles of bacterial genetics and genomics including mutations, horizontal gene transfer, and DNA sequencing technology.

APPENDIX F
Summary of Universal Precautions and Laboratory Safety Procedures

Precautions for Laboratories

1. All specimens should be considered potentially infectious and thus should be put in a well-constructed container with a secure lid to prevent leaking during transport. Care should be taken when collecting each specimen to avoid contaminating the outside of the container and of the laboratory form accompanying the specimen.
2. All persons processing blood and body-fluid specimens or microbial cultures should wear gloves. Masks and protective eyewear should be worn if mucous-membrane contact with blood or body fluids is anticipated. Gloves should be changed and hands washed after completion of specimen processing.
3. For routine procedures, such as histologic and pathologic studies or microbiologic culturing, a biological safety cabinet is not necessary. However, biological safety cabinets should be used whenever procedures are conducted that have a high potential for generating droplets. These include activities such as blending, sonicating, and vigorous mixing.
4. Mechanical pipetting devices should be used for manipulating all liquids in the laboratory. Mouth pipetting must not be done.
5. Use of needles and syringes should be limited to situations in which there is no alternative, and the recommendations for preventing injuries with needles outlined under universal precautions should be followed.
6. Laboratory work surfaces should be decontaminated with an appropriate chemical germicide after a spill of blood or other body fluids or microbial cultures and when work activities are completed.
7. Contaminated materials used in laboratory tests should be decontaminated before reprocessing or be placed in bags and disposed of in accordance with institutional policies for disposal of infective waste.
8. Scientific equipment that has been contaminated with blood or other body fluids or microbial cultures should be decontaminated and cleaned before being repaired in the laboratory or transported to the manufacturer.
9. All persons should wash their hands after completing laboratory activities and should remove protective clothing before leaving the laboratory.
10. There should be no eating, drinking, or smoking in the work area.

Use of Pathogens

Recently, there has been much discussion about the advisability of using opportunistic or pathogenic microorganisms in an introductory microbiology laboratory. Ideally, it would be desirable never to use pathogens with first-semester students. However, there are certain biochemical tests (e.g., coagulase) that cannot be performed without the use of pathogens. For this reason, their use has been minimized in this manual, but they continue to be used when necessary.

It is the belief of the authors that the most important lesson all introductory microbiology students can learn is that all microorganisms should be treated as potential pathogens. With careful emphasis on proper handling procedures and aseptic techniques, the use of microorganisms should not put the student at risk.

APPENDIX G
Biosafety Level (BSL)

When you order a microorganism from the American Type Culture Collection or other source, the accompanying paperwork usually specifies a biosafety level (BSL) for that microorganism.

The Centers for Disease Control and Prevention have recommended BSLs for most microorganisms. These BSLs are found in *BSL Recommendations for Select Agents—Supplement to CDC/NIH Biosafety in Microbiological and Biomedical Laboratories,* sixth edition (BMBL). This document is available online at https://www.cdc.gov/labs/BMBL.html.

Generally speaking, biosafety levels are a combination of laboratory practices and procedures, safety equipment, and laboratory facilities. The designation of an appropriate biosafety level for work with a particular microorganism is based on a risk assessment and includes a number of factors, such as infectivity or pathogenicity, biological stability, route of transmission, and communicability of the microorganism.

The CDC-recommended biosafety levels for the various microorganisms represent those conditions under which the microorganism can ordinarily be safely handled. There are four biosafety levels:

1. BSL1—work with microorganisms not known to cause disease in healthy adults; Standard Universal Precautions and Laboratory Safety Procedures apply; no safety equipment required; sinks required.
2. BSL2—work with microorganisms associated with human diseases; Standard Universal Precautions and Laboratory Safety Procedures apply plus limited access, biohazard signs, sharps precautions, and biosafety manual required; lab coats, gloves, face protection required; contaminated waste is autoclaved. The authors recommend using BSL2 conditions for all of the exercises in this lab manual to develop best practices among students new to microbiology.
3. BSL3—work with indigenous/exotic microorganisms which may have serious or lethal consequences and with potential for aerosol transmission; BSL2 practices plus controlled access; decontamination of all waste and lab clothing before laundering; determination of baseline serums; respiratory protection used as needed; physical separation from access corridors; double door access; negative airflow into lab; exhaust air not recirculated.
4. BSL4—work with dangerous/exotic microoganisms of life-threatening nature or unknown risk of transmission; BSL3 practices plus clothing change before entering lab; shower required for exit; all materials are decontaminated on exit; positive pressure personnel suit required for entry; separated/isolated building; dedicated air supply/exhaust and decontamination systems.

For more information on BSLs go to the American Biological Safety Association at http://www.absa.org.

Genus	Species	BSL Rating	Exercise
Acinetobacter	baylyi	1	52
Alcaligenes	denitrificans	1	11
Alcaligenes	faecalis	1	12, 19, 20, 26, 32
Aspergillus	niger	1	48
Bacillus	subtilis	1	2, 15, 16, 21, 24, 31
Bacillus	megaterium	1	10
Bacillus	macerans	1	10
Bacillus	circulans	1	10
Citrobacter	freundii	1	28
Clostridium	acetobutylicum	1	10
Clostridium	sporogenes	1	17
Enterobacter	aerogenes	1	23, 24, 28, 41, 49
Enterococcus	faecalis	2	25
Escherichia	coli	1	6–9, 13–24, 26, 27, 29–33, 35, 36, 41, 49, 51–54
Geobacillus	stearothermophilus	1	31
Halobacterium	salinarium	1	33
Klebsiella	pneumoniae	2	11, 23, 24, 27, 28, 35, 41
Micrococcus	roseus	1	5
Micrococcus	luteus	1	14, 25
Moraxella	nonliquefaciens	2	45
Mycobacterium	phlei	2	9
Mycobacterium	smegmatis	2	9
Neisseria	subflava biovar subflava	2	45
Neisseria	flavescens	2	45
Penicillium	notatum	1	48
Proteus	vulgaris	2	20, 21, 23, 24, 27–29
Pseudomonas	aeruginosa	2	2, 12, 17, 20, 24, 26, 31, 32, 34, 35
Pseudomonas	fluorescens	1	30
Rhizopus	stolonifer	1	48
Rhodotorula	rubrum	1	47
Saccharomyces	cerevisiae	1	6, 19, 32, 47, 49
Salmonella	enterica ser. Typhimurium	2	19, 27
Serratia	marcescens	1	5, 15, 16, 31, 49
Shigella	flexneri	2	20
Spirillum	volutans	1	2
Staphylococcus	aureus	2	6–8, 15, 19, 22, 25, 31, 33–35, 42, 44
Staphylococcus	epidermidis	1	30, 42
Staphylococcus	saprophyticus	1	42
Streptococcus	pyogenes	2	25, 44
Streptococcus	pneumoniae	2	43
Streptococcus	mitis	2	43
Streptococcus	equi	2	44
Streptococcus	bovis	2	44
Streptococcus	agalactiae	2	44
Vibrio	fischeri	1	40

APPENDIX H
pH and pH Indicators

pH is a measure of hydrogen ion (H$^+$) activity. In dilute solutions, the H$^+$ activity is essentially equal to the concentration. In such instances, pH = $-\log$ [H$^+$]. The pH scale ranges from 0 ([H$^+$] = 1.0^0 M) to 14 ([H$^+$] = 10^{-14} M).

A pH meter should be used for accurate pH determinations, observing the following precautions:

1. Adjust the temperature of the buffer used for pH meter standardization to the same temperature as the sample. Buffer pH changes with temperature; for example, the pH of standard phosphate buffer is 6.98 at 0°C, 6.88 at 20°C, and 6.85 at 37°C.
2. It is important to stir solutions while measuring their pH. If the sample is to be stirred with a magnetic mixer, stir the calibrating buffer in the same way.
3. Be sure that the electrodes used with tris buffers are recommended for such use by the manufacturer. This is necessary because some pH electrodes do not give accurate readings with tris (hydroxymethyl) aminomethane buffers.

In instances where precision is not required, such as in the preparation of routine media, the pH may be checked by the use of pH indicator solutions. By the proper selection, the pH can be estimated within ±0.2 pH units. Some common pH indicators and their useful pH ranges are listed in the following table.

pH Indicator	pH Range	Full Acidic Color	Full Basic Color
Brilliant green	0.0–2.6	Yellow	Green
Bromcresol green	3.8–5.4	Yellow	Blue-green
Bromcresol purple	5.2–6.8	Yellow	Purple
Bromophenol blue	3.0–4.6	Yellow	Blue
Bromothymol blue	6.0–7.6	Yellow	Blue
Congo red	3.0–5.0	Blue-violet	Red
Cresol red	2.3–8.8	Orange	Red
Cresolphthalein	8.2–9.8	Colorless	Red
2,4-dinitrophenol	2.8–4.0	Colorless	Red
Ethyl violet	0.0–2.4	Yellow	Blue
Litmus	4.5–8.3	Red	Blue
Malachite green	0.2–1.8	Yellow	Blue-green
Methyl green	0.2–1.8	Yellow	Blue
Methyl red	4.4–6.4	Red	Yellow
Neutral red	6.8–8.0	Red	Amber
Phenolphthalein	8.2–10.0	Colorless	Pink
Phenol red	6.8–8.4	Yellow	Red
Resazurin	3.8–6.4	Orange	Violet
Thymol blue	8.0–9.6	Yellow	Blue

Note: All of the above indicators can be made by (1) dissolving 0.04 g of indicator in 500 ml of 95% ethanol, (2) adding 500 ml of distilled water, and (3) filtering through Whatman No. 1 filter paper. Indicators should be stored in a dark, tightly closed bottle.

Index

A

Abbe condenser, 8
Abscesses, 210, 302
Absorbance, of medium, 134
Acetoin, 166, 168
Acid-alcohol, A–10, A–11, A–13
Acid buffer, A–12
Acid-fast staining, 67–69
   Kinyoun procedure of, A–11
   Ziehl–Neelsen procedure of, 68, 69, A–13
Acidic dyes, 43, 44
Acidity. *See* pH
Acidophiles, 225
Acid production
   in fermentation, 145–148
   in *Neisseria* identification, 325
   in *Staphylococcus* differentiation, 303
Acinetobacter baylyi
   biosafety level (BSL) of, A–28
Actidione (cycloheximide) agar, A–14
Actinomycetes, in soil, 273–275
Agar, 97, A–14–A–21. *See also specific types and procedures*
   bottom, A–14
   culture media preparation with, 97–99
   nutrient, A–18
   selective/differential, 110–112
   top, A–20
Agar cultures
   smear preparation from, 50
Agar deeps, 97
   anaerobic bacteria cultures in, 124, 127
Agaricus, 345–349
Agarose gel electrophoresis, 364, 366
Agar plates (petri plates), 97
   bioluminescence on, 288, 289
   Brewer's anaerobic, 125, 127
   disposal of, 100
   Mueller–Hinton II, 244–247
   Oxyrase systems for, 126
   pouring of, 97, 98, 99
   sea water, 287
   starch, 157

Agar slant, 97
   anaerobic bacteria cultures in, 124
   bile esculin, 317, 319
   Simmons citrate, 168, 170
   triple sugar iron agar test in, 151–154
   urea, 192
Alcaligenes denitrificans
   biosafety level (BSL) of, A–28
   capsule staining of, 81–83
Alcaligenes faecalis
   biosafety level (BSL) of, A–28
   fermentation (β-galactosidase) testing in, 145–148
   flagella staining of, 87–90
   oxidase test for, 185–187
   pH and growth of, 225–226
   triple sugar iron agar for, 151–154
Alcohol, A–10
   as decolorizer, 61
   disinfection, 238
   in DNA isolation, 356–359
   fermentation, 146–147
   safety of, 81, 103
   for slide cleaning, 83
Algae, 336
Alkaline buffer, A–12
Alkaline methylene blue
   in acid-fast staining, 67
Alkaliphiles, 225
α-Amylases, 157–158
Alpha-hemolysis, 302, 303, 310
American Biological Safety Association, A–27
American Public Health Association, 268
American Society for Microbiology, A–24
American Society for Microbiology (ASM) Style Manual,
   on dilution ratios, 134–135
American Type Culture Collection, A–22, A–27
Amino acids
   IMViC tests, 165–170
Ammonium sulfate API broth, A–14
Ampicillin, susceptibility testing of, 245
Amylopectin, 157
Amylose, 157

Anaerobic bacteria
  cooked meat medium for, 124
  cultivation of, 123–128
  GasPak for, 125, 126, 127
  pathogenic, 124
  simplified method of growing, 127–128
Anaerobic chamber, 126, 127
Anthony, E. E., Jr., 82
Anthony's capsule staining method, 82
Anthrax, 74, 81, 210
Antibiograms, 244
Antibiotics, 216
  discovery of, 243
  minimum inhibitory concentration of, 244
  resistance to
    mutation and, 370
  susceptibility testing of, 243–247
    Staphylococcus, 243–247, 301–306
    Streptococcus, 315–319
Antimicrobials, 237–239
  efficiency of, 238–239
  ingredients of commercial products, 239
Antisepsis, 96, 238
Antiseptics, 238
Arm of microscope, 8
Ascomycota, 337–339, 345–349
Ascospores, 338
Asepsis, 35
Aseptic technique, 33–37, A–24
Aspergillus
  identification of, 345–349
  in soil, 273–275
Aspergillus niger
  biosafety level (BSL) of, A–28
  identification of, 345–349
ATCC 23724 cell, 382
ATCC 23740 cell, 382
Autoclave, 96, 97, 99
Autoclaving procedure, 99
Auxotroph, 382
Azotobacter nitrogen-free broth, A–14

B

Bacillus
  catalase test for, 182
  in normal human flora, 332
Bacillus anthracis, 74, 81, 210
Bacillus circulans
  biosafety level (BSL) of, A–28
  endospore staining of, 73–76
Bacillus macerans
  biosafety level (BSL) of, A–28
  endospore staining of, 73–76
Bacillus megaterium
  biosafety level (BSL) of, A–28
  endospore staining of, 73–76

Bacillus subtilis, 173
  biosafety level (BSL) of, A–28
  casein hydrolysis test for, 173–176
  gelatin hydrolysis test for, 173–176
  motility of, 15–16
  negative staining of, 43–45
  pour plate, 117–119
  starch hydrolysis test for, 157–158
  streak plate, 109–112
  temperature sensitivity of, 217–219
Bacitracin sensitivity test, 317
Bacitracin susceptibility test, 245
Bacteremia, 210
Bacteria, 32, 354
  adherence to slide, 69
  biochemical activities of, 144
  chemolithotrophic, 210
  chemoorganoheterotrophic, 210
  colonies of, 103–104
    characteristics of, 104
    number (count) of, 134–137
    pour-plate technique for, 117–119
    spread-plate technique for, 103–106
    streak-plate technique for, 109–112
  first observations of, 2
  flagella of, 15–16
  lawn of, 244–245
  morphology and staining of, 32
  motility of, 15–16
    flagella and, 15–16, 88
    gliding, 16
    spirochete and, 16
    swarming, 16
  nonmotile, 16
  shapes of, 3, 5, 6
  in soil, 273–274
Bacterial cell biology, 32
Bacterial cell envelope, 59
Bacterial conjugation, 381–383
Bacterial count, 133–137
  medical application of, 133
  most probable number (MPN) test, 264, 266–269
  principles of, 134
  spectrophotometric (turbidimetric), 134, 137
  standard (viable) plate, 134, 135–137
  too few to count, 136
  too numerous to count, 136, 274
Bacterial growth, 251–255
  control of, 216
    antibiotic susceptibility testing of, 243–247
    chemical agents for, 216, 237–239
  environmental factors affecting, 216
  osmotic pressure and, 231–233
  pH and, 225–226
  phases of, 253
  temperature and, 217–219

Bacterial growth (*Continued*)
  turbidimetric analysis of, 134, 136–137, 251–253
Bacterial growth curve, 251–255
  turbidimetric analysis of, 251–253
Bacterial smear, 49–50
Bacterial transformation, 376
Bacteriological filter, 97
Bacteroides, as pathogenic anaerobe, 124
Bacteroides fragilis
  gelatin hydrolysis test for, 174
  as pathogenic anaerobe, 124
BactiCard Neisseria rapid test, 323, 326
BactiCard Strep rapid test, 315, 318
Bacto lipase reagent, 161
Baker's yeast. *See also Saccharomyces cerevisae*
Barritt's reagent, A–10
  in IMViC tests, 165
  safety considerations with, 165
  in Voges–Proskauer test, 166, 168, 170
Base of microscope, 8
Basic dyes, 50, 51
Basic Local Alignment Search Tool (BLAST), 364
Basidiomycota, 345–349, 348
Bauer, A., 244
Beggiatoa, 279–281, A–14
Benchtop incinerator, 37
β-Galactosidase activity, 145–148
Beta-hemolysis, 302, 303, 316
Bile esculin agar, 317, 319, A–14
Bile solubility test, 310, 311, A–10
Biochemical tests, 144, 166
Biocide, 238
Bioluminescence, 287–289
Biomass, 134, 251
Biosafety level (BSL), A–27–A–28
Bismark brown stain, 60
Blank tube, 136
Blood, pathogenic anaerobic bacteria of, 124
Blood agar, 304–305, 310–311, A–14
  hemolysis patterns, 302, 303
  Streptococcus culture on, 315–319
Blow-out pipette, 33
Boiling, bacterial sensitivity to, 218
Bordetella pertussis, 88
Bottom agar, A–14
Botulism, 74
Brain-heart infusion agar (broth), 254, A–14
Bread mold, 346, 349
Brewer's anaerobic agar (petri plate), 125, 127, A–15
Bright-field light microscope, 2, 3–8
  proper use of, 4–6
  student skill with, A–24
  troubleshooting, 6
Bright-phase-contrast microscopy, 26
Brilliant green bile lactose (2%) broth, A–15

Bromcresol purple, in decarboxylase test, 198–200
Broth media, A–14–A–21
  anaerobic bacteria cultivation in, 124, 127
  pour plate from, 117–119
  preparation of, 98, 99
  streak plate from, 109–112
Brownian movement, 16
Brucellosis, 293
BSL Recommendations for Select Agents–Supplement to CDC/NIH Biosafety in Microbiological and Biomedical Laboratories, A–27
Bud, 338
Budding, 338
Buffered water, A–12
Buffers, 226
2,3-Butanediol, 166, 167, 168
Butanediol fermenters, 166
Butanol, as fermentation product, 147
Butyric acid, as fermentation product, 147

C

Cadaverine, 199
CAMP factor, 317–319
Campylobacter
  milk contamination, 294
  water contamination, 265
Candida albicans, 337–339
Candidiasis, 338
Candle jar, for *Neisseria,* 325, 326
Capsule, 82
  composition of, 82
  medical significance of, 81
Capsule staining, 81–83
  Anthony's method of, 82
  India ink for, 81–83, A–11
  modified (Graham and Evans) method of, 83
Carbenicillin, susceptibility testing of, 245
Carbohydrates
  fermentation of, 145–148
  triple sugar iron agar test for, 151–154
Carbolfuchsin, A–11
  in acid-fast staining, 67
  in simple staining, 49–51
Carbuncles, 302
Cardinal temperature, 218
Casein, 174
Casein hydrolysis, 174, 175–176
Catalase, 182
Catalase test, 181–182
  for *Bacillus,* 182
  for *Clostridium,* 182
  for *Neisseria,* 326
C carbohydrate, 317
Centers for Disease Control and Prevention (CDC), A–27
CFUs. *See* Colony-forming units

Chemically defined media, 97, 99
Chemolithotrophic bacteria, 210
Chemoorganoheterotrophic bacteria, 210
Chemotherapy, 238
Chimeras, 354
Chloramphenicol, susceptibility testing of, 245
Chlorine dioxide, 96
Chlorobium, 279–281
Chloroform, safety considerations with, 387
Chocolate agar, 310–311, A–15
   Neisseria culture on, 325
   Staphylococcus aureus on, 302
Cholera, 88, 264
Chromatium, 279–281, A–15
Chromogen, 43–44, 50
Citizenry skills, A–25
Citrate, 166
Citrated rabbit plasma, 301
Citrate permease, 166, 168
Citrate utilization test, 166–168, 170
Citrobacter, monitoring water for, 265–270
Citrobacter freundii
   biosafety level (BSL) of, A–28
   decarboxylase tests of, 197–200
Clindamycin, susceptibility testing of, 245
Clinical microbiology laboratory, 300
Clostridium
   catalase test for, 182
   endospore staining of, 73–76
   medical significance of, 74
   as pathogenic anaerobe, 124
Clostridium acetobutylicum
   biosafety level (BSL) of, A–28
   medical significance of, 73–76
Clostridium botulinum, 74, 76
Clostridium difficile, 74
Clostridium perfringens
   endospores of, 74
   on gelatin medium, 174
   as pathogen, 124
Clostridium sporogenes
   biosafety level (BSL) of, A–28
   cultivation of, 123–128
Clostridium tertium, 88
Clostridium tetani, 74, 76
Coagulase test, 301, 303–304, 305
Coarse adjustment knob, 8
Cocci, 3
Coenocytic hyphae, 346
Cognitive processes, A–24
Coliform bacteria
   as indicator of fecal contamination, 264
   lactose fermentation test for, 264
   in milk, 294–296
   most probable number (MPN) test for, 264, 266–269
Coliform group, 264

Colistin, susceptibility testing of, 245
Colitis, pseudomembranous, 74
Colonies of bacteria, 103–104
   characteristics of, 104
   coliform, 267, 268
   number (count) of, 134–137
   pour-plate technique for, 117–119
   spread-plate technique for, 103–106
   streak-plate technique for, 109–112
Colonies of mold, 345
Colony counters, 136
Colony-forming units (CFUs), 118, 134, 136
Commercial antimicrobial products, 239
Communication skills, A–24
Competent bacterium, 375
Competent cells, preparation and transformation of, 376–377
Completed tests, for coliforms, 266, 267–268, 269–270
Complex media, 98
Confirmed test, for coliforms, 266, 267, 268, 269
Congo red, 44
Conidia, 346
Conjugation, 381–383
Conjugative plasmids, 382
Conjunctivitis, 309
Conklin, Marie E., 74
Contamination
   external, 35
   fecal, 264, 267 (*See also* Coliform bacteria)
   food (*See* Food microbiology)
   inoculating loop, 35
   pipette, 33–35
Cooked meat medium, 124, A–23
Copper sulfate, A–10
   in capsule staining, 81–82
Coprinus, 345, 348
Corynebacterium
   in normal human flora, 332
Counterstain, in Gram staining, 58, 60, 61
COVID-19 pandemic, 354
Cryptococcosis, 81
Cryptococcus neoformans, capsule staining of, 81
Cryptosporidium, water contamination, 265
Crystal violet
   in capsule staining, 81–83
   in flagella staining, 88
   in Gram staining, 57–58, 60–61
   simple staining of, 49–51
Crystal violet capsule stain, A–11
Culture media
   aseptic technique for, 35–36
   dispensing of, 97
   and equipment, 95–100
   maintaining sterility of, 98
   nonsynthetic, 98
   preparation of, 95–100

Culture media (*Continued*)
　sterilization methods, 96–97
　types of, 97–99, A–14–A–21
Culture transfer
　aseptic technique in, 35–36
　pipette use for, 35
　procedure for, 35–37
Cysteine, 169
　degradation of, 169
Cysteine desulfurase, 169
Cysteine-trypticase agar, A–22–A–23
Cystine tryptic agar, A–15
Cytochrome *c* oxidase, 185–186

D

Dairy products, lipolytic bacteria of, 161–162
Dark-field microscope, 2, 19–21
Dark-phase-contrast microscopy, 26
Death phase, of growth, 251, 252
Decarboxylase tests, 197–200
Decarboxylation, 198
Decline, of growth, 251, 252
Decolorizers, A–10
　in capsule staining, 82
　in Gram staining, 58, 60–61
Deoxyribonucleic acid. *See* DNA
Deoxyribo-nucleotide triphosphates (dNTPs), 364
Desulfovibrio, 279–281
Dextrose, fermentation of, 146–148
Difco Laboratories
　flagella staining method, 88
　SpotTest Nitrate Reagent C, 211
Differential interference contrast (DIC) microscope, 2
Differential media, 110–111
Differential staining, 57, 58
Differential tests, 144
Dilution(s)
　with sample problems, A–1–A–9
　serial, 136, 137, A–2
　　student skill with, A–24
　to specified volume, A–1–A–2
　to unspecified volume, A–1–A–2
Dilution factor, A–1
Dilution ratios, 134–135, A–2
Diphenylamine reagent, A–10
Disinfectants, 216, 237–239
Disinfection, chemical, 237–239
Disposal, of glass slides, 61
Distilled water, in culture media, 99
DNA, 354, 356
　bacterial transformation and, 376
　extraction from gram-negative bacterium, 357–359
　genomic
　　isolation of
　　　from *Escherichia coli,* 355–359
　purification of, 356–357
　structure of, 356
　sequencing chromatograph, 364, 366
DNase Agar, 303–305
DNase test, for *Neisseria,* 324–326
DNase test agar, A–15
DNase test agar with methyl green, A–15
DNA size ladder, 364
Donor bacterium, in transformation, 375–377
Doubling time, 253
Drinking (potable) water, 264
Dry-heat sterilization, 96
Dry Slide Oxidase, 186
Durham, Herbert Edward, 146
Durham tube, 145–148
　for coliform testing, 267, 270

E

Ehrlich, Paul, 68
Electrocompetent cells, preparation and transformation of, 377
Electroporation apparatus, 376
Endo agar, A–15
Endocarditis, 302, 316
Endospore(s), 74
　medical significance of, 74
　position of, 74
　sizes of, 73–74
　temperature-resistant, 218
Endospore staining, 73–76
　microwave procedure of, 75–76
　Schaeffer–Fulton procedure of, 74, 75
　Wirtz–Conklin procedure of, 74, 75
Enteric bacteria *(Enterobacteriaceae)*
　IMViC tests for, 165–170
　nitrate reduction test for, 209–212
　phenylamine deamination test for, 203–205
　triple sugar iron agar for, 151–154
Enterobacter, monitoring water for, 265–270
Enterobacter aerogenes, 173
　biosafety level (BSL) of, A–28
　casein hydrolysis test for, 173–176
　decarboxylase tests of, 197–200
　gelatin hydrolysis test for, 173–176
　IMViC tests for, 165–170
Enterococcus
　bile esculin reaction of, 317–318
　catalase test for, 181
　rapid detection of, 318–319
Enterococcus faecalis
　biosafety level (BSL) of, A–28
　catalase test for, 181–182
　diagnostic tests for, 318
Environmental microbiology, 264
　soil, 264, 273–275

Enzymes, 144
   catalase activity, 182
   nitrate reduction test for, 209–212
   pH and, 225
   phenylamine deamination test for, 203–205
   proteolytic, 174
Eosin, 43, 44
Eosin blue, A–10
Eosin methylene blue (EMB) agar, 111–112, A–15
   growth patterns on, 111
Erysipelas, 316
Erythromycin, susceptibility testing of, 245
Escherichia coli, 43–45, 61, 173
   acid-fast staining of, 67–69
   antibiotic-resistant
     bacterial conjugation and, 381–383
     safety considerations with, 381
   antibiotic susceptibility of, 243–247
   biosafety level (BSL) of, A–28
   carbohydrate fermentation by, 145–148
   casein hydrolysis test for, 173–176
   cultivation of, 123–128
   in culture media preparation, 95–100
   and equipment, 95–100
   gelatin hydrolysis test for, 173–176
   generation time of, 251
   gene recombination in, 355–359
   Gram stain of, 57–61
   growth of
     growth curve of, 251–255
     osmotic pressure (salt concentration) and, 231–233
     pH and, 225–226
     temperature and, 217–219
   lipase test for, 161–162
   lytic phage infection of, 388
   milk contamination, 293–296
   nitrate reduction test for, 209–212
   number (count) of, 133–137
   oxidase test for, 185–187
   phenylamine deamination test for, 203–205
   pour plate, 117–119
   selective/differential media for, 110–111
   simple staining of, 49–52
   spread plate, 103–106
   starch hydrolysis test for, 157–158
   streak plate, 109–112
   triple sugar iron agar for, 151–154
   urease activity test for, 191–193
   in water
     as fecal contamination indicator, 264, 267
     most probable number test for, 264, 266–269
Escherichia coli O157:H7, 294
Esculin reaction, in *Streptococcus* differentiation, 317–318
Ethanol. *See* Alcohol
Ethylene oxide, for sterilization, 96
Euglena, phase-contrast microscopy of, 26

Eugon agar, A–16
Eugon broth, A–16
Eukaryotic microbiology, 336. *See also specific types*
   medical significance of, 336
   phase-contrast microscopy of, 26
Eukaryotic microorganisms, 336
Exponential phase, of growth, 251, 252
External contamination, 35
Extreme halophiles, 232
Eyepiece, 4, 7
   insertion of ocular micrometer in, 7

F

Facultative anaerobes, 124, 125
Fastidious microorganisms, A–23
Fecal contamination, 264, 267. *See also* Coliform bacteria
Fermentation, 145–148
   citrate utilization test for, 166–168, 170
   defined, 146
   in food microbiology, 264
   IMViC tests for, 165–170
   methyl red test for, 165–167, 170
   in *Neisseria* identification, 324
   pathways of, 147
   triple sugar iron agar test for, 151–154
   Voges–Proskauer test for, 166, 168, 170
   yeast, 338–339
Ferric chloride reagent, A–10–A–11
Ferric chloride solution
   in phenylalanine deamination, 203–205
   safety considerations with, 203
Filtration, for sterilization, 96
Fine adjustment knob, 8
Flagella, 88
   Gray method of, 88
   Leifson method of, 88
   medical significance of, 88
   motility from, 15–16, 88
Flagellar motion, 16
Flagella staining, 87–90
   Difco method of, 88
   West method of, 88–89, A–12
Flagellation, patterns of, 88
Fluorescence microscopy, 2
Focal length, 4
Food intoxications, 264
Food microbiology, 264
   milk examination for bacteria, 295
   plate counts and quality assessment of milk, 293–296
Food poisonings, 210, 264, 302
Freeze-drying, A–22
Fuchsin, in flagella staining, 88
Fulton, MacDonald, 74
Fungal meningitis, 81

Fungi, 336
  commercially prepared slides of, 347–348
  isolation and culturing of, 337–339, 345–349
  safety issues with, 345
  in soil, 273–275
Furoxone-Tween-Oil Red (FTO) agar, 331, A–16
Fusobacterium, 124

G

Gas gangrene, 74
GasPak Anaerobic System, 125, 126, 127, 128
GasPak pouches, 126, 127
Gas production
  in fermentation, 145–148
  in triple sugar iron agar test, 151–154
Gastritis, 191
Gastroenteritis, 264
Gelatinase, 173–174, 305, 306
Gelatin hydrolysis, 173–174
Gelatin hydrolysis test, 173–176
  procedure of, 175
Gel diffusion agar, A–16
Generalized transduction, 388–389
Generation time, 251–255
Genetic engineering, 354
Genetics, 354
  genomic DNA isolation in, 355–359
  mutations in, 369–372
  16S rRNA in, 363–366
  transformation in, 375–377
Genomics, 354
Gentamicin, susceptibility testing of, 245
Geobacillus stearothermophilus
  biosafety level (BSL) of, A–28
  temperature sensitivity of, 217–219
Giardia, water contamination, 265
Glass rod, spooling DNA on, 356–357
Glassware, cleaning solution for, A–10
Gliding motility, 16
Glomerulonephritis, 316
Glucose fermentation, 146
  in methyl red test, 166–167, 170
  in *Neisseria* identification, 324, 325
  triple sugar iron agar for, 151–154
  in Voges–Proskauer test, 166, 168, 170
  yeast, 338–339
Glucose-minimal salts, A–16
Glucose oxidation, 165–170
Glycerol yeast agar, 274, 275
Gonococcal, 325
Gonococcal urethritis, 325
Gonorrhea. *See Neisseria gonorrhoeae*
Graham and Evans method, of capsule staining, 83
Gram-negative, 57–61
Gram-negative bacteria, 57–61
Gram-positive, 57–61

Gram-positive bacteria, 57–61
Gram's iodine, A–10–A–11
  for Gram stain, 57
  for starch hydrolysis test, 157–158
Gram stain, 32, 57–61, A–11
  control procedure in, 60–61
  decolorizers in, 58, 60–61, A–10
  mordant in, 58, 61
  in *Neissseria* identification, 325
  principles of, 58–60
  sources of errors in, 61
  traditional technique of, 60–61
Gram-variable, 60
Gram-variable bacteria, 60
Gray method, of flagella staining, 88
Green lactose broth, 267
Group A streptococci, 315–319
Group B streptococci, 315–319
Group C streptococci, 315–319
Group D streptococci, 315–319
Growth. *See* Microbial growth
Growth curve, 251–255
Growth optimum
  pH, 225–226
  temperature, 217–218
Growth range, 218
Growth rate constant, 253–255

H

Halobacterium salinarium
  biosafety level (BSL) of, A–28
  osmotic pressure (salt concentration) and, 231–233
Halophiles, 232
Hand washing, commercial products for, 239
Hanging drop slides, 15–16
Hawaii, culture restrictions of, A–22
Heat-fixing, 50
Helicobacter pylori
  electron micrograph of, 88
  urease activity of, 191
Hemolysis, 310–311
  alpha-hemolysis, 302, 303, 310
  beta-hemolysis, 302, 303, 316
  observation of, 316
  in *Staphylococcus* differentiation, 302
  in *Streptococcus* differentiation, 316–319
Hepatitis A, 265
Hfr strain, conjugation using, 383
High-dry objective lens, 4
High-dry objective lens (40x, 4mm), 4
"Hockey stick," 103, 104
Horizontal gene transfer, 382
Human flora, normal, 329–332
Human skin
  architecture of, 330, 331
  microorganisms associated with, 332

Hyaluronidase, 405
Hydrochloric acid, for cleaning slides, 90
Hydrogen peroxide, 182
   in catalase test, 181–182
   vaporized, 96–97
Hydrogen sulfide (H_2S), triple sugar iron agar for, 151–154
Hydrogen sulfide test, 169
Hydrolases, 157
Hydrolysis
   casein, 174
   defined, 157
   gelatin, 173–174, 173–176
   lipid, 161–162
   protein, 169
   starch, 157–158
Hyperthermophiles, 218
Hypertonic solution, 232
Hyphae, 346
Hypotonic solution, 232

I

Illuminator, 8
IMViC tests, 165–170
India ink, A–11
   in capsule staining, 81, 82, 83, A–11
   in negative staining, 43, 44, 45
Indole-negative bacteria, 166
Indole-positive bacteria, 166
Indole production, 166, 169–170
Indole test, 166–167, 169–170
Induced mutation, 370
Infectious microbes, 300
Inoculating loops, 35
   in culture transfer, 35–36
   helpful hints, 37, 52, 112
   in smear preparation, 49–50
   sterilization of, 35
   in streak-plate technique, 109–112
Inoculating needles, 35
   in culture transfer, 35–36
   helpful hints, 37
   sterilization of, 35
Interpersonal and citizenry skills, A–25
Iodine
   Gram's, A–10–A–11
      for Gram stain, 57
      for starch hydrolysis test, 157–158
Ionizing radiation, 96
Iris diaphragm, 4, 8
Iron agar
   Kligler, A–17
   lysine, 198–200, A–17
   triple sugar, 151–154, A–21

Isopropanol, as fermentation product, 147
Isotonic solution, 232

K

Kanamycin, susceptibility testing of, 245
KEY tablets
   in decarboxylase tests, 198–199
   in ONPG test, 146, 148
   in phenylalinine deaminase test, 203–204
KEY test strips
   in decarboxylase tests, 204
   in oxidase test, 186
KF streptococcus agar, A–16
Kinyoun acid-fast stain, A–11
Kinyoun carbolfuchsin, A–11
Kirby, William, 244
Kirby–Bauer method, 243–244
Klebsiella, monitoring water for, 265–270
Klebsiella pneumoniae
   antibiotic susceptibility of, 243–247
   biosafety level (BSL) of, A–28
   capsule staining of, 81–83
   casein hydrolysis test for, 173–176
   citrate fermentation by, 166
   decarboxylase tests of, 197–200
   gelatin hydrolysis test for, 173–176
   IMViC test for, 165–170
   as nitrate reducer, 210
   urease activity test for, 191–193
Kligler Iron Agar, A–17
Kovacs' reagent, 166, A–11
   helpful hints, 170
   in IMViC tests, 165, 166, 170
   in indole production test, 166–167, 169–170
   safety considerations with, 165

L

Laboratory Core Curriculum, A–24
Laboratory safety procedures, A–26. *See also specific material and procedures*
Laboratory skills, A–24–A–25
Laboratory thinking skills, A–24–A–25
Lactococcus lactis, 294
Lactose fermentation, 146
   in IMViC tests, 166
   triple sugar iron agar for, 151–154
Lactose fermentation broth, A–17
Lag phase, of growth, 251, 252
Lambda phage, 388
Lancefield classification, 316
Lauryl tryptose broth, A–17
   for most probable number test, 264, 267
Lawn, bacterial, 244–245

Leeuwenhoek, Antoni van, 2
Legal issues
   in *Neissseria gonorrhoeae* identification, 324
   in use of normal microbiota, 337
Leifson method, of flagella staining, 88
Leukocidins, 316
Levine EMB agar, A–17
Light intensity, difference in, 26
Light microscope
   bright-field, 2, 3–8
   dark-field, 2, 19–21
   parts and functions of, 8
   phase-contrast, 2, 25–28
Linear magnification, 4
Lipase(s), 161
Lipase test, 161–162
Lipid(s), 161–162
Lipid hydrolysis, 161–162
Lipolysis, 162
Liquid (broth) media, 97, A–14–A–21
   anaerobic bacteria cultivation in, 124, 127
   pour plate from, 117–119
   preparation of, 98, 99
   spread plate from, 103–106
   streak plate from, 109–112
Listeria, milk contamination, 294
Listeria monocytogenes, medical significance of, 88
Listeriosis, 293
Litmus milk, 174–175, A–17
   procedure, 176
   reactions, examples of, 175
   tube, coagulation in, 175
Living microbes, 32
Loeffler's alkaline methylene blue, A–11
Logarithmic phase, of growth, 251, 252
Löwenstein–Jensen medium, A–17
Low-power objective lens (10x, 16mm), 4
Lycoperdon, 348
Lyophilization, A–22
Lysine decarboxylase test, 197–200, 199
Lysine iron agar (LIA), 198, A–17
Lysol, 237–239

M

MacConkey's agar, A–18
Malachite green, A–11
   in endospore staining, 73–76
Mannitol salt agar, 111, 303, 304–305, A–17
   Staphylococcus aureus and *Staphylococcus epidermidis* on, 303, 304
Maximum growth temperature, 218
Mean generation time, 253
Mean growth rate constant *(k),* 254–255
Measurement, microscopic, 6–8

Medical microbiology, 300
   normal human flora, 329–332
M-Endo broth, A–17
Meningitis
   bacterial, 309, 316, 324
   fungal, 81
Meningoencephalitis, 88
Mesophiles, 218
Methicillin, susceptibility testing of, 245
Methicillin-resistant *Staphylococcus aureus,* 302
Methyl cellulose, 25, 26
Methylene blue, 49–51, 124, 294, A–11. *See also* Eosin methylene blue (EMB) agar
Methylene blue reductase test, 294–296
Methylene blue reduction time (MBRT), 294
Methyl red reagent, A–11
Methyl red test, 166–167, 170
M-FC broth, A–17
Microaerophiles, 124
Microbial biochemistry, 144
Microbial growth
   control of, 216
      antibiotic susceptibility testing of, 243–247
      chemical agents for, 216, 237–239
   environmental factors affecting, 216
   growth curve of, 251–255
   osmotic pressure and, 231–233
   pH and, 225–226
   phases of, 253
   temperature and, 217–219
Microbial mass, 134
Microbicidal efficiency, 238
Microbiostatic efficiency, 238
Micrococcus, normal human flora, 329–332
Micrococcus luteus
   biosafety level (BSL) of, A–28
   catalase test for, 181–182
   in normal human flora, 332
   spread plate, 103–106
Micrococcus roseus, 33–37
   aseptic technique of, 33–37
   biosafety level (BSL) of, A–28
Micrococcus sp., 332
Microscopes
   bright-field light, 2, 3–8
   cleaning of, 4
   dark-field light, 2, 19–21
   differential interference, 2
   fluorescence, 2
   helpful hints, 8
   parts and functions of, 8
   phase-contrast light, 2, 25–28
   proper use of, 4–6
   troubleshooting, 6
   types of, 2
Microscopic measurement, 6–8

Microwave procedure, of endospore staining, 75–76
Milk (dairy products)
   contamination of, sources of, 293
   lipolytic bacteria of, 161–162
   methylene blue reductase test of, 294–296
   microbial growth in, 264
   pasteurization of, 293–294
   plate counts and quality assessment of, 293–296
   quality assessment of, 293–296
Minimum growth temperature, 218
Minimum inhibitory concentration (MIC), 244
Mixed acid fermenters, 166
MM 1 Medium, A–17
MM 2 Medium, A–18
Mobiluncus, 124
Moderate halophiles, 232
Moeller's decarboxylase broth with ornithine, A–18
Molds, 336, 345–349
   classification of, 346
   colonies of, 345
   pH and, 225
Moraxella
   characteristics of, 324
   oxidase test for, 186
Moraxella nonliquefaciens, 323–326
   biosafety level (BSL) of, A–28
   characteristics of, 324
   Gram stain of, 325
Morchella, 348
Mordant, in Gram staining, 58, 61
Morganella
   medical significance of, 191, 203
Morganella morganii, as nitrate reducer, 210
Most probable number (MPN) test, for coliform bacteria, 264, 266–269
Motility of bacteria, 15–16, 88
   flagella and, 15–16, 88
   gliding, 16
   spirochete and, 16
   swarming, 16
Motility test media, A–18
MR-VP broth, A–18
Mucor, in soil, 273–275
Müeller–Hinton agar, A–18
Mueller–Hinton II agar plate, 244–247
MUG (4-methylumbelliferyl-β-D-glucuronide), 268, 270
Mushrooms, 348
Mutagen, 370
Mutations, 369–372
   bacterial, 370
   during DNA replication, 370
   induced, 370
   nonsense, 370
   silent, 370
   spontaneous, 370
Mycelium, 346

Mycobacterium, water contamination, 265
Mycobacterium leprae, acid-fast staining of, 68
Mycobacterium phlei
   acid-fast staining of, 67–69
   biosafety level (BSL) of, A–28
Mycobacterium smegmatis
   acid-fast staining of, 67–69
   biosafety level (BSL) of, A–28
Mycobacterium tuberculosis, acid-fast staining of, 67–69
Mycology, 345. *See also* Fungi

N

N-acyl homoserine lactone (AHL), 288
Naphthol, alpha, 165, A–11
Negative citrate test, 168
Negative methyl red test, 166
Negative staining, 43–45, 49, 50
Negative VP test, 166
Neisser, Albert Ludwig Sigesmund, 324
Neisseria, 323–326
   candle jar for, 325, 326
   catalase test for, 326
   characteristics of, 324
   culture on chocolate agar, 325
   culture on modified Thayer–Martin agar, 325
   DNase test for, 324–326
   fermentation patterns of, 324, 326
   Gram stain of, 325
   morphology of, 324, 325
   nitrate reduction test for, 324, 326
   oxidase test for, 186, 325
   rapid identification of, 326
   safety considerations with, 323
Neisseria flavescens, 323–326
   biosafety level (BSL) of, A–28
Neisseria gonorrhoeae, 323–326
   characteristics of, 324
   culture on modified Thayer–Martin agar, 325
   fermentation patterns of, 324, 326
   Gram stain of, 325
   morphology of, 324, 325
   nitrate reduction test for, 324, 326
   oxidase test for, 186
   rapid detection of, 324
Neisseria lactamica, 324
Neisseria meningitidis, 324
Neisseria sicca, 323–326
Neisseria subflava, 323–326
Neisseria subflava ser subflava, biosafety level (BSL) of, A–28
Neomycin, susceptibility testing of, 245
Nephelo culture flask, 255
Nessler's reagent, A–11
Neutrophiles, 225, 226
Nigrosin, 43, 44, 45, A–11–A–12
Nitrate agar slants, A–18

Nitrate broth, 210–211, A–18
Nitrate-free broth, A–18
Nitrate reductase, 210
Nitrate reduction test, 209–212
   for *Neisseria*, 324, 326
   for *Staphylococcus*, 302
Nitrite test reagents, A–12
Nocardia
   acid-fast staining of, 67–69
   medical significance of, 68
Nocardia asteroides, acid-fast staining of, 68
Nocardia brasiliensis, acid-fast staining of, 68
Non-acid fast microorganisms, 68
Nonmotile bacteria, 16
Nonsense mutations, 370
Nonsynthetic media, 98
Normal human flora, 329–332
Normal microbiota, 329
   fungi identification in, 337–339
   laboratory use, legal issues with, 337
Nosocomial infections, 210
Nostoc caeruleum, phase-contrast microscopy of, 26
Novobiocin, susceptibility testing of, 245
Novobiocin sensitivity, 305, 306
Nutrient agar, A–18
Nutrient broth, 97, A–18
Nutrient gelatin, A–18

O

Objective lens, 4, 8
   cleaning of, 4
   high-dry, 4
   low-power, 4
   ocular lens (eyepiece), 4, 7, 8
   oil immersion, 4, 5–6
   proper use of, 4–6
Obligate aerobes, 124
Obligate anaerobes, 124
Ocular lens (eyepiece), 4, 7, 8
Ocular micrometer, 6–8
   calibration of, 6–7
   insertion in eyepiece, 7, 8
Oil immersion objective lens (90x, 100x, 1.8mm), 4, 5–6
Oleandomycin, susceptibility testing of, 245
ONPG (o-nitro-phenyl-β-D-galactopyranoside), 146, A–12
   in β-galactosidase test, 146, 148
   in coliform testing, 268
Optical density, of medium, 134
Optimal growth pH, 226
Optimum growth temperature, 218
Optochin sensitivity, 311
Optochin test, 309–311
Ornithine decarboxylase test, 197–200
Osmotic pressure
   defined, 231–232
   growth effects of, 231–233
Otitis media, 309, 316
Oxidase test, 185–187
   for *Neisseria*, 186, 325
   for *Staphylococcus*, 302
Oxidase test reagent, A–12
Oxidation–reduction reaction, 124
OxyDish, 126
Oxygen, and bacterial growth, 124, 125, 126
OxyPlate, 123
Oxyrase for Agar, 123, 126
Oxyrase for Broth, 123, 126

P

Paramecium, 25–28
   dark-field microscopy of, 19–20
   phase-contrast microscopy of, 25, 26, 27
Paramecium caudatum, phase-contrast microscopy of, 26
Pararosaniline, in flagella staining, 88
Pasteurization, 293–294
Pathogen(s)
   biosafety level (BSL) of, A–27–A–28
   use of, A–26
Pathogenic bacteria, 124
Penicillin G, susceptibility testing of, 245
Penicillium
   identification of, 345–349
   in soil, 273–275
Penicillium notatum
   biosafety level (BSL) of, A–28
   identification of, 345–349
Peptic ulcer disease, 191
Peptone broth, A–18
Peptones, as buffers, 226
Peptostreptococcus, 124
Peritonitis, 309
Peroxidase, 182
Petri plates. *See* Agar plates
pH, A–29
   indicators
      in citrate utilization test, 166–168, 170
      in decarboxylase tests, 198–199
      in fermentation and bacteria identification, 146–147
      in lipid hydrolysis test, 161
      methyl red, 166–168, 170
      and mold growth, 346
      in triple sugar iron agar test, 152–154
   and microbial growth, 225–226
Phagocytosis, capsule and, 81
Pharyngitis, streptococcal, 316
Phase-contrast microscope, 2, 25–28
Phenol
   as disinfectant, 239
   safety considerations with, 237
Phenol coefficient (PC), 237–239

Phenol red
   in carbohydrate fermentation, 145–148, 152
   in decarboxylase tests, 199
   in urease activity test, 192
Phenol red broth, 146
Phenol red dextrose broth, A–18
Phenol red lactose broth, A–19
Phenol red sucrose (saccharose) broth, A–19
Phenylalanine deaminase (phenylalanine) agar, 203–204, A–19
Phenylalinine deamination (PAD) test, 203–205
pH indicators, A–29
Phosphate-buffered saline, A–12
Phosphate buffers, 226, A–12
Phospholipids, 161–162
Photobacterium, 287
Physiological saline, A–12
Pipettes, 33–37
   filling of, 34
   procedure for using, 35
   transferring liquid by, 34
Plasmid pMMB66, 376
Plasmids, 354, 375–376
   conjugative, 382
Plasmolysis, 232
Plate count agar, A–19
Pneumococci, 309–311. *See also* Streptococcus pneumoniae
Pneumonia
   Klebsiella, 81, 210
   Proteus, 88
   staphylococcal, 302
   streptococcal, 81
Polymerase chain reaction (PCR), 354, 364, 365
Polymyxin B, susceptibility testing of, 245
Polyporus, 348
Polysorbate 80, A–21
Pond water, phase-contrast microscopy of, 25–26
Positive methyl red test, 166
Positive staining, 50. *See also* Simple staining
Potable water, 264
Potato dextrose agar, 346, 347, 348, A–19
Pour-plate technique, 117–119
P1 phage, 388
P22 phage, 388
Presence-absence (P-A) broth, A–19
Presumptive test, for coliforms, 267–269
Primary stain, 58
Propionibacterium, normal human flora, 330
Propionic acid, as fermentation product, 147
Proteases, 174
Proteins, 174
   casein hydrolysis, 173–176, 174
   catalase activity, 181–182
   decarboxylase tests, 197–200
   gelatin hydrolysis, 173–176
   IMViC tests, 165–170
   nitrate reduction test, 209–212
   oxidase test, 185–187
   phenylamine deamination test for, 203–205
   urease activity test for, 191–193
Proteolysis, 174
Proteolytic enzymes, 174
Proteus
   antibiotic susceptibility of, 245
   casein hydrolysis test for, 173–176
   gelatin hydrolysis test for, 173–176
   selective/differential media for, 111
Proteus mirabilis
   medical significance of, 191
   as nitrate reducer, 210
Proteus vulgaris, 173
   biosafety level (BSL) of, A–28
   capsule of, 88
   casein hydrolysis test for, 173–176
   decarboxylase tests of, 197–200
   flagellar stain of, 88
   flagella staining of, 87–90
   gelatin hydrolysis test for, 173–176
   IMViC tests of, 165–170
   phenylamine deamination test for, 203–205
   starch hydrolysis test for, 157–158
   triple sugar iron agar for, 151–154
   urease activity test for, 191–193
Prototroph, 382
Protozoa, 336
Providencia, medical significance of, 191, 203
Pseudomembranous colitis, 74
Pseudomonas
   decarboxylase test for, 198
   selective/differential media for, 111
Pseudomonas aeruginosa, 173, 237
   antibiotic susceptibility of, 243–247
   biosafety level (BSL) of, A–28
   casein hydrolysis test for, 173–176
   cultivation of, 123–128
   disinfectants and, 237–239
   flagella staining of, 87–90
   gelatin hydrolysis test for, 173–176
   medical significance of, 88
   motility of, 15
   oxidase test for, 185–187
   pH and growth of, 225–226
   temperature sensitivity of, 217–219
   triple sugar iron agar for, 151–154
Pseudomonas fluorescens
   biosafety level (BSL) of, A–28
   nitrate reduction test for, 209–212
Psychrophiles, 218
Psychrotrophs, 218

Puccinia, 348
Pure (stock) cultures, 104
   Hawaii restrictions on, A–22
   maintenance of, A–22–A–23
   medical application of, 103, 109
   pour-plate technique for, 117–119
   sources for, A–22
   spread-plate technique for, 103–106
   streak-plate technique for, 109–112
Putrescine, 198

Q

Quick-freeze method, A–22
Quorum sensing, 288

R

Reactive oxygen species (ROS), 182
Reagents, A–10–A–13. *See also specific reagents*
Recipient bacterium, in transformation, 375–377
Recipient cell, 354
Rectal swabs, legal issues with, 315
Refractive index, 4
REMEL
   BactiCard Neisseria rapid test, 323, 326
   BactiCard Strep rapid test, 315, 318
   IDS RapidID Yeast Plus Panel, 337, 339
   RIM A.R.C. STREP A TEST, 318
Reproductive hyphae, 346
Resazurin, 124
Respiratory burst, 182
Rheumatic fever, 316
Rhizoidal hyphae, 346
Rhizopus, 348
   in soil, 273–275
Rhizopus stolonifer
   biosafety level (BSL) of, A–28
   identification of, 345–349
Rhodotorula
   isolation and culturing of, 337–339
Rhodotorula rubrum
   biosafety level (BSL) of, A–28
Ribonuclease, 356
Rifampin, susceptibility testing of, 245
Rods, 3
Rotating nosepiece, 8
Russell, F. F., 152

S

Sabouraud dextrose agar, 274, 275, 338–339, 345–347, A–19
Saccharolytic microorganisms, 152
Saccharomyces cerevisae, 43–45
   biosafety level (BSL) of, A–28
   fermentation activity of, 145–148
   life cycle of, 338
   Sabouraud dextrose agar for, 338–339
Safety, A–26–A–27. *See also specific material and procedures*
Safranin, A–11
   in capsule staining, 81–83
   in endospore staining, 73–76
   in Gram staining, 57–60
Saline
   phosphate-buffered, A–12
   physiological, A–12
Salmonella
   IMViC tests of, 165–170
   milk contamination, 294
   selective/differential media for, 111
   water contamination, 265
Salmonella enterica
   carbohydrate fermentation by, 145–148
   urease activity test for, 191–193
Salmonella enterica ser Typhimurium, biosafety level (BSL) of, A–28
Salt medium-Halobacterium, A–19
Salt tolerance, of bacteria, 231–233
Saprolegnia, 348
SARS-CoV-2 virus, 354
Scarlet fever, 316
Schaeffer, Alice B., 74
Schaeffer–Fulton procedure, 74, 75
Seafood contamination, 267
Sea water agar plate, 287, A–19
Sea water synthetic, A–19
Selection method, in genetic engineering, 354
Selective media, 110–111
Semisolid media, 97
Septate hyphae, 346
Septum, of molds, 346
Serial dilutions, 136, 137, A–2
   student skill with, A–24
Serological pipette, 33
Serratia marcescens, 33–37
   aseptic technique of, 33–37
   biosafety level (BSL) of, A–28
   pour plate, 117–119
   streak plate, 109–112
   temperature sensitivity of, 217–219
Sewage spills, 264
Sexual spores, 338
Shigella
   IMViC tests of, 165–170
   water contamination, 265
Shigella flexneri
   biosafety level (BSL) of, A–28
   triple sugar iron agar for, 151–154
Shigellosis, 293
Silent mutations, 370

Silver nitrate, 88, 89
SIM medium (agar), 169, A–19
Simmons citrate agar, A–20
Simmons citrate agar slants, 168, 170
Simple staining, 49–52
16S rRNA gene, 363–364
Skin, human
   architecture of, 330, 331
   microorganisms associated with, 332
Slant. *See* Agar slant
Slide culture, 346
Slides
   bacterial adherence to, 69
   cleaning of, 83, 90
   hanging drop, 15, 16
   student skill with, A–24
   Vaseline on, 16
   wet-mount, 4, 6
Smear
   bacterial, 49–50
   preparation of, 44, 45, 49, 50–51
   thin, 44, 45
Sodium chloride, bacterial tolerance to, 231–233
Sodium hypochlorite, 239
Soil
   habitat of, 274
   microorganisms in, 264, 273–275
Solid media, 97
Solute, 232
Solution(s), A–10–A–13
   hypertonic, 232
   hypotonic, 232
   isotonic, 232
   stock, A–1, A–12
Solvent, 232
Specialized transduction, 389
Spectrophotometer, 134, 136–137
Spectrophotometry, 134
   analysis with, 134, 136–137
Spirilla, 3
Spirillum volutans, 88
   biosafety level (BSL) of, A–28
   flagella staining of, 87–90
   motility of, 15
Spirit blue agar, lipase test in, 161–162
Spirit blue agar with 3% lipase, A–20
Spirochete motility, 16
Spizizen's salts, A–20
Spontaneous mutations, 370
Spores, mold, 346
Spread-plate technique, 103–106
SS agar, A–20
Stage micrometer, 6, 7
Stage of microscope, 8
Staining. *See also* specific methods and organisms
   acid-fast, 67–69
   capsule, 81–83
   differential, 58
   disadvantage of, 43
   endospore, 73–76
   flagella, 87–90
   Gram, 32, 57–61, A–11
   negative, 43–45, 49, 50
   positive, 50
   simple, 49–52
Standard methods agar, A–20
Standard plate count, 134, 135
Staphylococcus, 301–306
   antibiotic susceptibility testing of, 243–247
   catalase test for, 182
   characteristics of, 302, 303
   coagulase test for, 303, 304
   normal human flora, 330
   phenol coefficient test of, 237
Staphylococcus aureus, 43–45, 61, 301–306, 315, 317
   antibiotic susceptibility testing of, 243–247
   biosafety level (BSL) of, A–28
   catalase test for, 181–182
   characteristics of, 303
   on chocolate agar, 302
   disinfectants and, 237–239
   Gram stain of, 57–61
   lipase test for, 161–162
   mannitol salt agar on, 303, 304
   as nitrate reducer, 210
   in normal human flora, 332
   osmotic pressure (salt concentration) and, 231–233
   selective/differential media for, 110–111
   simple staining of, 49–52
   temperature sensitivity of, 217–219
Staphylococcus epidermidis, 301–306
   biosafety level (BSL) of, A–28
   characteristics of, 303
   mannitol salt agar on, 303, 304
   nitrate reduction test for, 209–212
   in normal human flora, 332
Staphylococcus saprophyticus, 301–306
   biosafety level (BSL) of, A–28
Starch, composition of, 157
Starch agar, 157, A–20
Starch agar plate, 157
Starch hydrolysis, 157–158
Stationary phase, of growth, 251, 252
Sterilization, 96–97, 238
Stock cultures. *See* Pure cultures
Stock solutions, A–1, A–12
Streak-plate technique, 109–112
Strep throat, 316
Streptococcus, 315–319
   bacitracin sensitivity test of, 317, 318

Streptococcus (*Continued*)
   bile esculin reaction of, 317–318, 319
   CAMP factor test of, 317, 318, 319
   catalase test for, 182
   characteristics of, 303
   culture on blood agar, 315–319
   group A, 315–319
   group B, 315–319
   group C, 315–319
   group D, 315–319
   hemolytic activity of, 315–319
   Lancefield classification of, 316
   legal considerations with, 315
   medical significance of, 316
   morphology of, 316
   rapid detection of, 318–319
   safety considerations with, 315
   SXT sensitivity of, 315–319
Streptococcus agalactiae, 315, 317–318
   biosafety level (BSL) of, A–28
Streptococcus bovis, 315–318
   biosafety level (BSL) of, A–28
Streptococcus equi, 315–318
   biosafety level (BSL) of, A–28
Streptococcus mitis, 309–311, 310, 318
   biosafety level (BSL) of, A–28
Streptococcus mutans, 316–318
   capsule staining of, 81
Streptococcus pneumoniae, 309–311, 310
   bile solubility test of, 310
   biosafety level (BSL) of, A–28
   capsule staining of, 81
   culture on blood agar, 310–311
   hemolysis of, 310
   medical significance of, 310
   morphology of, 310
Streptococcus pyogenes, 181, 315–318
   bacitracin sensitivity test of, 317, 318
   bile esculin reaction of, 317–319
   biosafety level (BSL) of, A–28
   CAMP factor test of, 317, 319
   classification of, 316
   culture on blood agar, 315–319
   hemolytic activity of, 316
   SXT sensitivity of, 315–319
Streptococcus salvarius, 316–318
Streptococcus viridans, 317–318, 332
   in normal human flora, 332
Streptodornase, 316
Streptokinase, 316
Streptolysin O, 316
Streptolysin S, 316
Streptomycin, susceptibility testing of, 245
Subculturing, 35
Substage condenser, 4
Sucrose, fermentation of, 146, 151–154

Sulfite, hydrogen sulfide production from, 169
Sulfonamides, susceptibility testing of, 245
Superoxide dismutase, 182
Susceptibility testing, of antibiotics, 243–247
Swarming motility, 16
SXT disks, 317, 318
Symbiotic relationships, of eukaryotes, 336

T

Tannic acid, in flagella staining, 88
T-broth, A–20
Temperature
   and food microbiology, 264
   and microbial growth, 217–219
Tergitol No. 4, 67
Tetanus, 74
Tetracycline, susceptibility testing of, 245
TFTC. *See* Too few to count
Thallus, 346
Thayer–Martin agar, modified, 325, A–20
Thermoduric bacteria, 218
Thermophiles, 218
Thermus aquaticus, 364
Thinking skills, A–24–A–25
Thin smear, 44, 45
Thioglycollate broth, A–20
   for anaerobic bacteria, 124, 127
Thiosulfate, hydrogen sulfide production from, 169
Throat swabs, legal issues with, 315
TNTC. *See* Too numerous to count
Todd–Hewitt broth, A–20
Tonsillitis, 316
Too few to count (TFTC), 136
Too numerous to count (TNTC), 136, 274
Tooth decay, 81
Top agar, A–20
Total coliform count, 264, 267
Toxic shock syndrome, 302
T phage, 388
Transformation, bacterial, 375–377
Treponema denticola, dark-field microscopy of, 19
Treponema pallidum
   dark-field microscopy of, 19, 21
   medical significance of, 19
Tributyrin agar, lipase test in, 161–162
Triglycerides, 161–162
Triple sugar iron agar, A–21
Triple sugar iron agar test, 151–154
Triton X-100 stock solution, 67, A–12
Trommsdorf's reagent, A–12
Tryptic nitrate broth, A–21
Tryptic soy agar (TSA), A–21
   for isolation of soil microorganisms, 274
   with lecithin and polysorbate 80, A–21

Tryptic soy agar (*Continued*)
   for *Neisseria,* 325, 326
Tryptic soy agar slants, A–23
Tryptic soy broth, A–21
   agar plate from, 99
   aseptic technique of, 33–37
   pour plate from, 117–119
   preparation of, 99
Tryptone agar, A–21
Tryptone broth, A–21
Tryptone glucose yeast agar, A–21
Tryptophan, 166
   hydrolysis of, 166
Tryptophanase, 166
TSA. *See* Tryptic soy agar
Tuberculosis, 293
Turbidimetric analysis, 134, 137
Typhoid fever, 264

U

Ultraviolet (UV) radiation, for sterilization, 96
Universal Precautions, A–26, A–27
Unpasteurized milk, quality assessment of, 293–296
Urea agar slants, 192
Urea broth, 192–193, A–21
Urea disks or tablets, 192–193
Urease, 192–193
Urease activity test, 191–193
Urinary tract infections, 88, 166, 191, 203, 210, 302
Use-dilution, 238

V

Vancomycin, susceptibility testing of, 245
van Leeuwenhoek, Antonie, 2
Vaporized hydrogen peroxide, 96–97
Vaseline, on slide, 16
Vaspar, A–12
Vegetative hyphae, 346
Viable (standard) plate count, 134
Vibrio, water contamination, 265
Vibrio cholerae, 88, 288
Vibrio fischeri, 287–289
   biolumniscence of, 287–289
   biosafety level (BSL) of, A–28
Vibrio parahemolyticus, 288
Violet red bile agar, A–21
Vogel–Johnson agar, 304–305, A–21
Voges–Proskauer test, 166, 168, 170
   for *Staphylococcus,* 302
Volume
   specified, diluting solution to, A–1–A–2
   unspecified, diluting solution to, A–1–A–2

Volvox, 25–28
   dark-field microscopy of, 21

W

Water
   buffered, A–12
   coliform bacteria in
      determination of, 264
      fecal contamination of, 264, 267
      most probable number (MPN) test for, 264, 266–269
      potable, 264
      total count of, 264
   distilled, in culture media, 99
   pond, phase-contrast microscopy of, 25–26
West stain, 88–89, A–12
Wet-mount slide
   measurement in, 6
   preparation of, 4
Whooping cough, 88
Wild-type, 370
Winogradsky columns, 279–281
Wirtz, Robert, 74
Wirtz–Conklin procedure, 74, 75
Working distance, 4
Wound infections, 88

Y

Yeast(s), 336, 337–339
   baker's (*See Saccharomyces cerevisae*)
   budding in, 338
   fermentation activity of, 145–148, 338
   genome of, 356
   in normal human flora, 332
   in normal microbiota, 338
   pH and, 225
   Sabouraud dextrose agar for, 338–339
Yersiniosis, 293
YM agar, A–21
YM broth, A–21

Z

Ziehl, Franz, 68
Ziehl–Neelsen acid-fast stain, 68, 69, A–13
Ziehl's carbolfuchsin, in acid-fast staining, 67
Zinc powder
   helpful hints, 212
   nitrate reduction test for, 209–212
   safety considerations, 209
Zone of inhibition, 244–245, 247
Zone of proteolysis, 174
Zygomycota, 345–349